AF761931

Assisted History Matching for Unconventional Reservoirs

Assisted History Matching for Unconventional Reservoirs

Sutthaporn Tripoppoom
Reservoir Engineer, PTT Exploration and Production, PLC, Bangkok, Thailand

Wei Yu
Chief Technology Officer, Sim Tech LLC and Research Associate, Hildebrand Department of Petroleum and Geosystems Engineering, The University of Texas at Austin, USA

Kamy Sepehrnoori
Professor, Hildebrand Department of Petroleum and Geosystems Engineering, The University of Texas at Austin, USA

Jijun Miao
CEO, Sim Tech LLC and Research Advisor, Hildebrand Department of Petroleum and Geosystems Engineering, The University of Texas at Austin, USA

ELSEVIER
Gulf Professional Publishing
An imprint of Elsevier

Gulf Professional Publishing is an imprint of Elsevier
50 Hampshire Street, 5th Floor, Cambridge, MA 02139, United States
The Boulevard, Langford Lane, Kidlington, Oxford, OX5 1GB, United Kingdom

Copyright © 2021 Elsevier Inc. All rights reserved.

No part of this publication may be reproduced or transmitted in any form or by any means, electronic or mechanical, including photocopying, recording, or any information storage and retrieval system, without permission in writing from the publisher. Details on how to seek permission, further information about the Publisher's permissions policies and our arrangements with organizations such as the Copyright Clearance Center and the Copyright Licensing Agency, can be found at our website: www.elsevier.com/permissions.

This book and the individual contributions contained in it are protected under copyright by the Publisher (other than as may be noted herein).

Notices

Knowledge and best practice in this field are constantly changing. As new research and experience broaden our understanding, changes in research methods, professional practices, or medical treatment may become necessary.

Practitioners and researchers must always rely on their own experience and knowledge in evaluating and using any information, methods, compounds, or experiments described herein. In using such information or methods they should be mindful of their own safety and the safety of others, including parties for whom they have a professional responsibility.

To the fullest extent of the law, neither the Publisher nor the authors, contributors, or editors, assume any liability for any injury and/or damage to persons or property as a matter of products liability, negligence or otherwise, or from any use or operation of any methods, products, instructions, or ideas contained in the material herein.

British Library Cataloguing-in-Publication Data

A catalogue record for this book is available from the British Library

Library of Congress Cataloging-in-Publication Data

A catalog record for this book is available from the Library of Congress

ISBN: 978-0-12-822242-3

For Information on all Gulf Professional Publishing publications
visit our website at https://www.elsevier.com/books-and-journals

Publisher: Joe Hayton
Senior Acquisitions Editor: Katie Hammon
Editorial Project Manager: Hilary Carr
Production Project Manager: Nirmala Arumugam
Cover Designer: Mark Rogers

Typeset by MPS Limited, Chennai, India

Working together to grow libraries in developing countries
www.elsevier.com • www.bookaid.org

Dedication

To my parents Sutisith and Supatsorn Tripoppoom and siblings Surisa and Suthaphong Tripoppoom

—Sutthaporn Tripoppoom

To my parents Zhongzhao Yu and Ruie Sun, in-laws Guilao Wu and Chunyun Tang, wife Kan Wu, daughter Emmy Yu, and son Evan Yu

—Wei Yu

To Professor Roy M. Knapp

—Kamy Sepehrnoori

To my parents Zhiyou Miao and Rongsheng Liu, wife Juan Zhou, son Zhouzi Miao, and daughter Vikki Miao

—Jijun Miao

Contents

About the authors

Mr. Sutthaporn Tripoppoom received his MS degree in the Hildebrand Department of Petroleum and Geosystems Engineering at the University of Texas at Austin in 2019. Currently, he is a reservoir engineer at Thai National Oil and Gas Company, PTT Exploration and Production Plc with 8 years of experience. His research interests include the development of Assisted History Matching (AHM) for naturally fractured reservoirs, hydraulically fractured reservoirs, and unconventional resources. He was the developer of an AHM workflow for unconventional reservoirs, and he also won the first place in SPE Regional Paper Contest (2019). He holds a BS degree in Petroleum Engineering from Chulalongkorn University in Thailand with First Class Honors and the outstanding student certificate. He is a member of SPE.

Dr. Wei Yu is the chief technology officer for Sim Tech LLC and a research associate in the Hildebrand Department of Petroleum and Geosystems Engineering at The University of Texas at Austin. He is an Associate Editor for *SPE Journal* and *Journal of Petroleum Science and Engineering*. His research interests include EDFM (Embedded Discrete Fracture Model) technology for modeling any complex fractures, shale gas and tight oil reservoir simulation, EDFM-AI for automatic history matching, and complex fracture characterization. Yu authored or coauthored more than 200 technical papers and holds five patents and two books ("*Shale Gas and Tight Oil Reservoir Simulation*" and "*Embedded Discrete Fracture Modeling and Application in Reservoir Simulation*"). He holds a PhD degree in petroleum engineering from The University of Texas at Austin.

Dr. Kamy Sepehrnoori is a professor in the Hildebrand Department of Petroleum and Geosystems Engineering at the University of Texas at Austin, where he holds the Texaco Centennial Chair in Petroleum Engineering. His research interest and teaching include computational methods, reservoir simulation, simulation of unconventional reservoirs, enhanced oil recovery modeling, flow assurance modeling, naturally fractured reservoirs, high-performance computing, and CO_2 sequestration. He has been teaching at the University of Texas for over 35 years and has graduated more than 230 MS and PhD students under his supervision working mainly in the areas of reservoir simulation and enhanced oil recovery modeling. For the last several years, he has been supervising a research group in simulation of unconventional reservoirs (shale gas and tight oil reservoirs). Sepehrnoori's research team along with his colleagues have been in charge of development of several compositional reservoir simulators (i.e., UTCOMPRS, UTCHEMRS, and UTGEL). Also, more recently, he supervised the development of a software package for embedded discrete fracture modeling for application in naturally and hydraulically fractured reservoirs. He has published over 650 articles in journals and conference proceedings in his research areas. He has also coauthored three books, which have been published by Elsevier. Sepehrnoori is the director of the Reservoir Simulation Joint Industry Project in the Center for Subsurface Energy and the Environment. He holds a PhD degree in petroleum engineering from the University of Texas at Austin.

Dr. Jijun Miao is the CEO at Sim Tech LLC and a research advisor in the Hildebrand Department of Petroleum and Geosystems Engineering at the University of Texas at Austin. His research interests include fracture characteristics, reservoir modeling and simulation, and EDFM development and application. Dr. Miao authored or coauthored more than 35 technical papers and holds 5 patents. He holds a PhD degree in petroleum exploration from the Research Institute of Petroleum Exploration & Development, PetroChina. Dr. Miao has years' hands-on experience in oilfield for oil and gas exploration and development and leads multidisciplinary research teams undertaking various major R&D programs.

Preface

Although fractures and reservoir properties are often expensive and sometimes difficult to be obtained in unconventional resources such as microseismic, log interpretation, and core data, production data are always available at no additional cost and can be beneficial to characterize hydraulic fracture and reservoir properties. The characterization of fractures and reservoir properties are important because they can calibrate fracturing propagation model to improve any future fracturing design and to optimize the well placement. However, history matching using production data is a time-consuming and challenging task traditionally for engineers, and after the historical data are matched, only one solution is available. With only one solution, it is insufficient to capture the geometry of hydraulic fractures and uncertainties of reservoir properties, and more importantly, it cannot handle the nonuniqueness of history-matching problem. Therefore this book was written to solve the problem by adopting the proposed assisted history-matching (AHM) workflow.

Two aspects are considered for history matching in unconventional reservoirs as shale formations including history-matching algorithm and fractures modeling. This book begins with a literature review of the current state-of-art AHM techniques and fracture modeling in Chapter 1, Introduction and literature review. Then the approach, which is practical, efficient, and accurate, used in this book is explained in Chapter 2, Methodology. For history-matching algorithm, we used multiple proxy-based Markov Chain Monte Carlo (proxy-based MCMC) algorithm in the proposed AHM workflow for shale reservoirs. The reason is that MCMC, one of the AHM algorithms in the Bayesian statistics, has benefits of quantifying uncertainty without bias or being trapped in any local minima. In addition, using MCMC with proxy model such as neural network unlocks the limitation of an infeasible number of simulation runs required by a traditional MCMC algorithm. For fractures modeling, the proposed AHM workflow utilized the benefits of embedded discrete fracture model to model fractures with a higher computational efficiency than a traditional local grid refinement method and more accuracy than the continuum approach.

Then, we performed the workflow validation for two scenarios of with and without natural fractures to validate the robustness and accuracy of the proposed workflow in Chapter 3, Validation of assisted history matching for a synthetic shale gas well. Then, we demonstrated the use of the workflow with many actual applications in each chapter including shale gas wells in Longmaxi shale in Chapter 4, Shale-gas well in longmaxi shale with bi-wing hydraulic fractures, and in Marcellus shale in Chapter 5, Shale-gas well in marcellus shale with bi-wing hydraulic fractures. We also compared the impact of using different proxy models: *k*-nearest neighbors and neural networks in terms of accuracy and computational efficiency in Chapter 6, Proxy comparison between neural network and k-nearest neighbors. We then introduced the concept of enhanced permeability area to represent complex fractures for history matching and investigating the effect from ignoring them during history matching and production forecast in Chapter 7, Shale-gas well with and without enhanced permeability area. The impact of natural fractures' existence was then investigated for shale gas wells and shale oil wells in Permian basin in Chapter 8, Shale-gas well with and without natural fractures, and Chapter 9, Shale-oil well with and without natural fractures, respectively. Subsequently, we emphasized the application of the workflow to investigate different production performances of three shale gas wells in Chapter 10, Investigation of different production performances in multiple shale-gas wells. Lastly, we concluded the book and recommended possible future works regarding the topic.

This book is intended for reservoir engineers and modeling engineers, who are responsible for modeling and simulation in unconventional reservoirs. For the readers new to the area of AHM and reservoir modeling in fractures and shale reservoirs, this book is not intended as a comprehensive theory book but instead provides the fundamental understanding of the topic with references and serves as an introductory book to this topic. For experienced researchers and engineers who have worked on this topic, this book can be used as a useful reference for actual applications. The approach can therefore be adopted to solve existing problems in the industry.

We would like to thank previous graduate students at The University of Texas at Austin who have worked on this topic including Marut Wantawin and Silpakorn Dachanuwattana. Without their solid work, this book would not be possible. We would like to extend our thanks to Chuxi Liu and Mauricio Fiallos who have helped us in various ways, direct and indirect, during the writing of this book. We also would like to thank Ms. Jessica Yeager for her careful editing of the manuscript.

Chapter 1

Introduction and literature review

1.1 Motivation

In shale reservoirs, the permeability of organic-rich mudrocks is very low, in the nanodarcy range, compared with the magnitude of 0.1–1000 md found in conventional reservoirs. This means oil and gas production from shale reservoirs using a traditional technique, such as is used in conventional reservoirs, is not achievable or economically viable. Fortunately, with the advancements of horizontal well drilling and multistage hydraulic fracturing in the recent decades, oil and gas production from shale reservoirs has been unlocked and has emerged to fulfill energy demand. Shale reservoir production has become one of the major contributors to oil and gas production worldwide.

Because the number of hydraulically fractured, shale reservoir wells has increased in the past decades, production data has become readily available at no additional cost and can be used to help characterize fracture geometry and reservoir properties. These fracture properties, such as effective fracture half-length, fracture height, and fracture conductivity, are not always identifiable by other methods. Although microseismic information is sometimes available, it does not provide effective fracture geometry or fracture conductivity, which we can obtain from production data. In addition, the matched model from history matching can be used further for production forecasting. However, history matching is an inverse problem, and the solutions are nonunique. Different sets of reservoir and fracture properties can mimic the same historical production data. In addition, traditionally, an engineer performs history matching manually, which is a tedious and time-consuming task. This labor-intensive task produces only one history matching solution. This causes concerns because a single-calibrated model is not capable of capturing reservoir and fracture uncertainties and history matching nonuniqueness. In addition, it is too risky to utilize for further production forecasts, estimation of reserve asset valuation, further fracturing design improvement, or well placement decisions. Assisted history matching (AHM) has been adopted to tackle these problems. The AHM tool developed in this book allows engineers to obtain multiple realizations of history matching in shale reservoirs by performing AHM and then performing

Assisted History Matching for Unconventional Reservoirs. DOI: https://doi.org/10.1016/B978-0-12-822242-3.00009-2
© 2021 Elsevier Inc. All rights reserved.

production forecasts probabilistically. The AHM tool features a workflow that is practical, efficient, and accurate.

1.2 Literature review

1.2.1 History matching algorithms

There are many AHM algorithms, but they can be grouped into three categories: optimization-based methods, ensemble-based methods, and Bayesian Markov chain Monte Carlo (MCMC) methods (Oliver and Chen, 2011).

Fig. 1.1 shows a comparison of each method based on accuracy and computational efficiency. Optimization-based algorithms are popular methods used to obtain multiple realizations due to their efficiency and relatively low computational efforts. Some algorithms rely on the gradient or derivatives of an objective function calculation, such as the gradient method (Ding, 2011) and the Gauss–Newton method (Gao et al., 2017; Chen et al., 2017). Others use a derivative-free method that includes particle swarm (Vazquez et al., 2015) and the evolutionary algorithm (Yin et al., 2011; Zhang and Fassihi, 2013; Nghiem et al., 2013; Xie et al., 2015). However, the optimization-based methods are not suitable for effectively quantifying uncertainty as a history matching problem. Moreover, the history matching solutions can be influenced by using different optimization algorithms (Goodwin, 2015). Erbas and Christie (2007) also demonstrated that different optimized algorithms (genetic algorithm and neighborhood algorithm) affect the history matching solution ensemble and production forecast.

On the contrary, the Bayesian inference methods such as the ensemble Kalman filter (EnKF) (Emerick and Reynolds, 2011) and the MCMC are specifically designed for uncertainty quantification due to Bayes' theorem. These Bayesian inference methods are based on using prior knowledge and the likelihood function to obtain a posterior distribution of history matching

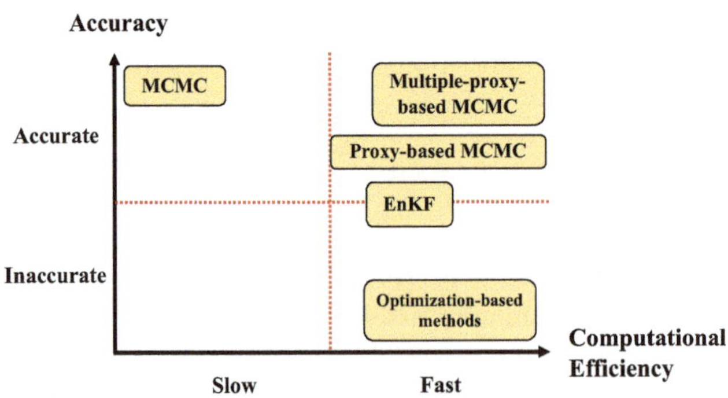

FIGURE 1.1 Cross plot comparing history matching algorithms in terms of accuracy and computational efficiency (Tripoppoom et al., 2020b).

solutions, including parameter values and their uncertainty. Ensemble-based methods such as EnKF (Emerick and Reynolds, 2011) and ensemble smoother (Chai et al., 2018; Chang and Zhang, 2018) are another category of AHM methods, which use covariance matrices to update an ensemble of parameters. Nevertheless, in most cases, history matching is a highly nonlinear problem, which is challenging for ensemble-based methods' assumption of a linear relationship between input and output parameters (Emerick and Reynolds, 2012; Goodwin, 2015). In addition, there is concern over whether EnKF can handle highly nonlinear cases because of its assumption of univariate Gaussian distribution.

The Bayesian inference method, MCMC, includes several types, such as Gibbs, Metropolis–Hasting, and Hamiltonian, which are used for history matching. In particular, these are used to capture history matching uncertainties. The advantage of MCMC is that it can quantify uncertainty without bias or being trapped in any local minima. However, due to the fact that MCMC is a sampling algorithm normally requiring more than 10^5-10^6 steps to reach a stable final distribution, it is limited by the infeasible computation effort traditional MCMC requires, as shown in Fig. 1.1. Therefore many studies have focused on overcoming this issue. Maucec et al. (2007) used a semianalytical streamlined model together with MCMC for history matching. Elsakout et al. (2015) proposed using a multilevel MCMC where a coarser grid is used before a finer grid with MCMC.

One prominent solution is to use a surrogate or proxy model to minimize the expensive reservoir simulation runs of MCMC steps (Slotte and Smorgrav, 2008; Maschio and Schiozer, 2014; Goodwin, 2015; Yeh et al., 2016). These methods are the proxy-based MCMC and multiple-proxy-based MCMC. The multiple-proxy-based MCMC method has more advantages over the proxy-based MCMC in terms of proxy model predictability. This is because using multiple proxy models can lessen the nonlinearity of a history matching problem instead of using a single proxy model for objective function values.

Various proxies are used together with MCMC in the AHM algorithm. Wantawin et al. (2017a,b) used a polynomial proxy with MCMC to perform history matching in shale gas wells. Yu et al. (2018a) and Tripoppoom et al. (2019) used the k-nearest neighbors (KNN) proxy with MCMC for history matching and probabilistic production forecast of shale gas wells.

Dachanuwattana et al. (2019) also compared different proxies, including the polynomial, KNN, and Kriging proxies, and found that Kriging is a relatively more predictive proxy than the rest, although it also requires more computational effort. They also found that KNN could be a balanced choice between model predictability and computational effort. Machine learning methods, such as neural networks (NNs), can be beneficial when used as the proxy model. Some studies used NNs with other history matching applications such as NN with the optimization-based method (Cullick et al., 2006), NN with pattern recognition technology for permeability field history matching

(Shahkarami et al., 2015), and NN in the proxy-based acceptance-rejection algorithm (Yang et al., 2015). Due to the flexibility of the NN formulation and training process to mitigate overfitting issues, NN can fit any highly nonlinear function better than the KNN proxy. Tripoppoom et al. (2020a) compared the performance of KNN and NN on shale gas wells. They found that KNN has the benefit of less running time while NN performs better than KNN in terms of predictability and requiring fewer simulation runs.

To sum up, to achieve practical application, efficiency, and accurate history matching results, one of the most suitable algorithms for AHM should be within the category of Bayesian MCMC used with proxy models to solve the issue of infeasible numbers of simulation runs, as found in traditional MCMC.

1.2.2 Fracture modeling

Another aspect to be considered for AHM in shale reservoirs is fracture modeling for both hydraulic fractures and natural fractures.

Firstly, analytical models such as rate transient analysis (RTA) have been used to characterize shale reservoirs and forecast production with various assumptions (Guo et al., 1994; Raghavan et al., 1997; Brown et al., 2009; Stalgorova and Mattar, 2013). Luo et al. (2014) combined RTA with Monte Carlo analysis to evaluate uncertainty in matrix and fracture properties leading to uncertainty in production forecast. Even though RTA is a fast method for characterizing reservoirs and forecasting production, the limitation of RTA is that there are many simplified assumptions made about fluid type, hydraulic fracture pattern, and heterogeneity. These assumptions constrain the application of RTA in real unconventional fields, which are complex. Moinfar et al. (2016) demonstrated the limitation of RTA assumptions when applied to real fields and emphasized the benefits of using a numerical model over an analytical model.

Since RTA is limited, several studies on numerical models in unconventional reservoirs have been undertaken. There are several modeling techniques used to mimic fracture flow behavior as shown in Fig. 1.2. One traditional method is a continuum approach such as the dual-porosity dual-permeability model (DPDK). However, the continuum approach has a limitation in modeling nonuniform and irregularly spaced fractures due to its assumption of uniformly distributed fractures. Rubin (2010) also found that the continuum approach cannot accurately capture the fracture–matrix connection in very low permeability reservoirs. This has led to the advent of the discrete fracture model (DFM). DFM using local grid refinement (LGR) (Cipolla et al., 2009) has been used to accurately model fractures; however, one major issue found with it is that it requires tremendous computational effort, especially when the number of fractures is high, and fracture systems become more complex. Other methods such as the logarithmically spaced, locally refined, and dual permeability grid (LS-LR-DK) by Rubin (2010) also attempt to improve model accuracy, but these are still tied to the concept of LGR.

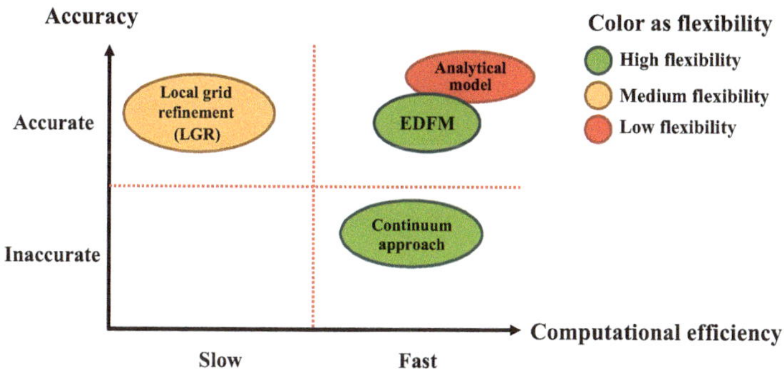

FIGURE 1.2 Comparison of different fracture modeling methods in terms of accuracy, computational efficiency, and flexibility (Tripoppoom et al., 2020b).

Therefore, a general three-dimensional (3D) embedded discrete fracture model (EDFM) has been developed to combine benefits from the continuum approach and LGR (Moinfar et al., 2014; Cavalcante Filho et al., 2015a; Shakiba and Sepehrnoori, 2015; Xu et al., 2017a,b; Yu et al., 2018b,d; Yu and Sepehrnoori, 2018; Sepehrnoori et al., 2020). The EDFM can model fracture flow behavior with an acceptable accuracy, on par with LGR, but with much lower computational effort (Moinfar et al., 2014; Xu et al., 2017a,b, 2018; Yu et al., 2018b). Many applications of EDFM have been implemented due to its computational efficiency, such as interwell fracture interference (Fiallos Torres et al., 2019a,b) and CO_2 Huff-n-Puff (Yu et al., 2019b). Cavalcante Filho et al. (2015b) used EDFM and the simplex optimization algorithm for AHM in fishbone wells. Eltahan et al. (2019) coupled EDFM with a commercial AHM tool to perform history matching in a shale oil well. Dachanuwattana et al. (2018a,b) used a proxy-based MCMC algorithm with EDFM in a shale oil and shale gas condensate well. Yu et al. (2018a) and Tripoppoom et al. (2019) used a multiple-proxy-based MCMC algorithm with EDFM for a shale gas reservoir.

In conclusion, as illustrated in Fig. 1.2, the analytical model provides the best accuracy and computational efficiency due to its exact solution nature; however, its lack of flexibility is a problem. Each analytical model is case-specific. Moreover, for some more complex cases, such as a complex fracture network or oil–gas–water multiphase flow, the availability and accuracy of the analytical model are limited. For a continuum approach such as the DPDK, the application is normally focused on uniform fractures such as natural fractures. However, it has a limitation when modeling nonuniform and irregularly spaced fractures accurately. It is not possible to interpret hydraulic fracture geometry from this method. For the LGR method, accurate results are produced due to grid refinement around fractures. Nevertheless, the LGR method requires relatively more computing time than other methods. This requirement will be more severe when more complex fractures or a greater number of fractures are involved. The EDFM method combines the benefits of the LGR method and the continuum approach. The EDFM concept is to embed fractures into matrix gridblocks and create two sets of grids, one for the matrix and one for the fractures. Then, each gridblock communicates through nonneighboring connection keywords and assumes a linear pressure drop within the gridblock. With this concept, the EDFM provides accuracy as acceptable as that of the LGR method but with a faster computing time than LGR. Even though the EDFM running time is not as fast as the analytical model, it has an advantage over the analytical model because of its flexibility to be applied to any case. A detailed EDFM calculation and its validation were discussed in Xu et al. (2017a, 2018, and 2019). Therefore, in this book, we used EDFM to model fractures during history matching.

1.2.3 Other challenges in unconventional reservoirs

To perform AHM in unconventional reservoirs, such as shale reservoirs, there are many challenges in addition to those found in conventional reservoirs. Those challenges include complicated fluid transport in nanopores, reservoir heterogeneity, the existence of natural fractures, and fracture modeling assumptions.

Firstly, complex fluid transportation in nanopores is illustrated by Javadpour et al. (2007, 2009) and Shabro et al. (2011, 2012). These are complicated mechanisms where traditional Darcy's law cannot be directly applied, such as slip flow, Knudsen diffusion, gas desorption, and gas diffusion in kerogen. Secondly, in shale reservoirs, very high heterogeneity in reservoir properties is normally observed, as demonstrated by Aderibigbe et al. (2016) and Amin et al. (2016). They performed a multimineral analysis using well-logging information, and the results indicated high heterogeneity in a vertical direction. Different rock qualities were observed, which is an important criterion in selecting production targets. Also, Yu et al. (2018b) demonstrated the effect of gas desorption, gas slippage and diffusion, and non-Darcy flow on the production forecast in shale gas reservoirs.

Another challenge in unconventional reservoirs is the existence of natural fractures in shale reservoirs. Natural fractures not only affect fracture propagation during hydraulic fracturing (Taleghani and Olson, 2013; Wu and Olson, 2016; Agrawal et al., 2019) but also have an effect on production performance in the long term. Yu et al. (2018c) investigated the effect of the existence of natural fractures as modeled by EDFM and found that the gas recovery could increase by 23.2%, in the case study, when natural fractures existed. Moreover, Yu et al. (2019a) showed the impact of natural fractures on the long-term production forecast of hydraulically fractured wells and well spacing decisions. Many studies have attempted to characterize natural fractures using information from core data, image logs, seismic data, and outcroppings. Shrivastava et al. (2018) used core data to characterize natural fractures and calibrate their fracture propagation model. However, information about natural fractures is challenging to obtain and often is not available. On the other hand, production data, which is always available, can be used to help reduce uncertainty regarding natural fractures via AHM. By using EDFM for fracture modeling, a higher computational efficiency when modeling natural fractures discretely can be achieved during AHM.

Lastly, to model fractures in conventional reservoirs, we model the hydraulic fractures with the bi-wing planar type fracture, which is an acceptable assumption. In shale reservoirs, however, there are cases where the bi-wing planar type of fracture cannot represent the actual hydraulic fracture propagation because the hydraulic fractures can interact with preexisting natural fractures and will induce a complex fracture network rather than bi-wing planar fractures. Fig. 1.3 (Warpinski et al., 2008) shows different fracture complexity from simple to extremely complex. Numerous results from microseismic data also support the complex-induced fracture network in shale reservoirs, as illustrated in Fig. 1.4 (Fisher et al., 2004).

To obtain complex fracture networks, a geomechanical model considering reservoir properties, fracturing operation, and interaction with preexisting natural fractures is utilized to predict the fracture propagation (Cipolla et al., 2012; Taleghani and Olson, 2013; Wu and Olson, 2015; Shrivastava et al., 2018). However, it is challenging to obtain accurate and reliable complex fracture geometries when generated from this fracture propagation model. This is because there are high levels of uncertainty and many assumptions in the model; input information is sometimes limited. Therefore, many studies

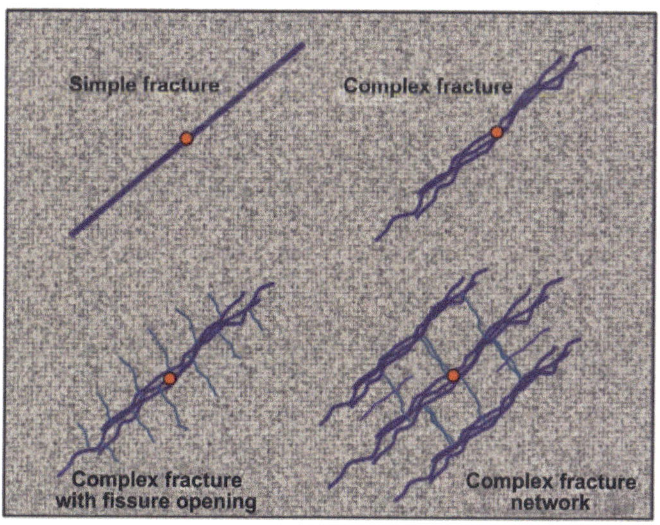

FIGURE 1.3 Different types of fracture complexity from simple fracture to complex fracture network (Warpinski et al., 2008).

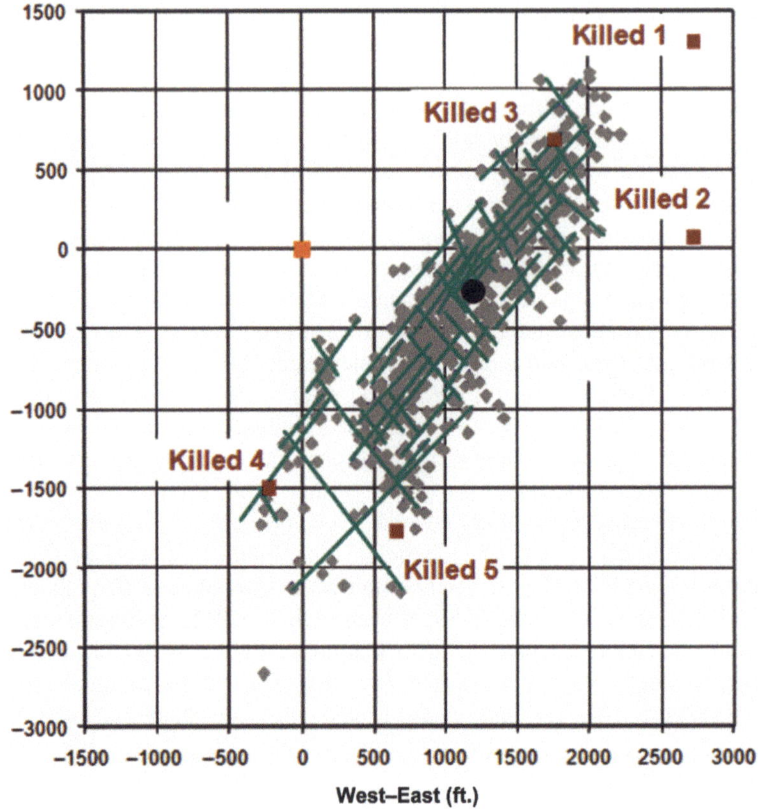

FIGURE 1.4 Microseismic events map in a vertical well, showing complex fracture network in Barnett (Fisher et al., 2004).

attempt to compromise by using the simpler concept of stimulated rock volume (SRV) rather than the fully complex fractures generated from the geomechanical model. The SRV is defined as the stimulated reservoir volume generated by hydraulic fractures and their interaction with preexisting fractures (Mayerhofer et al., 2010).

Du et al. (2010) adopted the dual-porosity model to represent fracture networks in shale gas wells. Chang and Zhang (2018) used dual permeability and dual porosity to model SRV together with the explicit hydraulic fractures for history matching in shale gas reservoirs. Rather than use the continuum approach alone, Rubin (2010) modeled the SRV area by -integrating the LGR concept and the dual permeability model, known as the LS-LR-DK. Yin et al. (2011) used the method of enhanced permeability area together with explicit hydraulic fractures to represent the SRV region. To estimate SRV, fracture extent, and orientation, several studies utilized the outputs from microseismic interpretation. Xie et al. (2015) utilized microseismic information and wire-mesh DFMs to represent SRV. They also integrated the analytical solution of calculating drainage volume by the fast marching method with a genetic algorithm to perform history matching of production data. Shakiba and Sepehrnoori (2015) used microseismic information to constrain the SRV modeling by using EDFM, which provides more running computational efficiency than traditional LGR.

Other studies also utilized microseismic interpretation to calibrate geomechanical models. Shaffner and Cheng (2011) used microseismic temporal and spatial events to confirm the completion diversion during each fracturing stage. They also used microseismic events mapping for fracture geometry over time by calibrating the fracture mechanical model, including aspects such as fracture half-length, fracture height, and azimuth. Xu et al. (2009) integrated microseismic interpretation with a semianalytical pseudo 3D-geomechanical model. They considered the mass conservation of injected fluid and the mechanic behavior interaction between fluid and fractures. Spatial and temporal microseismic events were utilized to calibrate the fracture propagation, fluid pressure, and fracture width in the fracture model.

Nevertheless, microseismic information is not always available for every well. Chang and Zhang (2018) show that history matching of SRV without microseismic interpretation is a challenging task. They investigated the nonuniqueness of SRV shape and volume by performing production history matching. This highlights the benefits of having microseismic interpretation as an initial constraint of fracture geometry and utilizing it for fracture model calibration.

One limitation of microseismic interpretation is that it cannot provide the effective fractures that contribute to production. In other words, the fracture conductivity and propped/unpropped fracture area cannot be extracted from microseismic data (Maxwell et al., 2013). This causes the fracture geometry, after calibrating with the actual production history or history matching, to be consistently less than the fracture geometry from microseismic data (Sharma and Gadde, 2004). Cipolla and Wallace (2014) also emphasized that SRV provided from microseismic data may not be a good indication of the effectiveness of fractures. They also supported using fracture modeling to estimate the propped and unpropped fracture area by performing production history matching. Mayerhofer et al. (2010) also used reservoir and fracture models to evaluate the effective fractures by varying fracture spacing for many cases with identical SRV from microseismic interpretation. Shakiba and Sepehrnoori (2015) demonstrated that microseismic interpretation could not capture the small fissures due to microseismic resolution because microseismic interpretation can only capture 0.1% of the fracturing events, most of which are aseismic processes.

In summation, the main limitation of using microseismic interpretation for fracture modeling is that microseismic interpretation cannot give any further information about the effective fracture volume. Simulation results will not be accurate if engineers only calibrate the fracture mechanical model or use fracture modeling techniques such as DPDK or discrete fracture geometry from microseismic interpretation alone. Also, not every well has microseismic information. Therefore, to determine the effective SRV, or effective fracture geometry, the need to calibrate with production historical data, which is always available at no additional cost, is crucial for later production forecasting, fracturing design, and future well spacing placement.

1.3 Assisted history matching in unconventional reservoirs

To conclude, by considering the advantages and disadvantages of each method when it comes to history matching algorithms, fracture modeling, and challenges in unconventional reservoirs, the use of a multiple-proxy-based MCMC algorithm with EDFM and numerical reservoir simulator could be a practical, efficient, and accurate AHM workflow for unconventional reservoirs. This could achieve the goals of obtaining multiple history matching solutions and performing production forecasts probabilistically. Details of the proposed AHM workflow are discussed in Chapter 2, Methodology.

References

Aderibigbe, A., Valdes, C., Heidari, Z., 2016. Integrated rock classification in the Wolfcamp shale based on reservoir quality and anisotropic stress profile estimated from well logs. Interpretation 4 (2), 1–18.

Agrawal, S., Shrivastava, K., Sharma, M.M., 2019. Effect of shear slippage on the interaction of hydraulic fractures with natural fractures. Paper SPE 194361, presented at the SPE Hydraulic Fracturing Technology Conference and Exhibition, The Woodlands, Texas, 5–7 February.

Amin, S., Wehner, M., Heidari, Z., Tice, M., 2016. Rock classification in the Eagle Ford Shale through integration of petrophysical, geological, and geochemical characterization. Paper SPWLA-2016-KKK, presented at the SPWLA 57th Annual Logging Symposium, Reykjavik, Iceland, 25–29 June.

Brown, M.L., Ozkan, E., Raghavan, R.S., Kazemi, H., 2009. Practical solutions for pressure transient responses of fractured horizontal wells in unconventional reservoirs. Paper SPE 125043, presented at the SPE Annual Technical Conference and Exhibition, New Orleans, Louisiana, 4–7 October.

Cavalcante Filho, J.S.D.A., Shakiba, M., Moinfar, A., Sepehrnoori, K., 2015a. Implementation of a preprocessor for embedded discrete fracture modeling in an IMPEC compositional reservoir simulator. Paper SPE 173289, presented at the SPE Reservoir Simulation Symposium, Houston, Texas, 23–25 February.

Cavalcante Filho, J.S.D.A., Xu, Y., Sepehrnoori, K., 2015b. Modeling fishbones using the embedded discrete fracture model formulation: sensitivity analysis and history matching. Paper SPE 175124, presented at the SPE Annual Technical Conference and Exhibition, Houston, Texas, 28–30 September.

Chai, Z., Tang, H., He, Y., Killough, J., Wang, Y., 2018. Uncertainty quantification of the fracture network with a novel fracture reservoir forward model. Paper SPE 191395, presented at the SPE Annual Technical Conference and Exhibition, Dallas, Texas, 24–26 February.

Chang, H., Zhang, D., 2018. History matching of stimulated reservoir volume of shale-gas reservoirs using an iterative ensemble smoother. SPE Journal 23 (2), 346–366.

Chen, C., Gao, G., Li, R., Cao, R., Chen, T., Vink, J.C., et al., 2017. Integration of distributed Gauss-Newton with randomized maximum likelihood method for uncertainty quantification of reservoir performance. Paper SPE 182639, presented at the SPE Reservoir Simulation Conference, Montgomery, Texas, 20–22 February.

Cipolla, C., Wallace, J., 2014. Stimulated reservoir volume: a misapplied concept? Paper SPE 168596, presented at the SPE Hydraulic Fracturing Technology Conference, The Woodlands, Texas, 4–6 February.

Cipolla, C.L., Lolon, E., Erdle, J.C., Rubin, B., 2009. Reservoir modeling in shale-gas reservoirs. Paper SPE 125530, presented at the SPE Eastern Regional Meeting, Charleston, West Virginia, 23–25 September.

Cipolla, C.L., Maxwell, S.C., Mack, M.G., 2012. Engineering guide to the application of microseismic interpretations. Paper SPE 152165, presented at the SPE Hydraulic Fracturing Technology Conference, The Woodlands, Texas, 6–8 February.

Cullick, A.S., Johnson, W.D., Shi, G., 2006. Improved and more rapid history matching with a nonlinear proxy and global optimization. Paper SPE 101933, presented at SPE Annual Technical Conference and Exhibition, San Antonio, Texas, 24–27 September.

Dachanuwattana, S., Jin, J., Zuloaga-Molero, P., Li, X., Xu, Y., Sepehrnoori, K., et al., 2018a. Application of proxy-based MCMC and EDFM to history match a Vaca Muerta shale oil well. Fuel 220, 490–502.

Dachanuwattana, S., Xia, Z., Yu, W., Qu, L., Wang, P., Liu, W., et al., 2018b. Application of proxy-based MCMC and EDFM to history match a shale gas condensate well. Journal of Petroleum Science and Engineering 167, 486–497.

Dachanuwattana, S., Yu, W., Sepehrnoori, K., 2019. An efficient MCMC history matching workflow using fit-for-purpose proxies applied in unconventional oil reservoirs. Journal of Petroleum Science and Engineering 176, 381–395.

Ding, Y.D., 2011. Development of a data-partition technique for gradient-based optimization methods in history matching. SPE Journal 16 (3), 582–593.

Du, C.M., Zhang, X., Zhan, L., Gu, H., Hay, B., Tushingham, K., et al., 2010. Modeling hydraulic fracturing induced fracture networks in shale gas reservoirs as a dual porosity system. Paper SPE 132180, presented at the International Oil & Gas Conference and Exhibition, Beijing, China, 8–10 June.

Elsakout, D., Christie, M., Lord, G., 2015. Multilevel Markov chain Monte Carlo (MLMCMC) for uncertainty quantification. Paper SPE 175870, presented at the SPE North Africa Technical Conference and Exhibition, Cairo, Egypt, 14–16 September.

Eltahan, E., Yu, W., Sepehrnoori, K., Kerr, E., Miao, J., Ambrose, R., 2019. Modeling naturally and hydraulically fractured reservoirs with artificial intelligence and assisted history matching methods using physics-based simulators. Paper SPE 195269, presented at SPE Western Regional Meeting, San Jose, California, 23–26 April.

Emerick, A.A., Reynolds, A.C., 2011. History matching a field case using the ensemble Kalman filter with covariance localization. SPE Reservoir Evaluation and Engineering 14 (4), 423–432.

Emerick, A.A., Reynolds, A.C., 2012. Combining the ensemble Kalman filter with Markov-chain Monte Carlo for improved history matching and uncertainty characterization. SPE Journal 17 (2), 418–440.

Erbas, D., Christie, M.A., 2007. Effect of sampling strategies on prediction uncertainty estimation. Paper SPE 106229, presented at the SPE Reservoir Simulation Symposium, Houston, Texas, 26–28 February.

Fiallos Torres, M.X., Yu, W., Ganjdanesh, R., Kerr, E., Sepehrnoori, K., Miao, J., et al., 2019a. Modeling interwell fracture interference and huff-n-puff pressure containment in Eagle Ford using EDFM. Paper SPE 195240, presented at SPE Oklahoma City Oil and Gas Symposium, Oklahoma City, Oklahoma, 9–10 April.

Fiallos Torres, M.X., Yu, W., Ganjdanesh, R., Kerr, E., Sepehrnoori, K., Miao, J., et al., 2019b. Modeling interwell interference due to complex fracture hits in Eagle Ford using EDFM. Paper IPTC 19468, presented at International Petroleum Technology Conference, Beijing, China, 26–28 March.

Fisher, M.K., Heinze, J.R., Harris, C.D., Davidson, B.M., Wright, C.A., Dunn, K.P., 2004. Optimizing horizontal completion techniques in the Barnett Shale using microseismic fracture mapping. Paper SPE 90051 presented at 2004 SPE Annual Technical Conference and Exhibition, Houston, Texas, 26–29 September.

Gao, G., Jiang, H., van Hagen, P., Vink, J.C., Wells, T., 2017. A Gauss-Newton trust-region solver for large-scale history-matching problems. SPE Journal 22 (6), 1999–2011.

Goodwin, N., 2015. Bridging the gap between deterministic and probabilistic uncertainty quantification using advanced proxy-based methods. Paper SPE 173301, presented at SPE Reservoir Simulation Symposium, Houston, Texas, 23–25 February.

Guo, G., Evans, R.D., Chang, M.M., 1994. Pressure-transient behavior for a horizontal well intersecting multiple random discrete fractures. Paper SPE 28390, presented at SPE Annual Technical Conference and Exhibition, New Orleans, Louisiana, 25–28 September.

Javadpour, F., 2009. Nanopores and apparent permeability of gas flow in mudrocks (shales and siltstone). Journal of Canadian Petroleum Technology 48 (8), 16–21.

Javadpour, F., Fisher, D., Unsworth, M., 2007. Nanoscale gas flow in shale gas sediments. Journal of Canadian Petroleum Technology 46 (10), 55–61.

Luo, H., Mahiya, G., Pannett, S., Benham, P.H., 2014. The use of rate-transient-analysis modeling to quantify uncertainties in commingled tight gas production-forecasting and decline-analysis parameters in the Alberta Deep Basin. SPE Reservoir Evaluation and Engineering 17 (2), 209–219.

Maschio, C., Schiozer, D.J., 2014. Bayesian history matching using artificial neural network and Markov chain Monte Carlo. Journal of Petroleum Science and Engineering 123, 62–71.

Maucec, M., Douma, S.G., Hohl, D., Lequijt, J., Jimenez, E., Datta-Gupta, A., 2007. Streamline-based history matching and uncertainty—Markov-chain Monte Carlo study of an offshore turbidite oil field. Paper SPE 109943, presented at SPE Annual Technical Conference and Exhibition, Anaheim, California, 11–14 November.

Maxwell, S.C., Weng, X., Kresse, O., Rutledge, J., 2013. Modeling microseismic hydraulic fracturing deformation. Paper SPE 166312, presented at SPE Annual Technical Conference and Exhibition, New Orleans, Louisiana, 30 September–2 October.

Mayerhofer, M.J., Lolon, E., Warpinski, N.R., Cipolla, C.L., Walser, D.W., Rightmire, C.M., 2010. What is stimulated reservoir volume? SPE Production & Operations 25 (1), 89–98.

Moinfar, A., Erdle, J.C., Patel, K., 2016. Comparison of numerical vs analytical models for EUR calculation and optimization in unconventional reservoirs. Paper SPE 180974, presented at SPE Argentina Exploration and Production of Unconventional Resources Symposium, Buenos Aires, Argentina, 1–3 June.

Moinfar, A., Varavei, A., Sepehrnoori, K., Johns, R.T., 2014. Development of an efficient embedded discrete fracture model for 3D compositional reservoir simulation in fractured reservoirs. SPE Journal 19 (2), 289–303.

Nghiem, L., Mirzabozorg, A., Yang, C., Chen, Z., 2013. Differential evolution for assisted history matching process: SAGD case study. Paper SPE 165491, presented at SPE Heavy Oil Conference-Canada—Calgary, Alberta, Canada, 11–13 June.

Oliver, D.S., Chen, Y., 2011. Recent progress on reservoir history matching: a review. Computational Geosciences 15 (1), 185–221.

Raghavan, R., Chen, C.C., Agarwal, B., 1997. An analysis of horizontal wells intercepted by multiple fractures. Paper PETSOC-HWC-94-39, presented at SPE/CIM/CANMET International Conference on Recent Advances in Horizontal Well Applications, Calgary, Canada, 20–23 March.

Rubin, B., 2010. Accurate simulation of non-Darcy flow in stimulated fractured shale reservoirs. Society of Petroleum Engineers. Paper SPE 132093, presented at the SPE Western Regional Meeting, Anaheim, California, 27–29 May.

Sepehrnoori, K., Xu, Y., Yu, W., 2020. Embedded Discrete Fracture Modeling and Application in Reservoir Simulation. Elsevier.

Shabro, V., Torres-Verdin, C., Javadpour, F., 2011. Numerical simulation of shale-gas production: from pore-scale modeling of slip-flow, Knudsen diffusion, and Langmuir desorption to reservoir modeling of compressible fluid. Paper SPE 144355, presented at North American Unconventional Gas Conference and Exhibition, The Woodlands, Texas, 14–16 June.

Shabro, V., Torres-Verdin, C., Sepehrnoori, K., 2012. Forecasting gas production in organic shale with the combined numerical simulation of gas diffusion in kerogen, Langmuir desorption from kerogen surfaces, and advection in nanopores. Paper SPE 159250, presented at SPE Annual Technical Conference and Exhibition, San Antonio, Texas, 8–10 October.

Shaffner, J.T., Cheng, A., 2011. The advantage of incorporating microseismic data into fracture models. Paper CSUG/SPE 148780 presented at the Canadian Unconventional Resources Conference, Calgary, Alberta, Canada, 15–17 November.

Shahkarami, A., Mohaghegh, S.D., Hajizadeh, Y., 2015. Assisted history matching using pattern recognition technology. Paper SPE 173405, presented at SPE Digital Energy Conference and Exhibition, The Woodlands, Texas, 3–5 March.

Shakiba, M., Sepehrnoori, K., 2015. Using embedded discrete fracture model (EDFM) and microseismic monitoring data to characterize the complex hydraulic fracture networks. Paper SPE 175142, presented at the SPE Annual Technical Conference and Exhibition, Houston, Texas, 28–30 September.

Sharma, M.M., Gadde, P.B., 2004. Slick water and hybrid fracs in the Bossier: some lessons learnt. Paper SPE 89876, presented at the SPE Annual Technical Conference and Exhibition, Houston, Texas, 26-29 September.

Shrivastava, K., Hwang, J., Sharma, M., 2018. Formation of complex fracture networks in the Wolfcamp Shale: calibrating model predictions with core measurements from the hydraulic fracturing test site. Paper SPE 191630, presented at the SPE Annual Technical Conference and Exhibition, Dallas, Texas, 24–26 September.

Slotte, P.A., Smorgrav, E., 2008. Response surface methodology approach for history matching and uncertainty assessment of reservoir simulation models. Paper SPE 113390, presented at Europec/EAGE Conference and Exhibition, Rome, Italy, 9–12 June.

Stalgorova, K., Mattar, L., 2013. Analytical model for unconventional multifractured composite systems. SPE Reservoir Evaluation and Engineering 16 (3), 246–256.

Taleghani, A.D., Olson, J.E., 2013. How natural fractures could affect hydraulic-fracture geometry. SPE Journal 19 (1), 161–171.

Tripoppoom, S., Yu, W., Huang, H., Sepehrnoori, K., Song, W., Dachanuwattana, S., 2019. A practical and efficient iterative history matching workflow for shale gas well coupling multiple objective functions, multiple proxy-based MCMC and EDFM. Journal of Petroleum Science and Engineering 176, 594–611.

Tripoppoom, S., Ma, X., Yong, R., Wu, J., Yu, W., Sepehrnoori, K., et al., 2020a. Assisted history matching in shale gas well using multiple-proxy-based Markov chain Monte Carlo algorithm: the comparison of K-nearest neighbors and neural networks as proxy model. Fuel 262, 116563.

Tripoppoom, S., Xie, J., Yong, R., Wu, J., Yu, W., Sepehrnoori, K., et al., 2020b. Investigation of different production performances in shale gas wells using assisted history matching: hydraulic fractures and reservoir characterization from production data. Fuel 267, 117097.

Vazquez, O., Young, C., Demyanov, V., Arnold, D., Fisher, A., MacMillan, A., et al., 2015. Produced-water-chemistry history matching in the Janice field. SPE Reservoir Evaluation & Engineering 18 (4), 564–576.

Wantawin, M., Yu, W., Dachanuwattana, S., Sepehrnoori, K., 2017a. An iterative response-surface methodology by use of high-degree-polynomial proxy models for integrated history matching and probabilistic forecasting applied to shale-gas reservoirs. SPE Journal 22 (6), 2012–2031.

Wantawin, M., Yu, W., Sepehrnoori, K., 2017b. An iterative work flow for history matching by use of design of experiment, response-surface methodology, and Markov chain Monte Carlo algorithm applied to tight oil reservoirs. SPE Reservation Evaluation and Engineering 20 (3), 613–626.

Warpinski, N.R., Mayerhofer, M.J., Vincent, M.C., Cipolla, C.L., Lolon, E., 2008. Stimulating unconventional reservoirs: maximizing network growth while optimizing fracture conductivity. Paper SPE 114173 presented at 2008 SPE Unconventional Reservoirs Conference, Keystone, Colorado, 10–12 February.

Wu, K., Olson, J.E., 2015. Simultaneous multifracture treatments: fully coupled fluid flow and fracture mechanics for horizontal wells. SPE Journal 20 (2), 337–346.

Wu, K., Olson, J.E., 2016. Numerical investigation of complex hydraulic-fracture development in naturally fractured reservoirs. SPE Production & Operations 31 (4), 300–309.

Xie, J., Yang, C., Gupta, N., King, M.J., Datta-Gupta, A., 2015. Integration of shale-gas-production data and microseismic for fracture and reservoir properties with the fast marching method. SPE Journal 20 (2), 347–359.

Xu, W., Le Calvez, J.H., Thiercelin, M.J., 2009. Characterization of hydraulically-induced fracture network using treatment and microseismic data in a tight-gas formation: a geomechanical approach. Paper SPE 125237 presented at the SPE Tight Gas Completions Conference, San Antonio, Texas, 15–17 June.

Xu, Y., Cavalcante Filho, J.S.A., Yu, W., Sepehrnoori, K., 2017a. Discrete-fracture modeling of complex hydraulic-fracture geometries in reservoir simulators. SPE Reservoir Evaluation and Engineering 20 (2), 403–422.

Xu, Y., Yu, W., Sepehrnoori, K., 2017b. Modeling dynamic behaviors of complex fractures in conventional reservoir simulators. Paper URTeC 2670513, presented at the SPE/AAPG/SEG Unconventional Resources Technology Conference, Austin, Texas, 24–26 July.

Xu, Y., Yu, W., Li, N., Lolon, E., Sepehrnoori, K., 2018. Modeling well performance in Piceance Basin Niobrara Formation using embedded discrete fracture model. Paper URTeC 2901327, presented at the SPE/AAPG/SEG Unconventional Resources Technology Conference, Houston, Texas, 23–25 July.

Xu, Y., Yu, W., Sepehrnoori, K., 2019. Modeling dynamic behaviors of complex fractures in conventional reservoir simulators. SPE Reservation Evaluation and Engineering 22 (3), 1110–1130.

Yang, C., Nghiem, L., Erdle, J., Moinfar, A., Fedutenko, E., Li, H., et al., 2015. An efficient and practical workflow for probabilistic forecasting of brown fields constrained by historical data. Paper SPE 175122, presented at the SPE Annual Technical Conference and Exhibition, Houston, Texas, 28–30 September.

Yeh, T., Uvieghara, T., Jennings, J.W., Chen, C., Alpak, F.O., Tendo, F., 2016. A practical workflow for probabilistic history matching and forecast uncertainty quantification: application to a deepwater West Africa reservoir. Paper SPE 181639, presented at the SPE Annual Technical Conference and Exhibition, Dubai, UAE, 25–28 September.

Yin, J., Xie, J., Datta-Gupta, A., Hill, A.D., 2011. Improved characterization and performance assessment of shale gas wells by integrating stimulated reservoir volume and production data. Paper SPE 148969, presented at the SPE Eastern Regional Meeting, Columbus, Ohio, 17–19 August.

Yu, W., Sepehrnoori, K., 2018. Shale Gas and Tight Oil Reservoir Simulation. Gulf Professional Publishing, Elsevier.

Yu, W., Tripoppoom, S., Sepehrnoori, K., Miao, J., 2018a. An automatic history-matching workflow for unconventional reservoirs coupling MCMC and non-intrusive EDFM methods. Paper SPE-191473-MS, presented at the SPE Annual Technical Conference and Exhibition, Dallas, Texas, 24–26 September.

Yu, W., Wu, K., Liu, M., Sepehrnoori, K., Miao, J., 2018b. Production forecasting for shale gas reservoirs with nanopores and complex fracture geometries using an innovative non-intrusive edfm method. Paper SPE 191666, presented at SPE Annual Technical Conference and Exhibition, Dallas, Texas, 24–26 September.

Yu, W., Xu, Y., Liu, M., Wu, K., Sepehrnoori, K., 2018c. Simulation of shale gas transport and production with complex fractures using embedded discrete fracture model. AIChE Journal 64 (6), 2251–2264.

Yu, W., Xu, Y., Weijermars, R., Wu, K., Sepehrnoori, K., 2018d. A numerical model for simulating pressure response of well interference and well performance in tight oil reservoirs with complex-fracture geometries using the fast embedded-discrete-fracture-model method. SPE Reservoir Evaluation and Engineering 2 (21), 489–502.

Yu, W., Hu, X., Liu, M., Wang, W., 2019a. Investigation of the effect of natural fractures on multiple shale-gas well performance using non-intrusive EDFM technology. Energies 12 (5), 932.

Yu, W., Zhang, Y., Varavei, A., Sepehrnoori, K., Zhang, T., Wu, K., et al., 2019b. Compositional simulation of CO_2 huff'n'puff in Eagle Ford tight oil reservoirs with CO_2 molecular diffusion, nanopore confinement, and complex natural fractures. SPE Reservoir Evaluation and Engineering 22 (2), 492–508.

Zhang, Z., Fassihi, M.R., 2013. Uncertainty analysis and assisted history matching workflow in shale oil reservoirs. Paper URTeC 1581398, presented at the Unconventional Resources Technology Conference, Denver, Colorado, 12–14 August.

Chapter 2

Methodology

2.1 Assisted history-matching framework

The assisted history-matching (AHM) framework consists of four key components: embedded discrete fracture model (EDFM) preprocessor, reservoir simulator, proxy model, and proxy-based Markov chain Monte Carlo (MCMC) algorithm, as shown in Fig. 2.1. First, the fractures are modeled by the EDFM preprocessor to precalculate the nonneighboring connections (NNCs). Then, the outputs from the EDFM preprocessor are coupled to the reservoir simulator DATA file. Next, the proxy is constructed to be used for the history matching (HM) algorithm. The proxy-based MCMC algorithm is used as the HM algorithm in the AHM workflow. This workflow repeats until the proxy model converges or the maximum number of simulations is exceeded.

2.2 Embedded discrete fracture model

The EDFM method combines the benefits from both the continuum approach and the local grid refinement (LGR) method. The EDFM concept creates two sets of grids: matrix and fractures. An example with three matrix grid

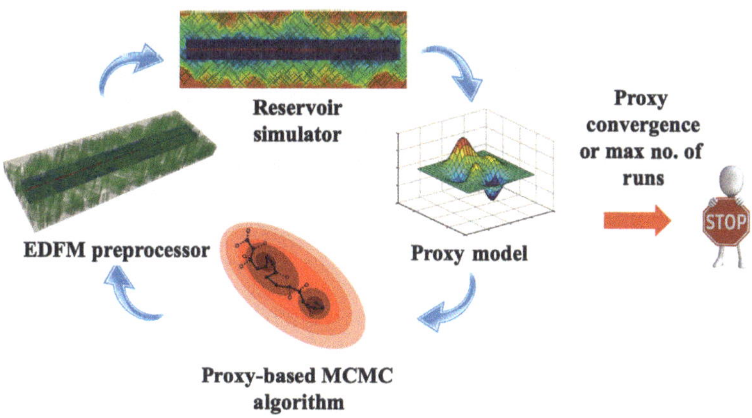

FIGURE 2.1 Integrated framework of AHM workflow (Tripoppoom, 2019). *AHM*, Assisted history matching.

Assisted History Matching for Unconventional Reservoirs. DOI: https://doi.org/10.1016/B978-0-12-822242-3.00007-9
© 2021 Elsevier Inc. All rights reserved.

blocks, two fractures, and wellbore is shown in Fig. 2.2. The fracture grid blocks are generated by the intersection between fracture and matrix grid blocks. Then, each grid block communicates through NNC keywords.

Because fractures are superimposed into the matrix grid and linear pressure gradient is assumed, there is no need to discretize a lot of small grid blocks, as in the traditional LGR. With this lower number of grid blocks, EDFM can provide accuracy as acceptable as LGR but with a lower computational time than LGR. This improvement becomes more evident when fracture networks become more complex (Xu et al., 2017a,b, 2018).

As illustrated by Xu et al. (2017a), EDFM creates additional cells to represent fractures. Each additional cell assigned to a fractured segment is

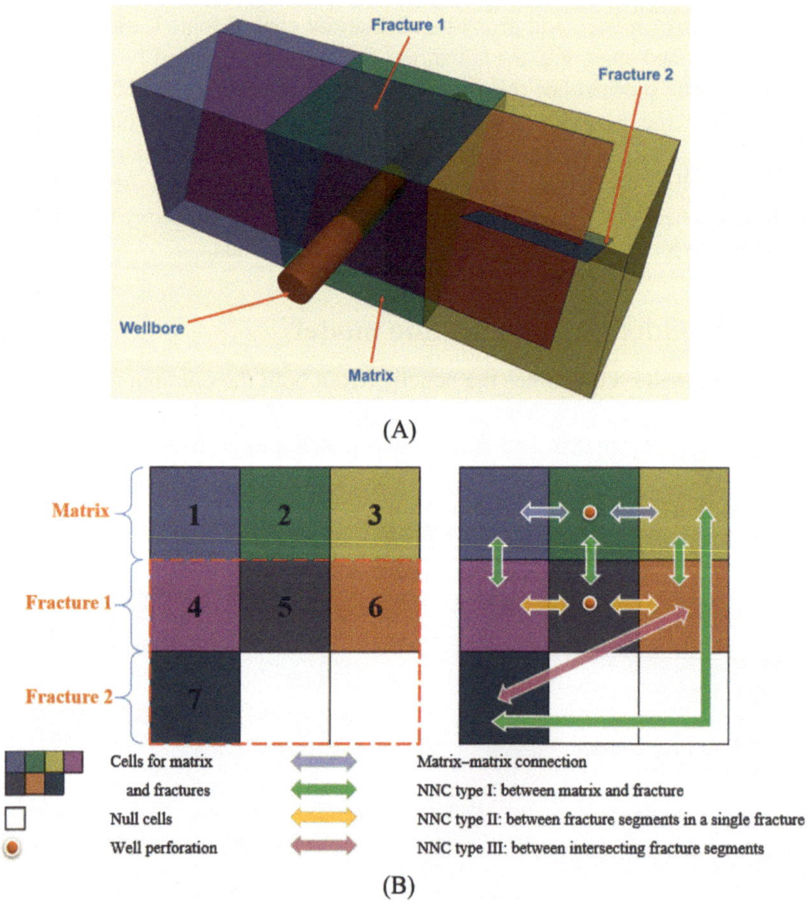

FIGURE 2.2 Concept of EDFM using a simple grid to model 3D complex fractures (Xu et al., 2017a): (A) physical domain and (B) computational domain. *EDFM*, Embedded discrete fracture model.

discretized by matrix cell boundaries. The effective porosity for each additional fracture cell, ϕ_f, is calculated by Eq. (2.1)

$$\phi_f = \frac{S_{\text{seg}} w_f}{V_b}, \tag{2.1}$$

where S_{seg} is the area of the fracture segment perpendicular to the fracture aperture, w_f the fracture width, and V_b the bulk volume of the cell assigned for the fracture segment.

Then, NNCs are defined to allow transport phenomena between cells. There are three types of NNCs: matrix-fracture, fracture-fracture in an individual fracture, and fracture-fracture of two different fractures and an effective well index of well-fracture connection, as shown in Fig. 2.3.

Xu et al. (2017a) explained the calculation of NNCs in detail, but in short, the flow between two cells is dictated by transmissibility factors.

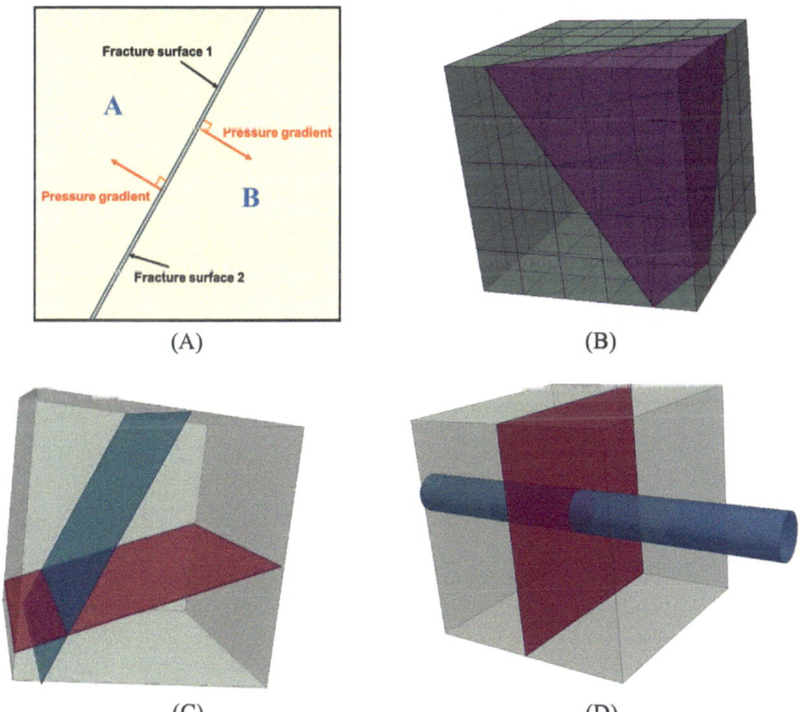

FIGURE 2.3 Three types of NNC and an effective well index in the EDFM method: (A) fluid flow between matrix grid blocks and fracture segments; (B) fluid flow between fracture segments within an individual fracture; (C) fluid flow between intersecting fracture segments within different fractures; and (D) fluid flow between fracture segments and wellbore (Xu et al., 2017a). *EDFM*, Embedded discrete fracture model; *NNC*, nonneighboring connection.

The volume flow rate of phase l between two cells in an NNC pair is calculated by Eq. (2.2)

$$q = \lambda T_{\mathrm{NNC}} \Delta P, \tag{2.2}$$

where λ_l is the relative mobility of phase l, T_{NNC} the NNC transmissibility factor, and ΔP the potential difference between two points.

The transmissibility calculation is different depending on the type of NNCs. For the matrix-fracture connection the NNC transmissibility depends on the matrix permeability and fracture geometry and is calculated by Eq. (2.3)

$$T_{\mathrm{NNC}} = \frac{k_{\mathrm{NNC}} A_{\mathrm{NNC}}}{d_{\mathrm{NNC}}}, \tag{2.3}$$

where k_{NNC} is the matrix permeability in the normal direction to the fracture plane, A_{NNC} the contact area of the fracture plane inside the matrix block, and d_{NNC} the average normal distance from matrix block to fracture plane. For the fracture-fracture connection the NNC transmissibility is calculated by using a two-point flux approximation scheme by Eq. (2.4)

$$T_{\mathrm{NNC}} = \frac{T_1 T_2}{T_1 + T_2}. \tag{2.4}$$

The detailed formulations to calculate each fracture segment transmissibility (T_1 and T_2) depend on whether the two fracture segments are in the same individual fracture discretized by the matrix cell boundary or if they are from two different fractures. For the well-fracture connection an effective well index is assigned to model each fracture segment that intersects the well trajectory. The well index formulation, WI_f in Eq. (2.5) and (2.6) (Moinfar et al., 2013), is derived from the formula proposed by Peaceman (1983):

$$\mathrm{WI}_f = \frac{2\pi k_f w_f}{\ln\left(\frac{r_e}{r_w}\right)}, \tag{2.5}$$

$$r_e = 0.14\sqrt{L_s^2 + H_s^2}, \tag{2.6}$$

where k_f is the fracture permeability, r_e the effective wellbore radius, r_w the wellbore radius, L_s the length of the fracture segment, and H_s the height of the fracture segment.

2.3 Reservoir simulator

For this study, after the outputs from the EDFM preprocessor are obtained, these outputs will be coupled with a reservoir simulator. Our developed code in the proposed AHM workflow will generate input to the EDFM preprocessor and output to edit the reservoir simulation data file automatically.

2.3.1 Multiphase flow governing equations

For the reservoir simulator the governing equation is the 3D multiphase flow black oil model. Starting with mass conservation, mass in minus mass out equals the net change in mass content. The continuity equation is shown in Eq. (2.7). Then, Darcy's law in Eq. (2.8) shows the relationship between the potential difference and velocity of phase *f*. Next, the flow equation is derived from the continuity equation, Darcy's law, and accounts for the well, as shown in Eqs. (2.9) and (2.10) (Aziz and Settari, 1979). In addition, we have auxiliary equations where capillary pressure relates to each phase pressure; the saturation of all phases must be equal to 1, as shown in Eqs. (2.11)–(2.13).

Continuity equation:

$$-\nabla\cdot(\rho_f \vec{v}_f) = \frac{\partial}{\partial t}(\phi s_f \rho_f), \tag{2.7}$$

Darcy's law:

$$v_{fs} = -\left(\frac{k_s k_{rf}}{\mu_f}\right)\frac{\partial \Phi}{\partial t}, \tag{2.8}$$

Flow equations:

$$\frac{\partial}{\partial x}\left(\frac{A_x k_x k_{rf}}{\mu B_f}\frac{\partial \Phi_f}{\partial x}\right)\Delta x + \frac{\partial}{\partial y}\left(\frac{A_y k_y k_{rf}}{\mu B_f}\frac{\partial \Phi_f}{\partial y}\right)\Delta y + \frac{\partial}{\partial z}\left(\frac{A_z k_z k_{rf}}{\mu B_f}\frac{\partial \Phi_f}{\partial z}\right)\Delta z + q_{fsc} = \frac{v_b}{5.615}\frac{\partial}{\partial t}\left(\frac{\phi s_f}{B_f}\right), \tag{2.9}$$

$$q_{fsc} = -\mathrm{PI}\left(\Phi_{\text{block}} - \Phi_{sf,\text{bottomhole}}\right), \tag{2.10}$$

where ρ_f is the density of phase *f*, $\vec{v}_f$ the velocity vector of phase *f*, ϕ the porosity, s_f the saturation of phase *f*, k_s the permeability, k_{rf} the relative permeability of phase *f*, μ_f the viscosity of phase *f*, Φ the potential equal to $P_f - \rho_f g$, *A* the area, B_f the formation volume factor of phase *f*, v_b the bulk volume, q_{fsc} the flow rate of phase *f* at standard condition, and PI the productivity index.

Auxiliary equations:

$$P_{\text{cow}}(S_w) = P_o - P_w, \tag{2.11}$$

$$P_{\text{cgo}}(S_g) = P_g - P_o, \tag{2.12}$$

$$S_o + S_w + S_g = 1, \tag{2.13}$$

where P_{cow} is the capillary pressure between oil and water and P_{cgo} the capillary pressure between gas and oil.

The boundary and initial conditions are as follows:

- Neumann boundary condition
  - No flow boundary: $\nabla P = 0$

- Initial condition
  - Initial pressure (P) = reservoir pressure
  - Other parameters as initial condition inputs are HM parameters or uncertain parameters

The simulation is constrained by either bottom-hole pressure (BHP) or main fluid flow rate (oil flow rate and gas flow rate).

2.4 Proxy model

A proxy model is a fast approximation of the original or actual model, which is normally more complicated and time-consuming to obtain the values for. Depending on the purposes, the term proxy model is interchangeably used with several other terms including surrogate model and response surface model, but the main objective of each is the same. In this study, two types of proxy models are utilized, including *k*-nearest neighbors (KNNs) and neural networks (NNs).

2.4.1 *k*-nearest neighbors

The concept behind KNN is to estimate the unmeasured point value by using information from measured neighboring points. To construct the KNN proxy model, each uncertain parameter is normalized to eliminate the effect of different value ranges. Then, the value of the unmeasured point is estimated by averaging the values of *k*-nearest measured points and giving more weight to the nearer points, as shown in Eq. (2.14). The inverse distance weighted averaging method is used, as shown in Eq. (2.15).

$$\hat{z}(\vec{\theta}_0) = \sum_{i=1}^{k} \lambda_{i0} z(\vec{\theta}_i), \tag{2.14}$$

where $\vec{\theta}_0$ is a vector of independent variables (uncertain parameters), $\hat{z}(\vec{\theta}_0)$ the estimated response (response parameters of unmeasured points), $\vec{\theta}_i$ a vector of the nearest k measured points, $z(\vec{\theta}_i)$ the response of measured points, and λ_{i0} the weight function between $\vec{\theta}_0$ and $\vec{\theta}_i$ defined as

$$\lambda_{i0} = \frac{\sum_{i=1}^{k} \left|\vec{\theta}_i - \vec{\theta}_0\right|}{\left|\vec{\theta}_i - \vec{\theta}_0\right|}. \tag{2.15}$$

As shown in Fig. 2.4, the predicted value is a function of two parameters, and we would like to predict the value at the central red point by having measured points around the red point available. However, the problem is how many neighboring points should be used. To answer this question, *k*-fold cross validation is performed, evaluating from $k = 1-1.5$ times the

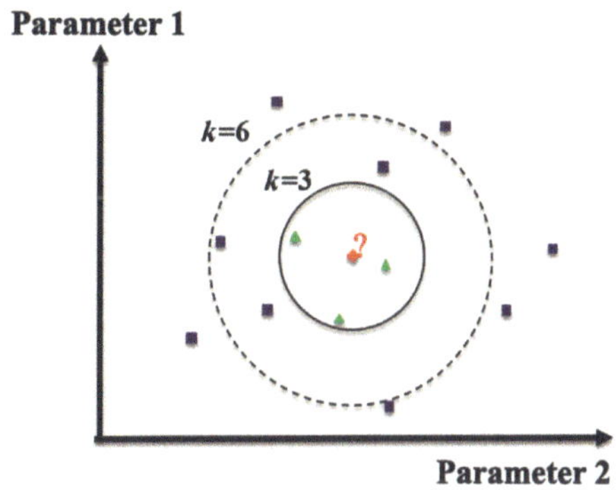

FIGURE 2.4 Concept of *k*-nearest neighbors (Tripoppoom, 2019).

square root of the number of measured points. For example, we have 50 measured points, so we evaluate *k* points from 1 to 10. Then, we generate fivefold from these 50 points. Each fold consists of different sets of 40 training points and 10 validating points. After that, we evaluate each *k* point value from the error between the estimated values from KNN and the actual values of the validating points. The error values from fivefold are averaged to represent the final error value of each *k* point. Then, we can select the nearest *k* points for each response-parameter proxy based on the one that gives the lowest root mean square error; in other words, the highest *R*-squared. As illustrated in Fig. 2.4, if *k* of 3 gives the lowest value, the red point values will be estimated from the nearest 3 points surrounding it, shown as green (triangular) points.

KNN has several benefits as a proxy model. First, KNN is an exact proxy. In other words, the value at the available measured point is the same as the actual value. This is different from some other proxy models such as the polynomial equation proxy model, where it is necessary to fit the polynomial equation coefficients with all available measured points. This could cause an overfitting issue. Another reason is that KNN requires low computing resources.

2.4.2 Neural networks

NNs are adopted from biological neurons in the human brain. As shown in Fig. 2.5, neurons are grouped into different layers including input, hidden, and output layers. Each neuron in the input layer represents different uncertain parameters for HM. And in the output layer, only one output neuron is used to represent the response-parameter value such as BHP or the water–gas ratio at a specific date in production history. In this study, we adopted multiple proxy points to lessen the nonlinearity of the problem. Therefore, NNs are trained to predict each response-parameter value or each selected proxy point. From the input to output layer, there are connections between each neuron. In this book, a fully connected neuron network architecture

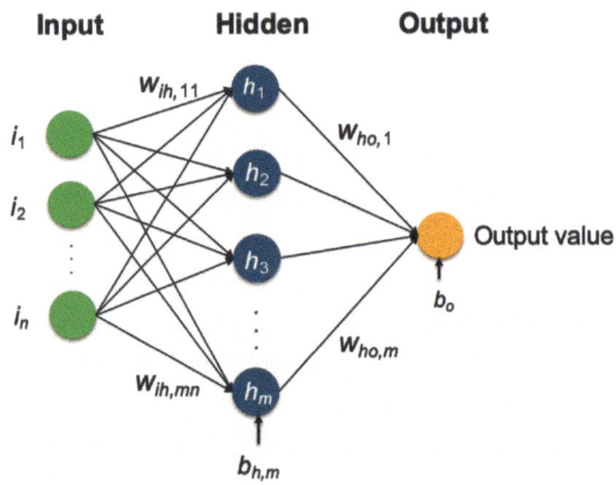

FIGURE 2.5 Structure of neural networks consisting of input, hidden, and output layers (Tripoppoom et al., 2020).

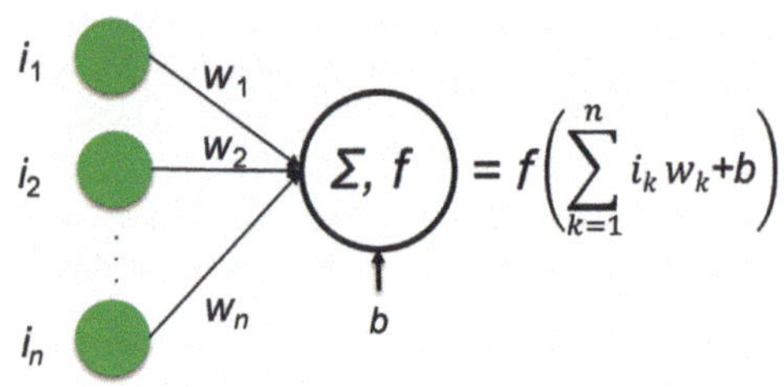

FIGURE 2.6 Components of simple input–output neuron networks including the input, weights, bias, and activation function and output (Tripoppoom et al., 2020).

with one hidden layer of 10 neurons is used. To calculate the output neuron value, Fig. 2.6 shows a simple input–output NN. The output value is calculated from the sum of input values, i_k, multiplied with the weight assigned to each connection, w_k, plus bias term, b. After that, an activation function, f, is evaluated to determine whether to activate the output neuron or not. When considering more complex NNs in Fig. 2.5, the hidden layer neurons, h_k, are first treated at output neurons calculated by Eqs. (2.16) and (2.17). Then, the hidden layer neurons are used as input to calculate the output value, as shown in Eq. (2.18). Where H is the $m \times 1$ matrix representing hidden layer neurons, W_{ih} the $m \times n$ matrix representing weights between input and hidden layer, I the $n \times 1$ matrix representing input layer neurons, B the $m \times 1$ bias term matrix, f the activation function, n the number of input layer neurons, and m the number of hidden layer neurons.

$$H = f(W_{ih}I + B), \tag{2.16}$$

$$\begin{bmatrix} h_1 \\ \cdots \\ h_k \\ h_m \end{bmatrix} = f\left(\begin{bmatrix} w_{ih,11} & \cdots & w_{ih,1j} & w_{ih,1n} \\ \cdots & \cdots & \cdots & \cdots \\ w_{ih,k1} & \cdots & w_{ih,kj} & w_{ih,kn} \\ w_{ih,m1} & \cdots & w_{ih,mj} & w_{ih,mn} \end{bmatrix}\begin{bmatrix} i_1 \\ \cdots \\ i_j \\ i_n \end{bmatrix} + \begin{bmatrix} b_{h,1} \\ \cdots \\ b_{h,k} \\ b_{h,m} \end{bmatrix}\right), \tag{2.17}$$

where h_k is the value of hidden layer neuron k, $w_{ih,kj}$ the weight between input layer neuron j and hidden layer neuron k, i_k the value of input layer neuron k, $b_{h,k}$ the bias term of hidden layer k, and f the activation function.

$$\text{Output value} = \begin{bmatrix} w_{ho,1} & \cdots & w_{ho,k} & w_{ho,m} \end{bmatrix}\begin{bmatrix} h_1 \\ \cdots \\ h_k \\ h_m \end{bmatrix} + b_o, \tag{2.18}$$

where $w_{ho,k}$ is the weight between hidden layer neuron k and output neuron and b_o the bias term for output neuron.

The process of training neuron networks starts with the step of preparing the data set. The data set is divided into three groups consisting of training, calibration, and blind-test data sets (Mohaghegh, 2017). First, the training data set is used to train NNs by minimizing the discrepancy between actual output values and predicted values from NNs. In this book, the Levenberg–Marquardt algorithm (Hagan and Menhaj, 1994), a supervised training method, is adopted for a back-propagation feedback algorithm to train the NNs by adjusting the weight values of each connection ($w_{ih,kj}$ and $w_{ho,k}$) and bias term ($b_{h,k}$ and b_o). Even though, Levenberg–Marquardt does not provide as exact a second-order convergence rate as Newton's method, it is similar to a quasi-Newtonian method where the algorithm maintains close to the second-order convergence rate without calculating the Hessian matrix. Next, the calibration data set is checked to ensure that the overfitting issue is mitigated. The error between the actual values and what the NNs predicted for the calibration data set is evaluated. The training of the NNs stops when this error in the calibration data set does not further reduce or starts to increase. Lastly, the blind data set has no effect during the training but is used as a tracker for users. In this book, the percentage of 70:15:15 is partitioned for training, calibration, and blind-test data sets, respectively.

2.5 Proxy-based Markov chain Monte Carlo algorithm

2.5.1 Markov chain Monte Carlo

The fundamental concept of all HM algorithms is to find the parameter domain that can provide low objective function value or HM solutions. Nevertheless, not all methods, such as optimization-based methods, are applicable for uncertainty quantification. MCMC, one of the Bayesian inference methods, is a suitable method to quantify uncertainty in HM. While a traditional MC is a random sampling algorithm, MCMC can be viewed as the MC algorithm with memory. This means the next sampling point is proposed around the current point and the decision of whether to move to the next point is based on the current condition. This is a reason why this method is called a Markov chain—the chain of sampling points is formed by the algorithm. In this book, Metropolis–Hastings (MH) MCMC is adopted. The MH MCMC, proposed by Metropolis et al. (1953) and Hastings (1970), is a random walk with probability. The concept is that if the next step gives a lower objective function value than the current state, there is a higher probability, close to one, that the algorithm will step to the next point rather than staying at the current state. On the other hand, if the next step gives a higher objective function value than the current state, there is a lower probability, closer to zero, that the algorithm will step to the next point and may prefer staying in the current state. In this manner, after sufficient step numbers (normally 10^5-10^6 steps), the MCMC algorithm will reach a stable distribution. In addition, with a proper setting of MCMC such as number of steps and a walking distance, this algorithm will not be trapped in any local minima and can explore the entire domain.

The formulation and detailed steps of MH MCMC are summarized in Fig. 2.7. First, the Markov chain is initialized at the realization that has the lowest objective function. Then, the proposed uncertain parameters realization, θ^*, for the next step of Markov chain is calculated from a random walking distance, not exceeding the specified maximum distance from the previous points as

$$\theta^* = \theta + \delta, \tag{2.19}$$

where θ is the uncertain parameter realization at the current step and δ the walking distance.

Then, the global error is estimated from the proxy. Normally, when the proxy is not used together with MCMC algorithm, the simulation run is needed to get the global error value of the proposed case. However, by using the proxy-based MCMC algorithm, the global error can be estimated from the proxy model. Next, the acceptance probability ratio, α^*, is calculated, defined by Eq. (2.20) with the assumption of Gaussian distribution.

$$\alpha^* = \min\left\{1, \frac{p(\theta^*|x)}{p(\theta|x)}\right\} = \min\left\{1, \exp\left[-\frac{\varepsilon^{*2} - \varepsilon^2}{2\sigma^2}\right]\right\} \tag{2.20}$$

where $p(\theta|x)$ is the posterior distribution of the uncertain parameters, θ, given the measured production history x. The initial value of variance σ^2 is calculated based on the variance of global error of the known samples (cases with actual simulation results). ε^{*2} and ε^2 are the global errors calculated from multiple response-parameter proxies of the proposed uncertain parameter realization and the current step, respectively.

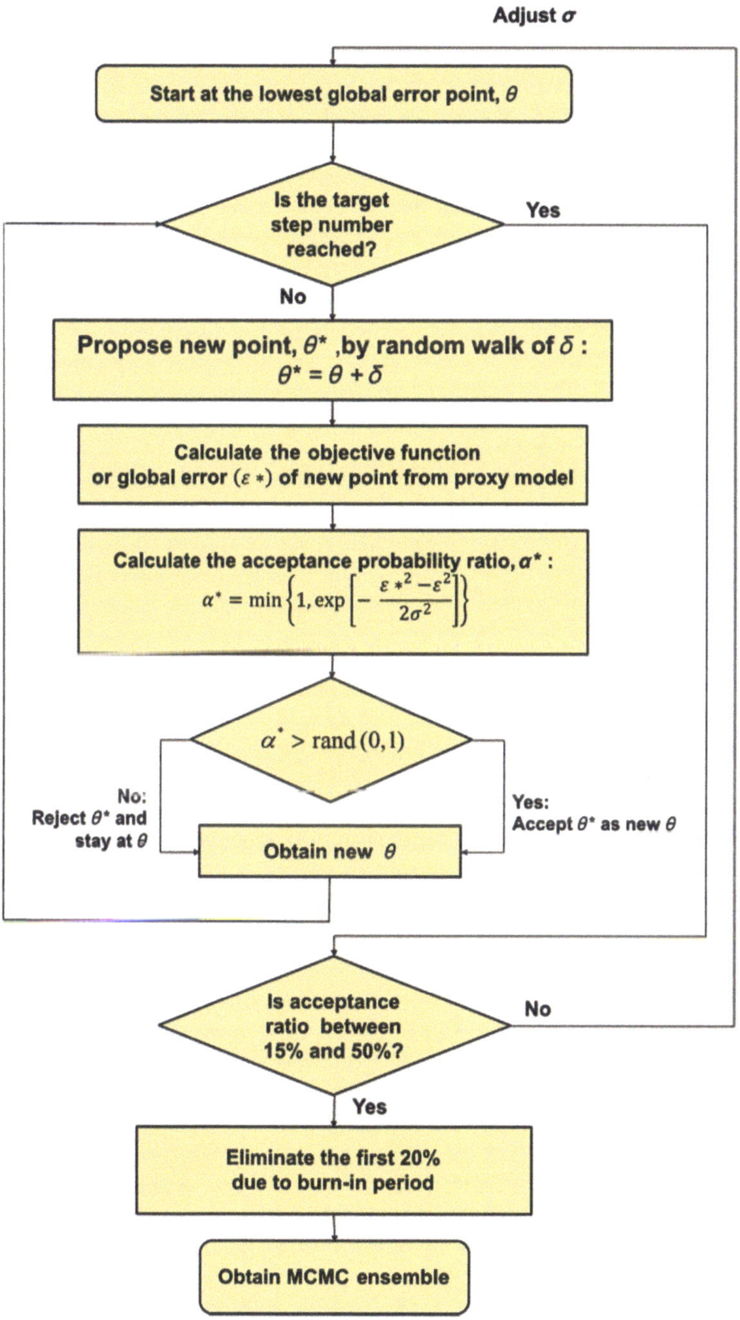

FIGURE 2.7 Steps and formulation in MH MCMC algorithm (Tripoppoom, 2019). *MCMC*, Markov chain Monte Carlo; *MH*, Metropolis–Hastings.

The MH algorithm will accept the proposed θ^* with probability of α^* and reject it with probability $1 - \alpha^*$. In other words, the Markov chain will move to the next point if α^* is higher than the random number between 0 and 1. Otherwise, the Markov chain will remain at the current point. The MH algorithm continues until the Markov chain is converged. Normally, a number of steps between 10^5 and 10^6 generate the converged Markov chain. In addition, the acceptance ratio should be in the range of 15%–50%. The variance, σ^2, will be adjusted automatically in the workflow until the acceptance ratio is within this range.

The initial 20% of the Markov chain should be removed due to a burn-in period before the chain converges to the target distribution (Slotte and Smorgrav, 2008). Lastly, the remaining MCMC ensemble is obtained for the later step of filtering for cases to be validated with the reservoir simulator.

2.5.2 Proxy-based MCMC algorithm and stopping criteria

One issue with traditional MCMC is that it requires simulation runs for each of the 10^5–10^6 steps, which is not feasible for a practical use, such as HM. Therefore the proxy model is utilized together with MCMC, as shown in Fig. 2.8. Either KNNs or NNs are constructed as the proxy model to minimize the number of actual reservoir simulation runs. While HM is a highly nonlinear problem, depending on one proxy model to predict objective function values may present a challenge in handling the nonlinearity of the problem. Therefore, multiple proxies are constructed to represent each production data point, such as gas rate and water–gas ratio at different selected dates. Then, these multiple points are calculated to obtain a global objective function value. There are some rules when constructing multiple proxies. First, the number of proxies should be higher than the number of input parameters. Second, constructing more proxy models should give more information to the global objective function calculation, such as a different production trend (Goodwin, 2015).

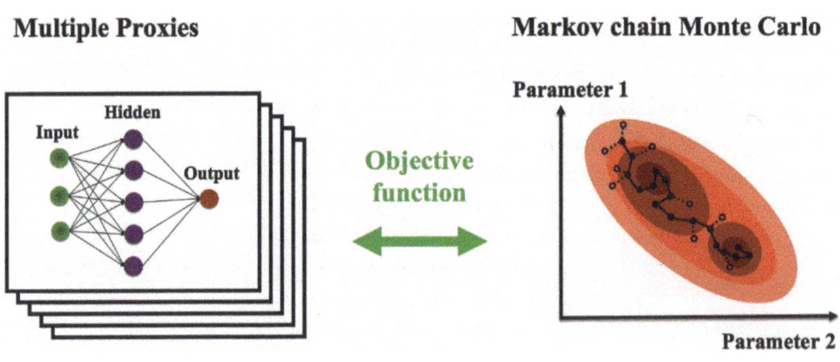

FIGURE 2.8 Proxy-based MCMC algorithm illustration. *MCMC*, Markov chain Monte Carlo.

Using a proxy-based MCMC algorithm, HM can be performed automatically in an iterative manner, and at the end of each iteration, new proposed cases are filtered from the MCMC ensemble obtained from the algorithm to be validated with reservoir simulator.

The first filtering scheme for the MCMC ensemble is the lowest 10% of the proxy-estimated global error criteria. The reason is that more points in a low global error area are needed to improve proxy quality (adding more measured points from actual simulation runs in the possible HM solution domain). More accuracy in the low global error domain is needed to better estimate the objection function value from proxy. The remaining ensembles are further filtered by two criteria. For example, if the final number of cases to be validated with the reservoir simulator is 25, then the first half of the set of 25 is filtered by criteria 1 of the lowest objective function value and the latter half is filtered by criteria 2 of the farthest points from all previously measured points. The reason we include two criteria is, first, to find the HM solutions based on proxy-estimated values (criteria 1) and, second, to improve proxy quality by scattering the points over the uncertain parameters domain at the same time (criteria 2). This means the workflow would find more possible HM solutions in other uncertain parameter domains. Finally, the candidate cases obtained at the end of each iteration are validated with the reservoir simulator.

Then, the new simulation results are included in the data set for further re-training of the proxy models. This algorithm continues iteratively until the stopping criteria is triggered. In this book, the maximum number of simulations and the proxy convergence are used. The reason that proxy convergence is used is because no additional resources should be utilized if the proxy model improvement is minimal. For the calculation of proxy convergence, we evaluated whether the current proxy negligibly changes from the previous proxy. 100,000 random uncertain parameter realizations are generated. Next, proxy-estimated global errors from both current and previous proxy models are calculated and compared. The percentage of relative discrepancies of global error between the two proxies (current and previous) of all 100,000 random cases is averaged and evaluated to determine whether it reaches the specified proxy convergence threshold. In this workflow, 5% relative discrepancy was used as the threshold.

2.6 Steps in assisted history-matching workflow

There are three main steps in the AHM workflow including (1) parameter identification and screening, (2) HM, and (3) probabilistic forecasting. The main steps of the AHM workflow are illustrated in Fig. 2.9. The significant parameters and objective function formulation should be determined at the end of the first step: parameter identification and screening. Then, at the

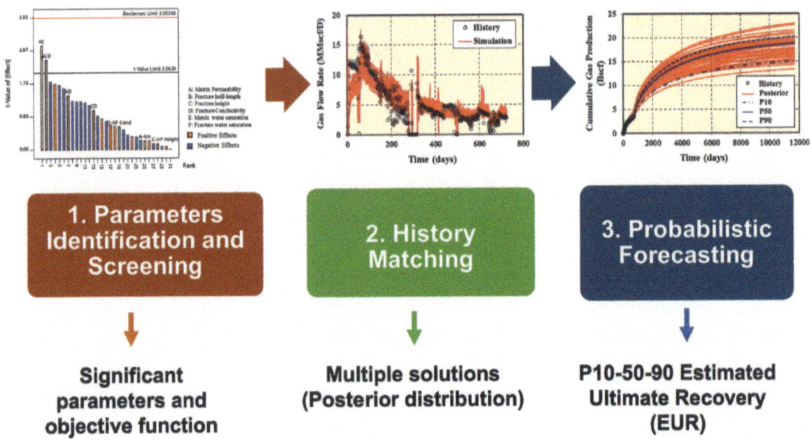

FIGURE 2.9 Main steps in AHM workflow (Tripoppoom, 2019). *AHM*, Assisted history matching.

end of the HM step, multiple solutions and posterior distribution of uncertain parameters can be obtained. Lastly, at the end of probabilistic forecasting, P10–P50–P90 estimated ultimate recovery (EUR) should be available based on production prediction. The detailed AHM workflow is also shown in Fig. 2.10.

2.6.1 Parameters identification and screening

There are two types of parameters: uncertain parameters and response parameters. First, uncertain parameters should be identified. These are subsurface parameters that have uncertainty and need to be characterized by dynamic data, for example, matrix permeability, fracture conductivity, and fracture geometry. Then, response parameters are determined, which are flowing BHP and historical production data used in HM, for example, oil, gas, and water flow rate.

2.6.1.1 Multiple objective functions

First, the objective function is defined. Multiple objective functions are used and defined either as Eq. (2.21) or Eq. (2.22).

$$\text{Objective function} = \text{global error}(\varepsilon) = \frac{\sum_{j=1}^{m}\sum_{i=1}^{n}\left|\frac{x_{ij,\text{model}} - x_{ij,\text{history}}}{x_{ij,\text{history}}} \times 100\right| \times w_{ij}}{\sum_{j=1}^{m}\sum_{i=1}^{n} w_{ij}}, \tag{2.21}$$

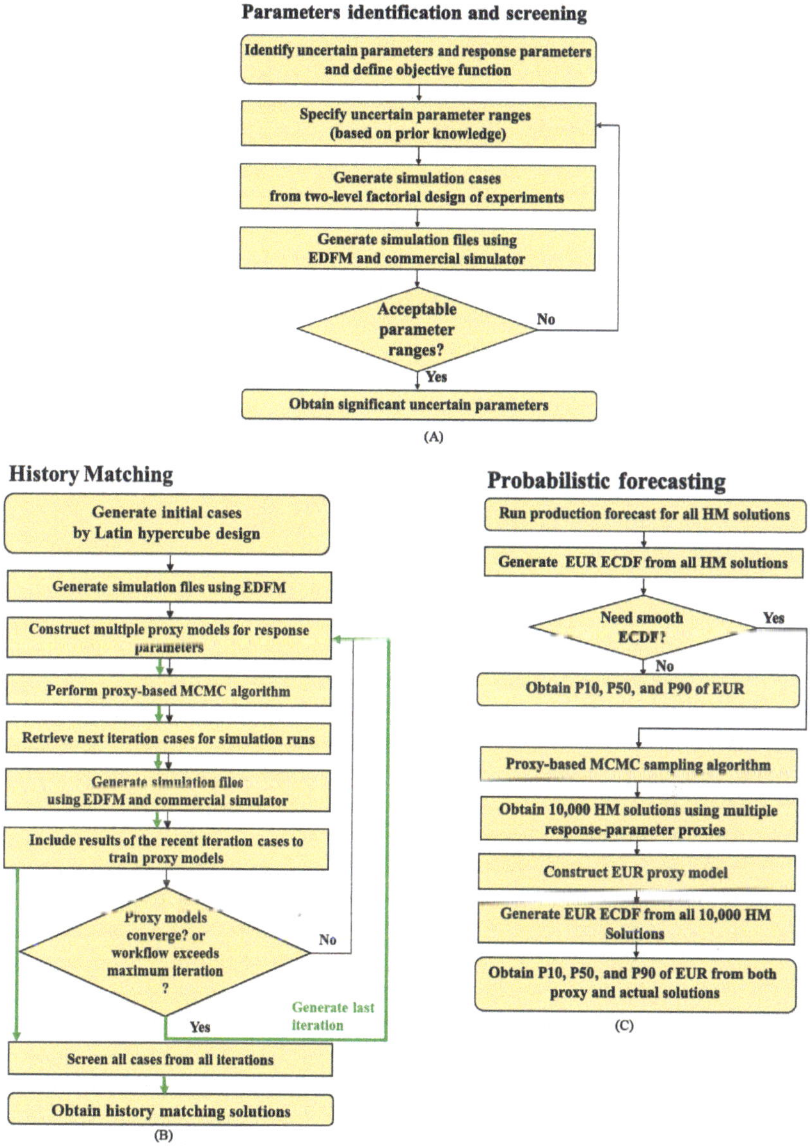

FIGURE 2.10 Steps in AHM workflow: (A) parameters identification and screening, (B) history matching, and (C) probabilistic forecasting (Tripoppoom, 2019). *AHM*, Assisted history matching.

where i is an index of data point, j an index of response parameter, n a number of data points, m a number of response parameters, $x_{ij,\text{model}}$ the value from model (simulation run or proxy model) at index i of response parameter j, $x_{ij,\text{history}}$ the historical data at index i of response parameter j, and w_{ij} the weight of data points i of response parameter j.

$$\text{Objective function} = \text{global error}(\varepsilon) = \frac{\sum_{j=1}^{m}\sum_{i=1}^{n}\left[\frac{x_{ij,\text{model}} - x_{ij,\text{history}}}{\text{Normalized Factor}_j} \times 100\right]^2 \times w_{ij}}{\sum_{j=1}^{m}\sum_{i=1}^{n} w_{ij}}, \tag{2.22}$$

where i is an index of data point, j an index of response parameter, n the number of data points, m the number of response parameters, $x_{ij,\text{model}}$ the value from the model (simulation run or proxy model) at index i of response parameter j, $x_{ij,\text{history}}$ the historical data at index i of response parameter j, Normalized Factor$_j$ defined as the maximum of data value minus the minimum of data value of response parameter j, and w_{ij} the weight of data points i of response parameter j.

The main difference between Eq. (2.21) and Eq. (2.22) is that the latter magnifies the high relative error case to have a higher error than those of the previous one due to its squared formula. This means if we would like to distinguish the low and high error dramatically, Eq. (2.22) should be used.

The concept of using global error as the objective function is to have multiple objective functions in one objective function, normalized by some factor to eliminate the different values and ranges of each response parameter. In addition, there is a weight factor to be applied to each response parameter and, further, to each data point as needed (Christie et al., 2013). In sum, the formulations can represent multiple objective functions in one value as the global error. Many response parameters such as oil flow rate and water cut at different production dates can be used.

After the objection function is defined, the effective ranges of each uncertain parameter are specified based on prior knowledge. Next, the two-level factorial design of experiments is performed to screen significant uncertain parameters for HM. After obtaining the response values of two-level factorial design, in this case global error, we evaluated whether any transformation was needed for the analysis by using Box–Cox plot (Box and Cox, 1964). Then, the significant uncertain parameters can be identified from the Pareto chart of t-value. Uncertain parameters with a higher t-value will have more effect on the objective function. Ones above the Bonferroni limit should be selected due to their certain importance to objective function. Ones above t-value limit are possibly important and will be dependent on whether the user decides to include them or not. The rest below t-value limit are insignificant parameters and should be selected only to support hierarchy.

In addition, a check is needed to determine whether the specified parameter ranges generate the simulated production results covering all the historical data range or not. If they fail, the parameter ranges have to be adjusted and the workflow in this step should be run again.

The benefit of having the parameter-screening step is that the number of simulation runs can be optimized in the HM step by eliminating the insignificant uncertain parameters.

2.6.2 History matching

After identifying the significant uncertain parameters, the initial simulation cases are generated by Latin hypercube design. The concept of Latin hypercube sampling (LHS) is to divide the domain into squares in cases of two parameters or cubes for three parameters. The same concept is also applicable to any higher dimension. Then, the LHS evenly samples points all over each square or cube as shown in Fig. 2.11, unlike MC sampling that will sample randomly in the domain. This ensures that Latin hypercube design generates a good spatial location of initial points for proxy modeling by scattering initial points all over the parameter domain (Goodwin, 2015; Yang et al., 2015) unlike MC or a two-level factorial design that evaluates at the maximum and minimum values only. There is no strict rule for the number of initial points because it depends on the complexity of the problem and number of uncertain parameters. Based on experience and empirical observation, we used a number of initial points around five times the number of uncertain parameters. However, the concept is that we would not like to waste too many simulation points in the initial iteration in a domain which has high-objective function values—the non-HM solution domain. More points should be added into the possible HM solution domain (Wantawin et al., 2017).

Then, the multiple response-parameter proxies will be constructed based on a training data set of the initial iteration cases. The proxy-based MCMC algorithm is performed automatically.

The algorithm continues iteratively until the stop criterion is triggered. The last iteration is generated but with a different filtering scheme to obtain the cases for the next iteration. The remaining MCMC ensemble will be

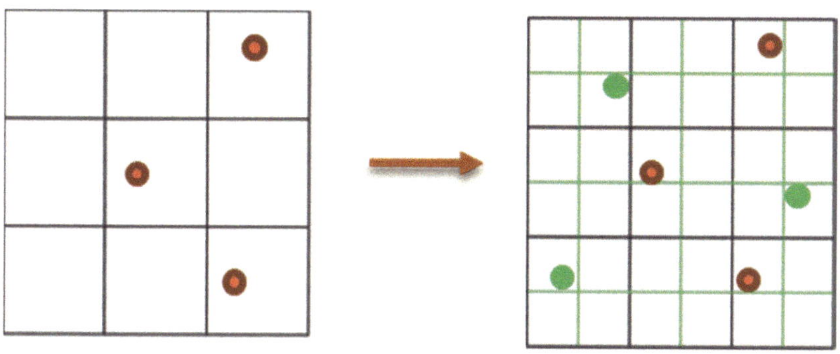

FIGURE 2.11 LHS concept (Tripoppoom, 2019). *LHS*, Latin hypercube sampling.

screened only with criteria 1 of the lowest objective function value for the last iteration. The reason is that the final proxy is only used to obtain another set of possible HM solutions and not to consider improving the proxy accuracy using criteria 2 of the farthest points from all previously measured points, as in the previous iterations. Finally, all simulation runs from all iterations are screened by HM tolerance to obtain HM solutions.

2.6.3 Probabilistic forecasting

After multiple HM solutions are retrieved from the HM step, production forecasts based on these solutions can be performed. Then, the forecasted EUR of all solutions are retrieved as the empirical cumulative distribution function (ECDF). The P10, P50, and P90 of EUR are quantified. In some cases the number of solutions is insufficient to create a smooth EUR ECDF. The MH MCMC algorithm can be used to generate 10,000 different solutions, as an example, from the latest multiple response-parameter proxies. This is to collect the realizations where the proxy-estimated global error is lower than the HM solution tolerance until 10,000 solutions are reached. Another proxy for EUR is created from the actual simulation EUR retrieved previously. The EUR of 10,000 solutions will be estimated from the EUR proxy. Finally, a smoother ECDF and P10–P50–P90 of EUR for proxy solutions can be obtained and compared to the ECDF from the actual solutions.

References

Aziz, K., Settari, A., 1979. Petroleum Reservoir Simulation. Elsevier Applied Science Publishers, London.

Box, G.E.P., Cox, D.R., 1964. An analysis of transformations. Journal of the Royal Statistical Society, Series B 26 (2), 211–243.

Christie, M., Eydinov, D., Demyanov, V., Talbot, J., Arnold, D., Shelkov, V., 2013. Use of multi-objective algorithms in history matching of a real field. In: Paper SPE 163580, Presented at SPE Reservoir Simulation Symposium, The Woodlands, TX, February 18–20, 2013.

Goodwin, N., 2015. Bridging the gap between deterministic and probabilistic uncertainty quantification using advanced proxy-based methods. In: Paper SPE 173301, Presented at SPE Reservoir Simulation Symposium, Houston, TX, February 23–25, 2015.

Hagan, M.T., Menhaj, M.B., 1994. Training feedforward networks with the Marquardt algorithm. IEEE Transactions on Neural Networks 5 (6), 989–993.

Hastings, W.K., 1970. Monte Carlo sampling methods using Markov chains and their applications. Biometricka 57 (1), 97–109.

Metropolis, N., Rosenbluth, A.M., Rosenbluth, M.N., Teller, A.H., 1953. Equation of state calculations by fast computer machines. The Journal of Chemical Physics 21 (6), 1087–1092.

Mohaghegh, S.D., 2017. Shale Analytics Data-Driven Analytics in Unconventional Resources. Springer International Publishing.

Moinfar, A., Varavei, A., Sepehrnoori, K., Johns, R.T., 2013. Development of a coupled dual continuum and discrete fracture model for the simulation of unconventional reservoirs.

In: Paper SPE 163647, Presented at SPE Reservoir Simulation Symposium, The Woodlands, TX, February 18–20, 2013.

Peaceman, D.W., 1983. Interpretation of well-block pressures in numerical reservoir simulation with nonsquare grid blocks and anisotropic permeability. SPE Journal 23 (3), 531–543.

Slotte, P.A., Smorgrav, E., 2008. Response surface methodology approach for history matching and uncertainty assessment of reservoir simulation models. In: Paper SPE 113390, Presented at EUROPEC/EAGE Conference and Exhibition, Rome, Italy, June 9–12, 2008.

Tripoppoom, S., 2019. Assisted History Matching Workflow for Unconventional Reservoirs (M.S. thesis). The University of Texas at Austin.

Tripoppoom, S., Ma, X., Yong, R., Wu, J., Yu, W., Sepehrnoori, K., et al., 2020. Assisted history matching in shale gas well using multiple-proxy-based Markov chain Monte Carlo algorithm: the comparison of *K*-nearest neighbors and neural networks as proxy model. Fuel 262, 116563.

Wantawin, M., Yu, W., Sepehrnoori, K., 2017. An iterative response-surface methodology by use of high-degree-polynomial proxy models for integrated history matching and probabilistic forecasting applied to shale-gas reservoirs. SPE Journal 22 (6), 2012–2031.

Xu, Y., Cavalcante Filho, J.S.A., Yu, W., Sepehrnoori, K., 2017a. Discrete-fracture modeling of complex hydraulic-fracture geometries in reservoir simulators. SPE Reservoir Evaluation & Engineering 20 (2), 403–422.

Xu, Y., Yu, W., Sepehrnoori, K., 2017b. Modeling dynamic behaviors of complex fractures in conventional reservoir simulators. In: Paper URTeC 2670513, Presented at SPE/AAPG/SEG Unconventional Resources Technology Conference, Austin, TX, July 24–26, 2017.

Xu, Y., Yu, W., Li, N., Lolon, E., Sepehrnoori, K., 2018. Modeling well performance in Piceance Basin Niobrara Formation using embedded discrete fracture model. In: Paper URTeC 2901327, Presented at SPE/AAPG/SEG Unconventional Resources Technology Conference, Houston, TX, July 23–25, 2018.

Yang, C., Nghiem, L., Erdle, J., Moinfar, A., Fedutenko, E., Li, H., et al., 2015. An efficient and practical workflow for probabilistic forecasting of brown fields constrained by historical data. In: Paper SPE 175122, Presented at the SPE Annual Technical Conference and Exhibition, Houston, TX, September 28–30, 2015.

Chapter 3

Validation of assisted history matching for a synthetic shale gas well

3.1 Introduction

As history matching is a nonunique solution inverse problem, in the actual application, we never know the exact true solution of the problem. Only posterior distributions of uncertain parameters with their P10–P50–P90 values are obtained. In this chapter, we aim to validate the proposed assisted history-matching (AHM) workflow and neural network Markov chain Monte Carlo (NN MCMC) algorithm. Therefore, we set up a synthetic shale gas model to validate the accuracy of AHM workflow. By knowing the true solution, we can validate the AHM workflow by comparing the posterior distributions of each parameter, that is, P50 values with true solution values.

A synthetic model or the true solution model was created using arbitrary but reasonable reservoir and fracture properties that we normally observe in shale gas reservoirs. The production history of this synthetic well was generated for 110 days including gas flow rate, water flow rate, and bottomhole pressure. A fluctuating gas production performance was generated to mimic a realistic production performance observed in an actual shale gas well.

In this study, we investigated two scenarios including cases with only hydraulic fractures and with both hydraulic and natural fractures. This is to demonstrate the validation of the AHM workflow for both scenarios with and without natural fractures.

3.2 Case 1: hydraulic fractures only

3.2.1 Reservoir model

We built a reservoir model with a model size (x, y, z) of $1800 \times 500 \times 20$ m. The grid cell dimension is $10 \times 10 \times 20$ m. A two-phase flow regime with gas and water was simulated. The initial reservoir pressure is 50 MPa and the reservoir depth is approximately 2700 m.

Assisted History Matching for Unconventional Reservoirs. DOI: https://doi.org/10.1016/B978-0-12-822242-3.00011-0
© 2021 Elsevier Inc. All rights reserved.

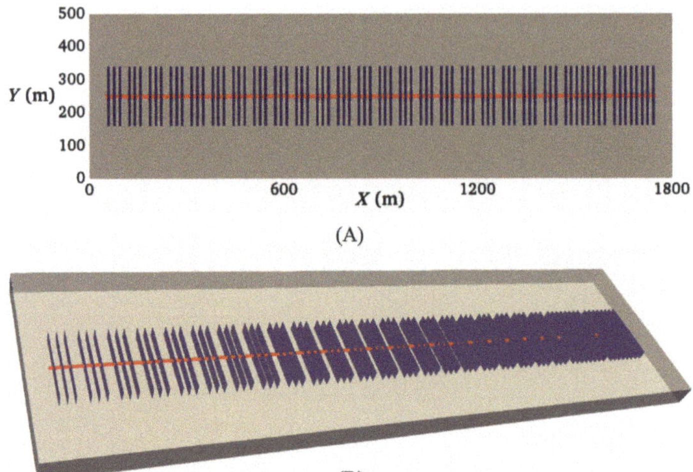

FIGURE 3.1 Synthetic reservoir model including 82 hydraulic fractures (blue) modeled by EDFM in a horizontal well (red): (A) top view; and (B) 3D view. *EDFM*, Embedded discrete fracture model.

As shown in Fig. 3.1, the well is a horizontal well with a lateral length of 1700 m and is completed with 24 stages of hydraulic fractures with a total cluster number of 82. We also considered the pressure dependency for fracture permeability, which is depicted in Fig. 3.2. Other key reservoir and fracture descriptions are summarized in Table 3.1.

We assumed hydraulic fractures are symmetrical around the wellbore. The model was also designed to capture a nonuniform cluster number for some of the toe side stages. This illustrated the workflow's ability to model different fracking designs.

3.2.2 History matching

The 110-day synthetic production history utilized for history matching and gas flow rate was used as a simulation constraint. Eight significant parameters, including matrix permeability, fracture height, fracture half-length, fracture water saturation, fracture conductivity, exponent of gas relative permeability, exponent of water relative permeability, and endpoint of water relative permeability, were selected to be uncertain parameters for history matching, as shown in Table 3.2.

Parameter values of a selected synthetic model or the so-called true values are summarized in Table 3.2. The true values of each parameter are specified based on the reasonable values from field experience in shale gas reservoirs. Then, the range of each uncertain parameter is specified as shown in Table 3.2. One can observe that we designed the true value model of each

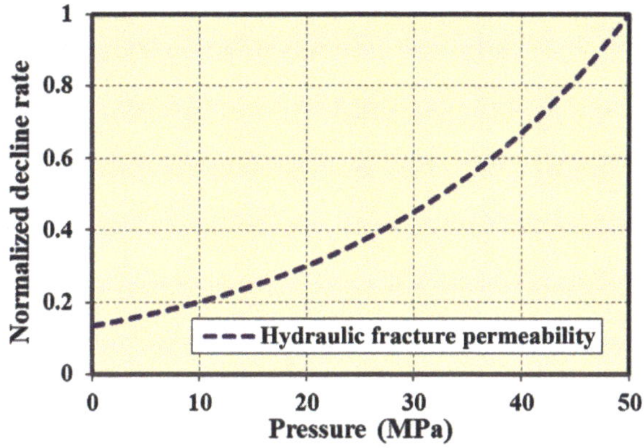

FIGURE 3.2 Pressure-dependent permeability curve for hydraulic fractures.

TABLE 3.1 Summary of basic reservoir properties used for validation of the automatic history-matching workflow.

Key parameters	Value	Unit
Model dimension ($x \times y \times z$)	$1800 \times 500 \times 20$	m
Number of grid blocks ($x \times y \times z$)	$180 \times 50 \times 1$	–
Grid cell dimension ($x \times y \times z$)	$10 \times 10 \times 20$	m
Reservoir top depth	2700	m
Initial reservoir pressure	50	MPa
Rock compressibility	4.35×10^{-7}	kPa^{-1}
Reservoir temperature	115	°C
Matrix porosity	6	%
Matrix water saturation	20	%
Residual water saturation	20	%
Residual gas saturation	10	%
Well length	1700	m
Number of stages	24	–
Stage spacing	30	m
Number of clusters	82	–
Cluster spacing	17	m
Langmuir pressure	7	MPa
Langmuir adsorption volume	3.5	m^3/t
Rock density	2.54	g/cm^3

TABLE 3.2 Summary of eight uncertain parameters for validation of the assisted history-matching workflow of the case with only hydraulic fractures.

Uncertain parameters	True value	Minimum	Maximum	Unit
Matrix permeability	0.0001	0.00001	0.0003	md
Fracture height	10	5	20	m
Fracture half-length	90	50	180	m
Fracture conductivity	50	20	70	md m
Fracture water saturation	70	60	90	%
Endpoint of water relative permeability	0.8	0.5	1.0	–
Exponent of water relative permeability	3	2.5	4	–
Exponent of gas relative permeability	2	1	2.5	–

parameter not to be situated at the average value. For example, a true value of fracture half-length is 90 m, while the average value is 115 m. The reason behind this is to avoid random convergence to the average value. Therefore, the posterior distributions of each uncertain parameter compared with the true value could validate the robustness of the AHM workflow and NN–MCMC algorithm used in the study, which is the objective of this chapter.

Next, we selected the location of proxy prediction points to calculate the global objective function. Each proxy point was used to construct a proxy model, in this case, NNs. These NNs were coupled with the MCMC algorithm to form the NN–MCMC algorithm for history matching. In this study, as shown in Fig. 3.3, 20 proxy points were selected for the proxy modeling including 10 proxies for bottomhole pressure at days 1, 8, 15, 24, 35, 50, 65, 80, 95, and 110 (Fig. 3.3A) and 10 proxies for water flow rate at days 2, 8, 15, 24, 35, 50, 65, 80, 95, and 110 (Fig. 3.3B).

Then, we generated 50 initial cases by using the Latin hypercube sampling method. The simulation results of 50 initial cases are shown in Fig. 3.4. The simulation results of these 50 initial cases cover the synthetic historical production data. This means the designed ranges are reasonable and the history-matching algorithm will be able to find the solutions from the specified ranges of uncertain parameters.

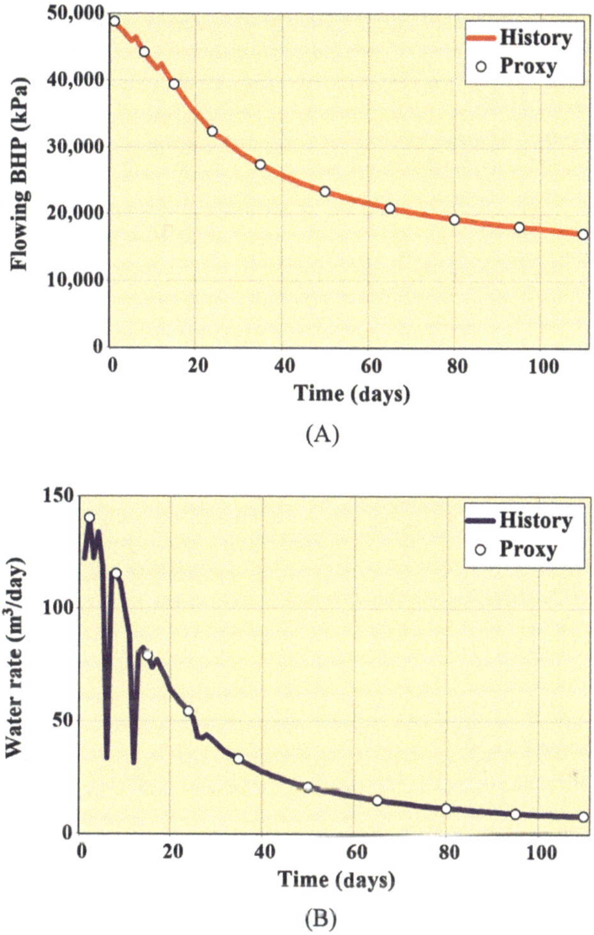

FIGURE 3.3 Proxy location set up for different synthetic response variables of the case with only hydraulic fractures: (A) flowing BHP; and (B) water flow rate. *BHP*, Bottomhole pressure.

For workflow termination criteria, we used a 5% discrepancy to check the proxy convergence or a specified maximum iteration number of 12. For this study, the workflow terminated at a maximum of 12 iterations before reaching a proxy convergence of 5%. After simulations were completed, we used the criteria of 6% error for the objective functions of both bottomhole pressure and water flow rate to obtain multiple history-matching solutions. Fig. 3.5 shows the number of history-matching solutions versus iteration number. Sixty-six solutions are filtered from the total of 325 simulation runs, which is around 20%. The number of solutions generally increases as iteration progresses, and the average solution number is higher after iteration 6.

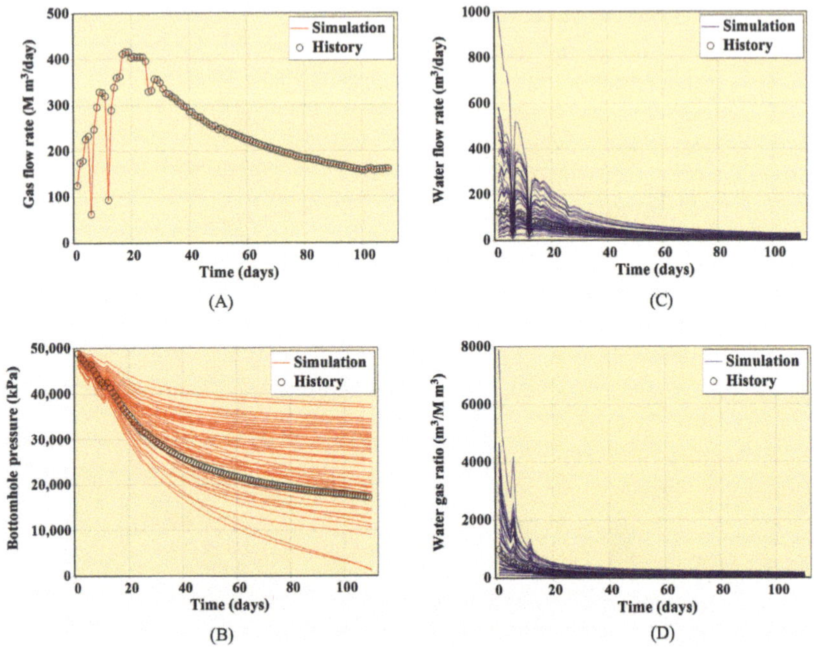

FIGURE 3.4 Comparison between simulation results and historical data of the cases from Latin hypercube design (initial iteration) of the case with only hydraulic fractures: (A) gas flow rate as constraint; (B) bottomhole pressure; (C) water flow rate; and (D) water–gas ratio.

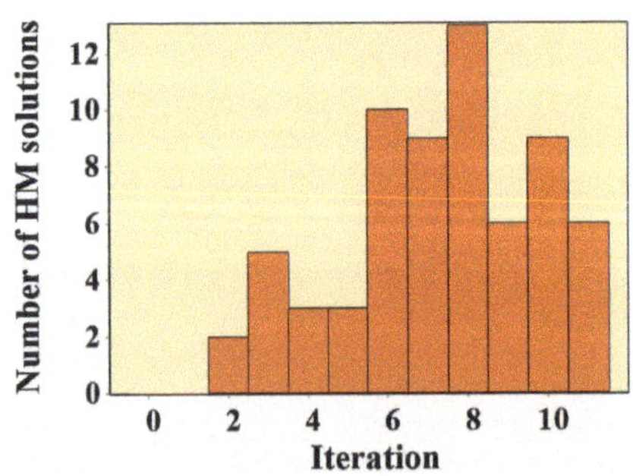

FIGURE 3.5 Number of history-matching solutions versus iteration number for the scenario with only hydraulic fractures.

All simulation results of the 66 history-matching solutions are shown in Fig. 3.6, including gas, water flow rates, and water–gas ratio. In addition, results of the best history-matching case (case with the lowest global objective function value) are shown in Fig. 3.7. The matching quality is excellent, and the history-matching solutions can capture the synthetic bottomhole

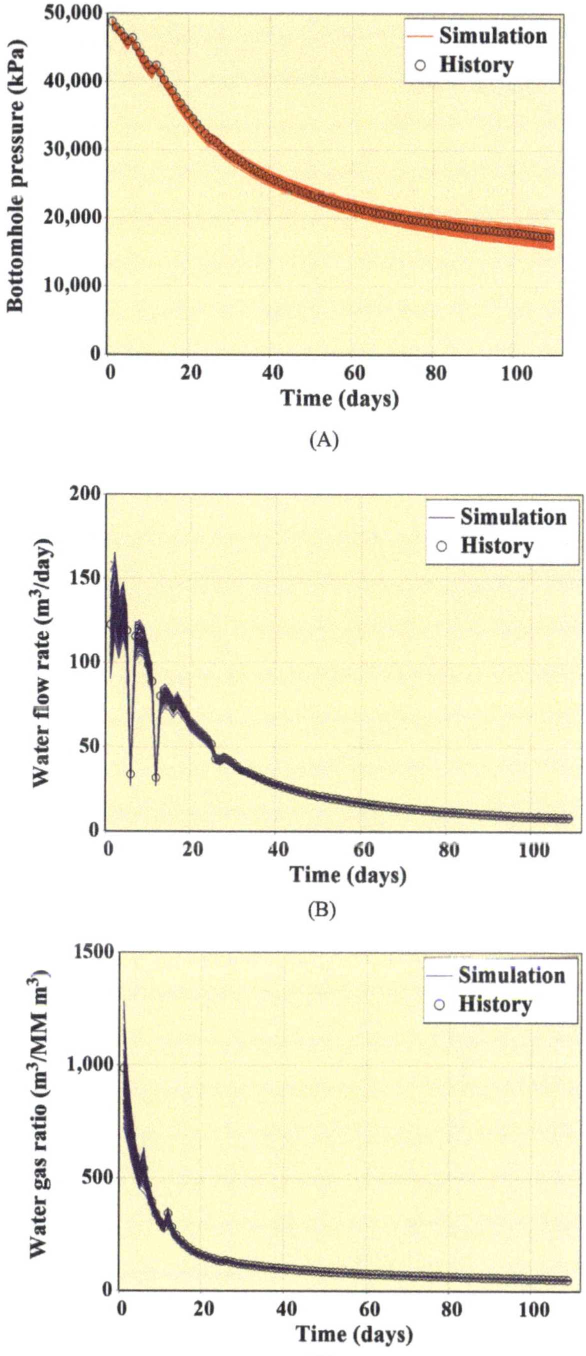

FIGURE 3.6 Comparison between simulation results of 66 history-matching solutions and synthetic history of the case with only hydraulic fractures: (A) bottomhole pressure; (B) water flow rate; and (C) water–gas ratio.

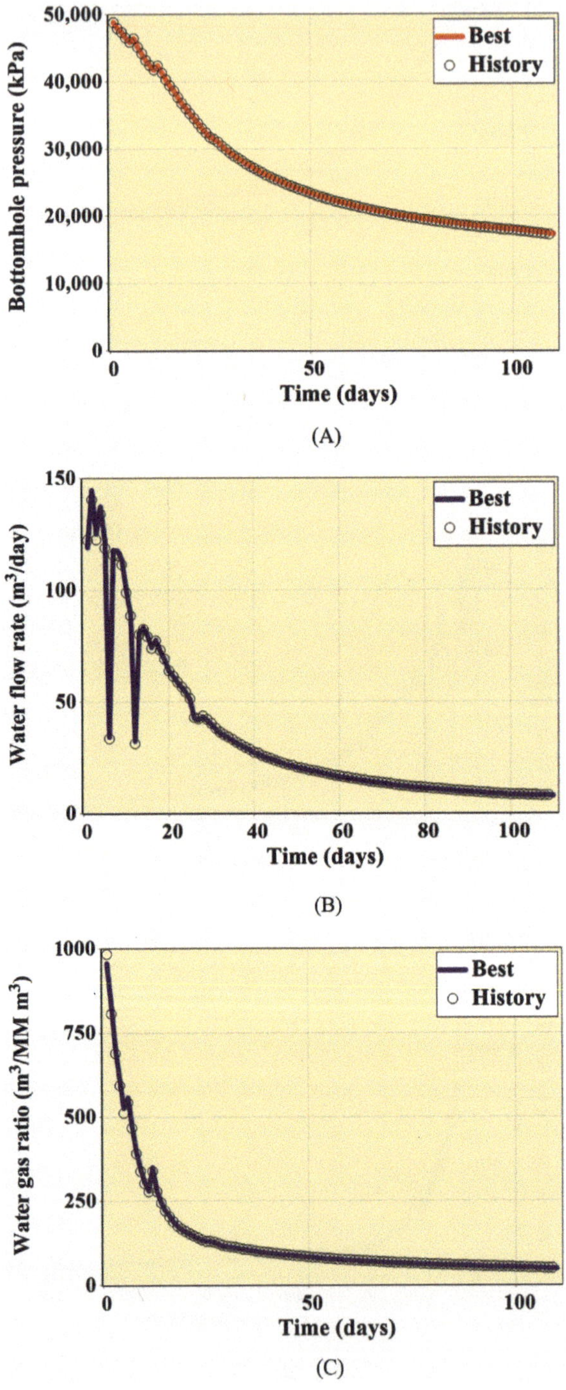

FIGURE 3.7 Comparison between simulation results of the best match and synthetic history of the case with only hydraulic fractures: (A) bottomhole pressure; (B) water flow rate; and (C) water–gas ratio.

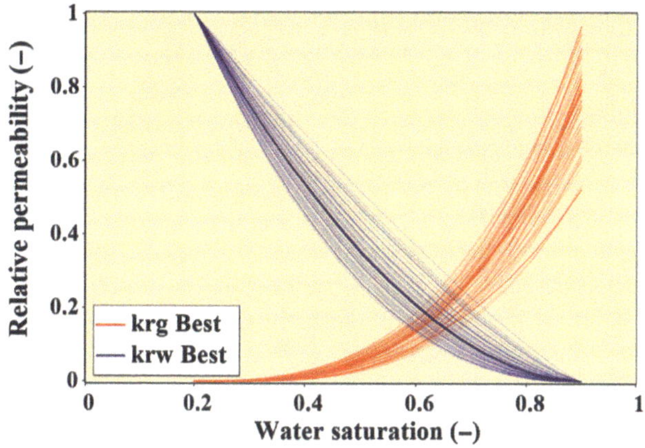

FIGURE 3.8 Relative permeability curves for history-matching solutions and the best match model of the case with only hydraulic fractures.

pressure and water flow rate history. In addition, Fig. 3.8 shows the relative permeability curves obtained from the AHM workflow.

3.2.3 Posterior distribution

The posterior distributions of uncertain parameters from 66 history-matching solutions are shown in Fig. 3.9. The prior uniform distributions (in a light gray color) we specified at the beginning transformed into the posterior distributions when the history-matching process finished. Parameters such as matrix permeability and fracture height have changed significantly from prior distribution to posterior distribution. As we observed, there are no history-matching solutions found for matrix permeability above 0.0002 md, fracture height above 14 m, and fracture half-length above 120 m. This means the ranges of these uncertain parameters could be further narrowed down for better efficiency of the history-matching process.

For other uncertain parameters, we found that any values in these ranges with a combination of other parameters could still satisfy the filtering criteria. This implies these uncertain parameters, including fracture conductivity of 20–70 md m, exponent of gas relative permeability curve of 1–2.5, exponent of water relative permeability curve of 2.5–4, and endpoint of water relative permeability of 0.5–1, did not as significantly impact objective function as much as matrix permeability, fracture height, and fracture half-length.

After we obtained the posterior distributions of the eight uncertain parameters, P50 value of each uncertain parameter was compared with the corresponding true value from the synthetic model. The comparison between true values and the posterior P50 values is summarized in Table 3.3. Parameters

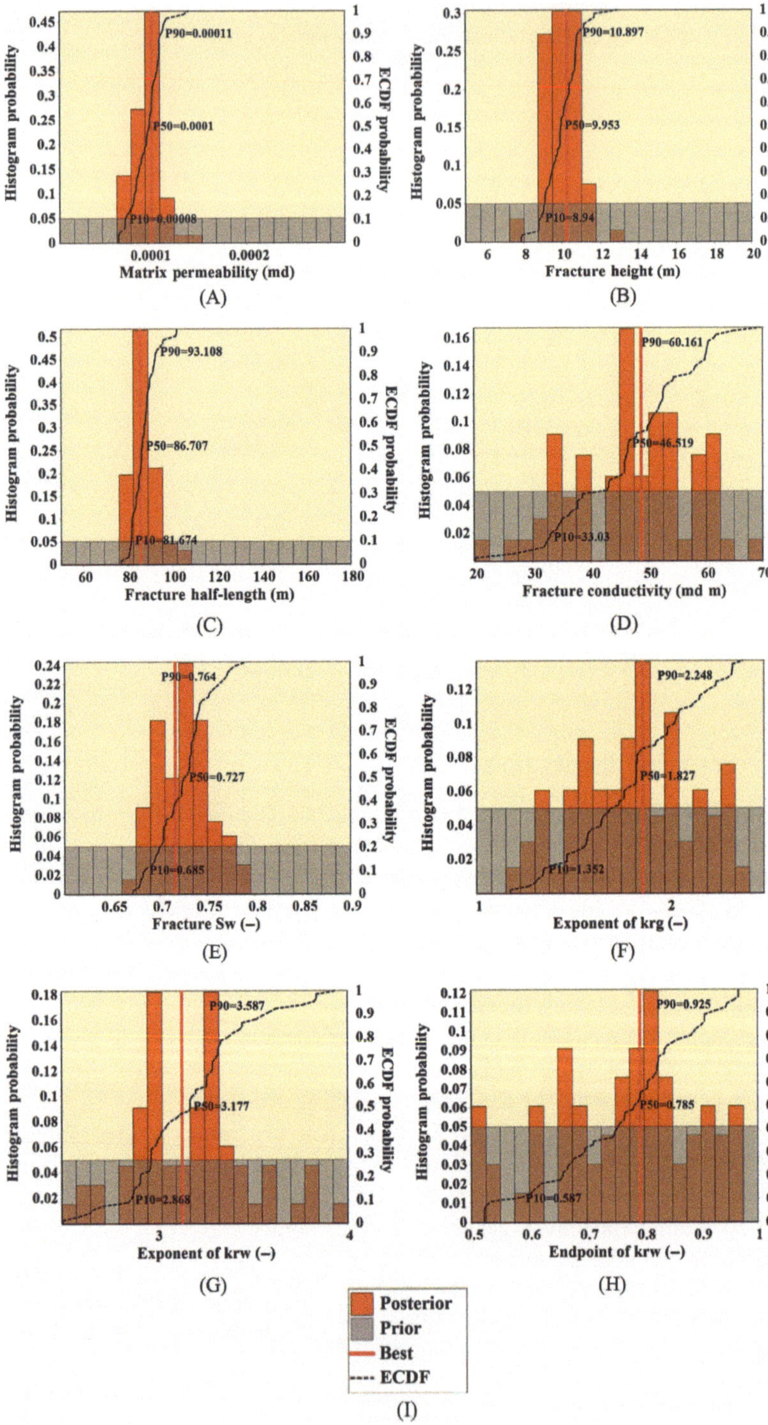

FIGURE 3.9 Comparison between prior and posterior distribution of uncertain parameters (66 solutions) of the case with only hydraulic fractures (ECDF represents the empirical cumulative distribution function): (A) matrix permeability; (B) fracture height; (C) fracture half-length; (D) fracture conductivity; (E) fracture water saturation; (F) exponent of k_{rg}; (G) exponent of k_{rw}; (H) endpoint of k_{rw}; and (I) figure legend.

TABLE 3.3 Summary of the comparisons between posterior P50 values and synthetic model's true values for the case with only hydraulic fractures.

Uncertain parameters	True value	Posterior P50	Unit	% Error
Matrix permeability	0.0001	0.000102	md	+2.0
Fracture height	10	9.95	m	−0.5
Fracture half-length	90	86.71	m	−3.7
Fracture conductivity	50	46.52	md m	−7.0
Fracture water saturation	70	72.7	%	+3.9
Exponent of gas relative permeability	2	1.83	–	−8.5
Exponent of water relative permeability	3	3.18	–	+6.0
Water endpoint relative permeability	0.8	0.785	–	−1.9

such as fracture conductivity, exponent of water relative permeability, and exponent of gas relative permeability have a relatively larger deviation from the true values, at 7.0%, 6.0%, and 8.5% error, respectively. Nevertheless, lower percentage errors (below 5%) were observed for other parameters. The average of absolute relative error from all parameters is 4.2%. In sum, this confirms the robustness and accuracy of the AHM workflow used in the study.

3.2.4 Production forecast

After the history-matching step, a 20-year production forecast was performed to check whether the estimated ultimate recovery (EUR) of the best match model obtained from the AHM workflow is consistent with the EUR of the true solution model or not. First, a 20-year production forecast of the 66 history-matching solutions is run with the bottomhole pressure constrained at 1.5 MPa (Fig. 3.10A). The simulation results are shown in Fig. 3.10. The predicted gas flow rate, water flow rate, cumulative gas production, and cumulative water production are shown in Fig. 3.10B, C, D, and E, respectively. The predicted 20-year gas EUR ranges approximately from 0.125 to 0.142 billion cubic meters and the 20-year water EUR ranges roughly from 5.7 to 6.5 thousand cubic meters. The EUR information from the

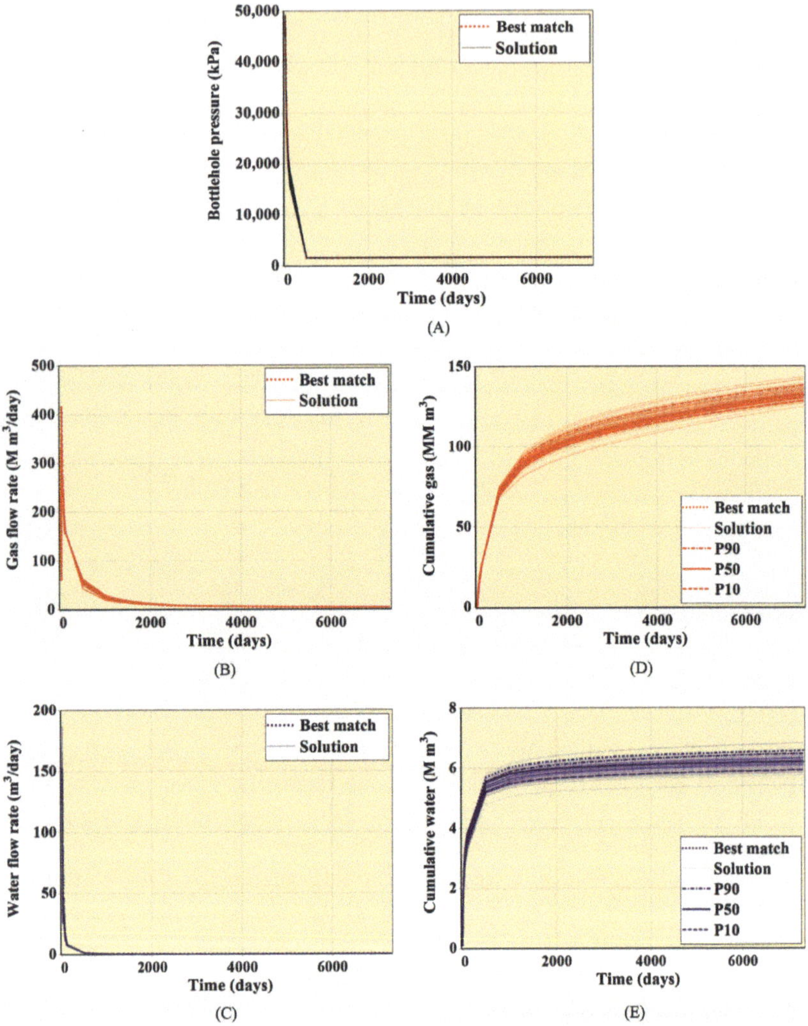

FIGURE 3.10 20-year production forecast of the 66 solutions for the case with only hydraulic fractures: (A) flowing BHP; (B) gas flow rate; (C) water flow rate; (D) cumulative gas production; and (E) cumulative water production. *BHP*, Bottomhole pressure.

history-matching solutions, best match, and synthetic model are summarized in Table 3.4, and the production comparison between the best match model and the true solution from synthetic model is shown in Fig. 3.11. We found that the P50 EUR of the history-matching solutions, best match EUR, and the true solution synthetic model EUR are in line with one another. Therefore, this alignment ensures the accuracy and robustness of the AHM workflow used in the study.

TABLE 3.4 Summary of 20-year gas estimated ultimate recovery (EUR) and water EUR of history-matching solutions, the best match model, and true value of synthetic model (case 1 only hydraulic fractures).

EUR statistics	Cumulative gas production	Cumulative water production
Unit	Million cubic meters	Thousand cubic meters
P10	129	5.94
P50	132	6.20
P90	137	6.44
Best match	132	6.23
True value of synthetic model	132	6.25

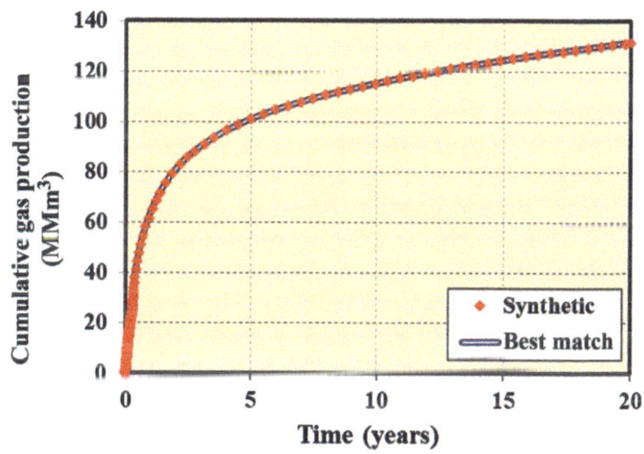

FIGURE 3.11 20-year production forecast of cumulative gas production profile of the synthetic model and best match model for the case with only hydraulic fractures.

3.2.5 Pressure visualization

We conclude this case with the pressure visualizations of the matrix blocks and the fracture segments at the end of production history and year 20. These visualizations for both the synthetic model and the best match model are compared in Fig. 3.12 (matrix pressure) and Fig. 3.13 (fracture pressure). For matrix pressure, only a negligible difference can be noticed between the synthetic model and best match model. For fracture pressure at the end of history, there is a negligible discrepancy between two models.

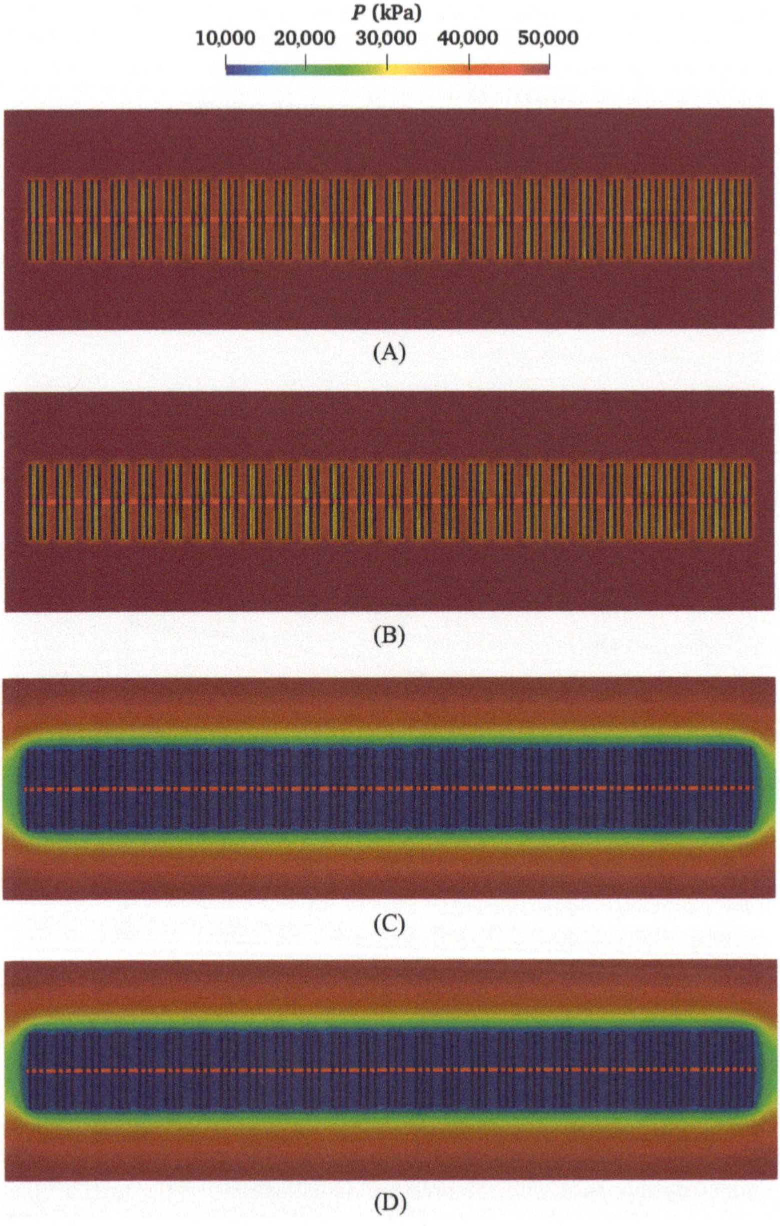

FIGURE 3.12 Matrix pressure visualization for the case with only hydraulic fractures: (A) synthetic model at 110 days; (B) best match model at 110 days; (C) synthetic model at 20 years; and (D) best match model at 20 years.

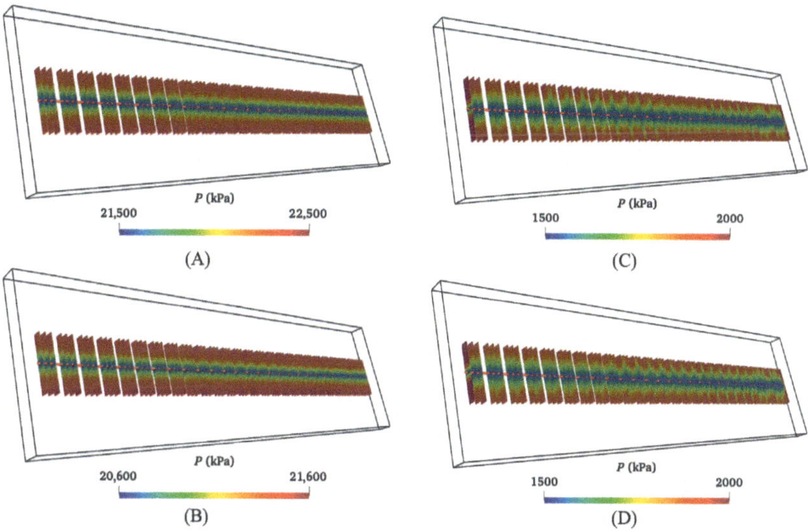

FIGURE 3.13 Fracture pressure visualization for the case with only hydraulic fractures: (A) synthetic model at 110 days; (B) best match model at 110 days; (C) synthetic model at 20 years; and (D) best match model at 20 years.

3.3 Case 2: hydraulic fractures and natural fractures

3.3.1 Reservoir model

We then studied the effect of natural fracture existence in the model on the robustness of the history-matching workflow. Two sets of natural fractures are included. All the natural fractures have dip angles of 90 degrees, and the strike angles for the two sets are 45 and 135 degrees, respectively. The natural fractures are designed to have heights equal to the reservoir thickness.

To focus more on the effects of natural fractures, we considered relative permeability parameters to be fixed but included the three most important parameters associated with natural fractures: total number of natural fractures, length of natural fractures, and conductivity of natural fractures. These three parameters have true values of 300, 100 m, and 5 md m, respectively, to be used in the new synthetic model.

Furthermore, as we discussed regarding the posterior distributions obtained for case 1, we also updated the range for matrix permeability, fracture half-length, and fracture conductivity by narrowing the range to be 0.00005–0.0002 md, 60–150 m, and 40–70 md m, respectively. All information regarding the uncertain and fixed parameters are summarized in Table 3.5. The new synthetic model with 300 natural fractures is shown in Fig. 3.14. Maximum and minimum cases for both total natural fracture number and length were demonstrated in Fig. 3.15 for a better visualization of the extreme cases.

TABLE 3.5 Summary of the uncertain/fixed parameters used in the scenario with hydraulic fractures and natural fractures.

Parameters	True value	Minimum	Maximum	Unit
Matrix permeability	0.0001	0.00005	0.0002	md
Fracture height	10	5	20	m
Fracture half-length	90	60	150	m
Fracture conductivity	50	40	70	md m
Fracture water saturation	70	60	90	%
k_{rg} exponent	1.7	Fixed		–
k_{rw} endpoint	0.778	Fixed		–
k_{rw} exponent	1.8	Fixed		–
Number of NF sets	2	Fixed		–
Azimuth of NF sets	45/135	Fixed		°
Dip of NF sets	90	Fixed		°
Total number of NF	300	200	450	–
NF length	100	70	150	m
NF conductivity	5	2	10	md m

NF, Natural fractures.

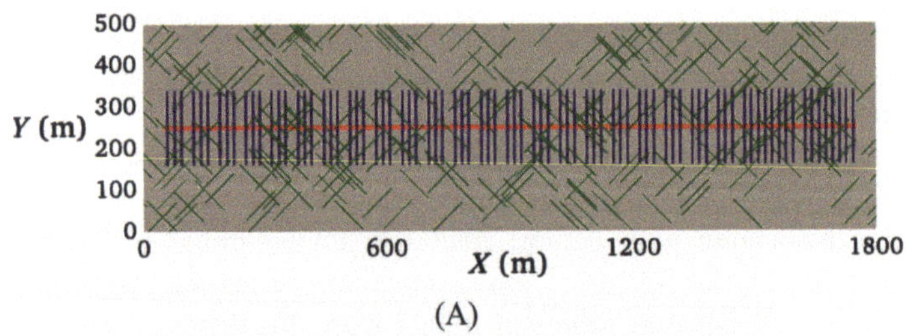

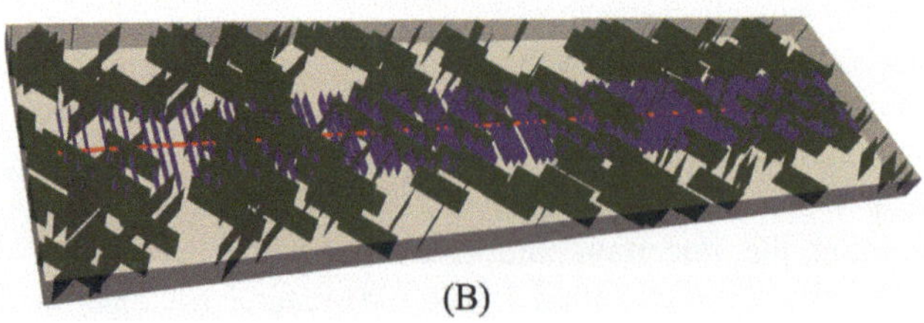

FIGURE 3.14 Synthetic reservoir model including 82 hydraulic fractures (blue) and 300 natural fractures (green) modeled by EDFM in a horizontal well (red): (A) top view; and (B) 3D view. *EDFM*, Embedded discrete fracture model.

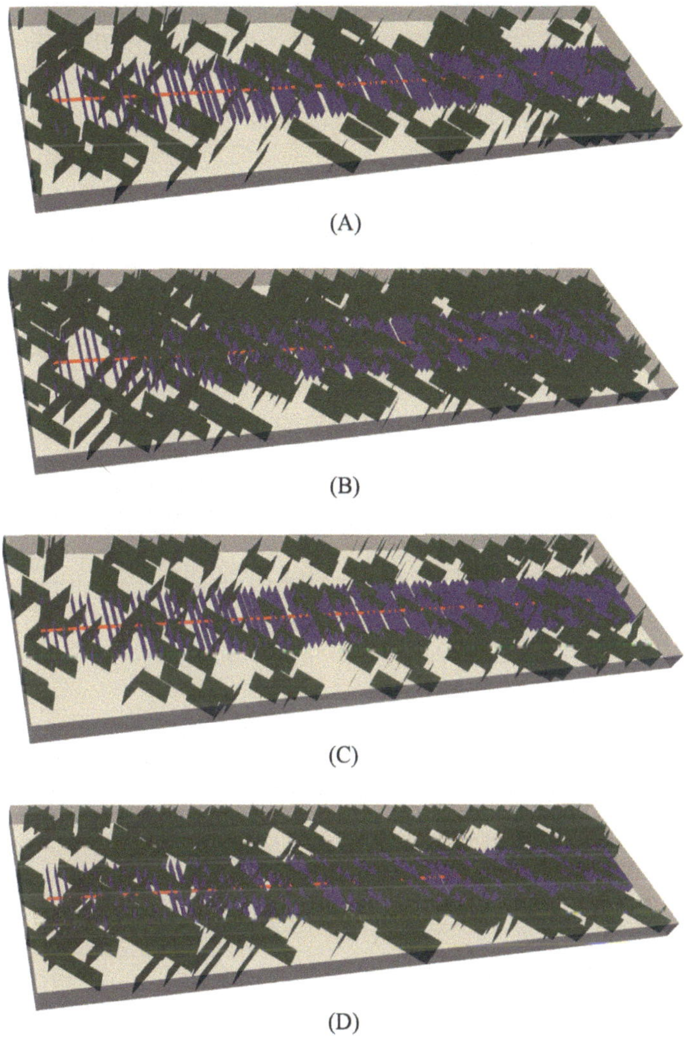

FIGURE 3.15 3D visualization of extreme cases of natural fracture spatial distributions in the reservoir model: (A) 200 natural fractures (minimum number); (B) 450 natural fractures (maximum number); (C) 70 m natural fractures (minimum length); and (D) 150 m natural fractures (maximum length).

3.3.2 History matching

The history-matching workflow progressed as same as the case with only hydraulic fractures. Fig. 3.16 shows the proxy points of synthetic production data. These included 10 proxies for bottomhole pressure at days 1, 8, 15, 24, 35, 50, 65, 80, 95, and 110 (Fig. 3.16A) and 10 proxies for water flow rate at days 2, 9, 17, 24, 35, 50, 65, 80, 95, and 110 (Fig. 3.16B).

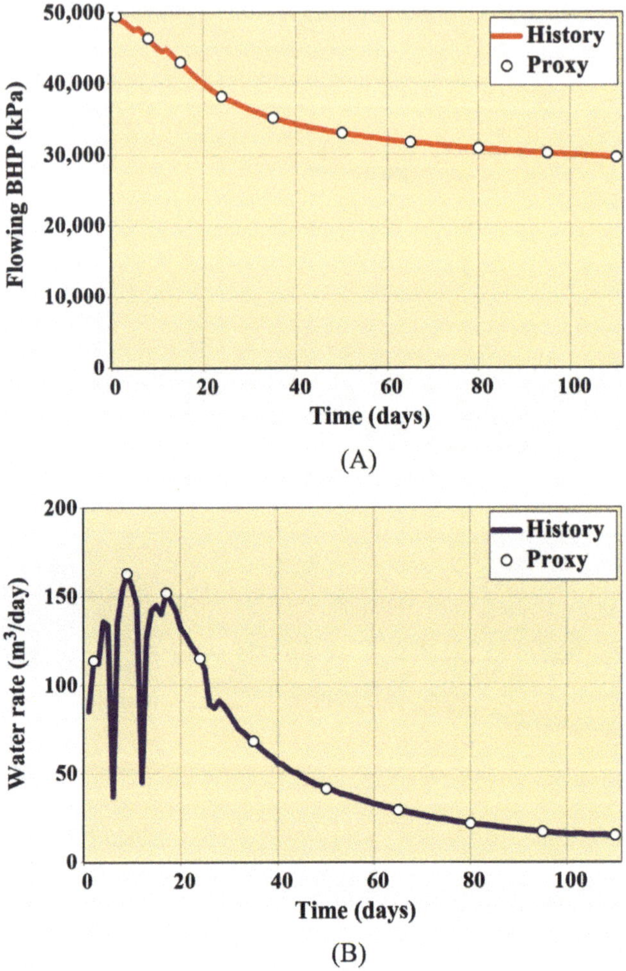

FIGURE 3.16 Proxy location set up for different synthetic response variables of the case with hydraulic and natural fractures: (A) flowing BHP; and (B) water flow rate. *BHP*, Bottomhole pressure.

Initial iteration cases by Latin hypercube design were run similarly and simulation results of the initial iteration are shown in Fig. 3.17. The simulation results cover the synthetic production data, implying history-matching solutions could be found in the specified uncertain parameter ranges.

Since the number of uncertain parameters of case 2 is same as case 1, we still used the same workflow termination criteria: either a maximum iteration of 12 (total simulation cases of 325) or proxy convergence of 5% discrepancy. For case 2, the workflow was terminated by reaching the specified maximum iteration number of 12. History-matching criteria of 3% relative

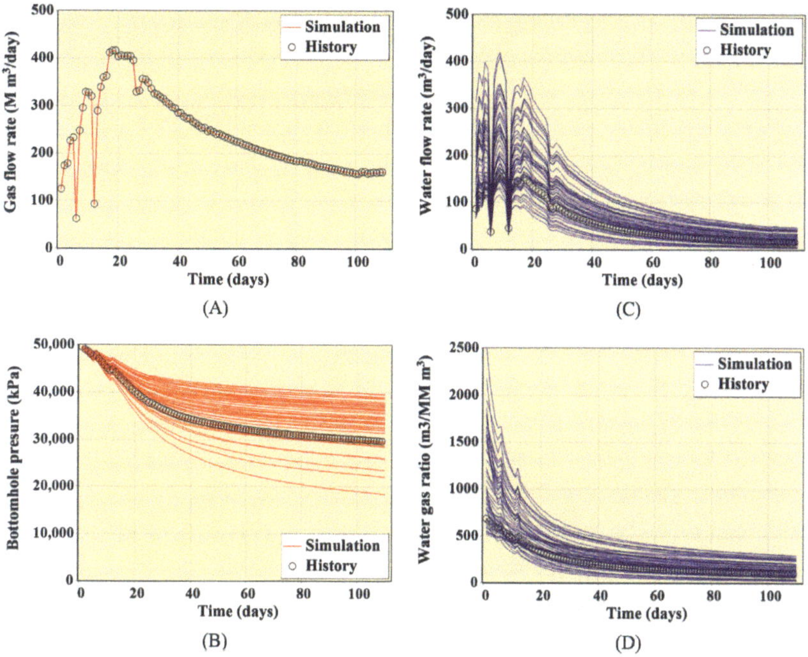

FIGURE 3.17 Comparison between simulation results and historical data of the cases from Latin hypercube design (initial iteration) of the case with hydraulic and natural fractures: (A) gas flow rate as constraint; (B) bottomhole pressure; (C) water flow rate; and (D) water–gas ratio.

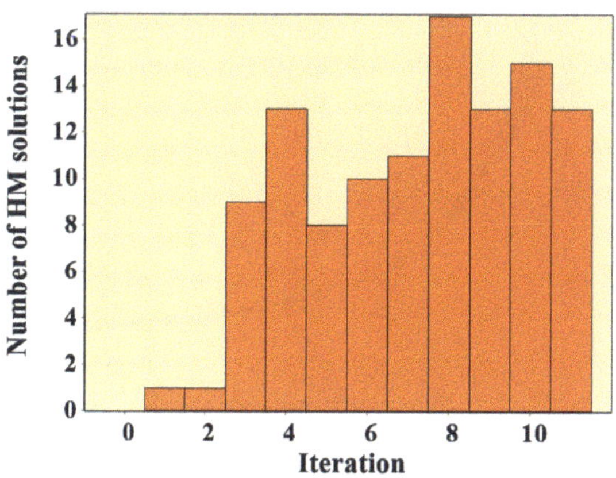

FIGURE 3.18 Number of history-matching solutions versus iteration number for the case with hydraulic and natural fractures.

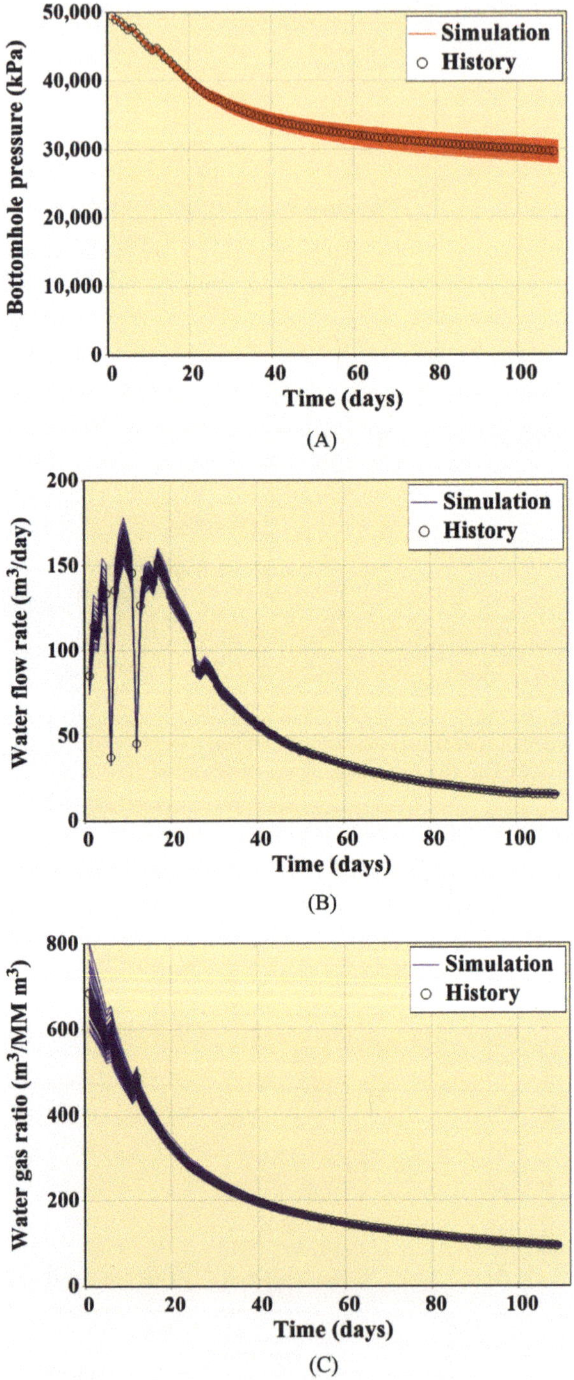

FIGURE 3.19 Comparison between simulation results of 111 history-matching solutions and the synthetic history for the case with hydraulic and natural fractures: (A) bottomhole pressure; (B) water flow rate; and (C) water–gas ratio.

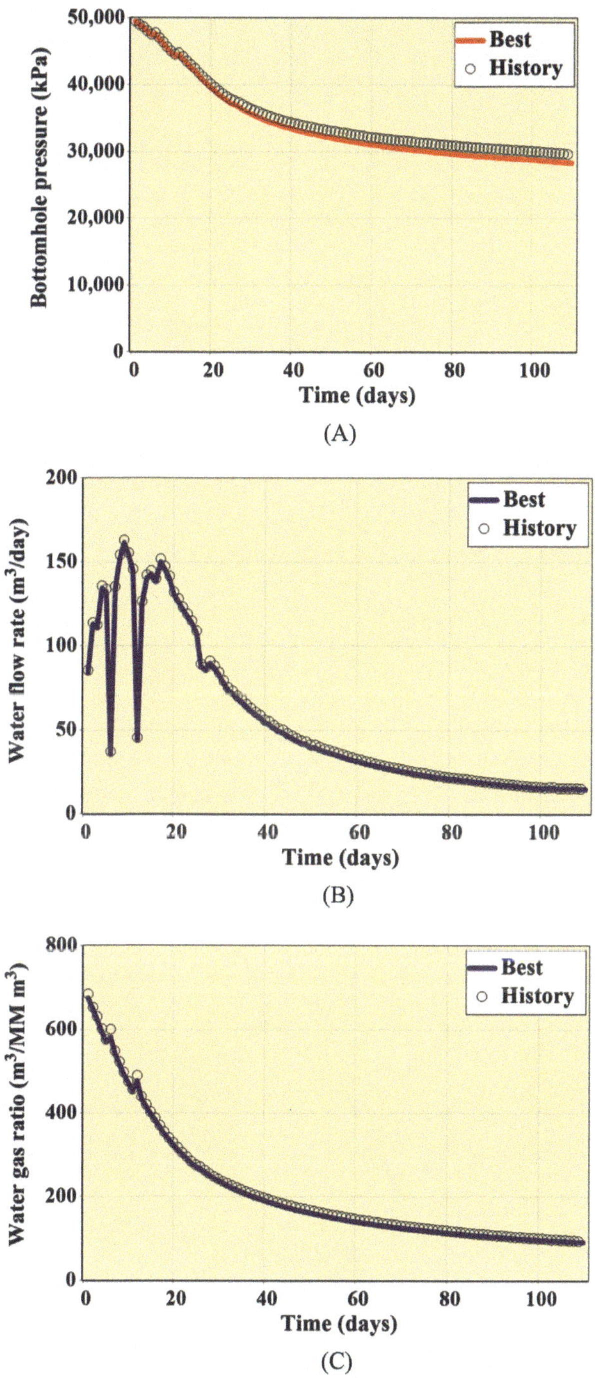

FIGURE 3.20 Comparison between simulation results of the best match and synthetic history of the case with hydraulic and natural fractures: (A) bottomhole pressure; (B) water flow rate; and (C) water gas ratio.

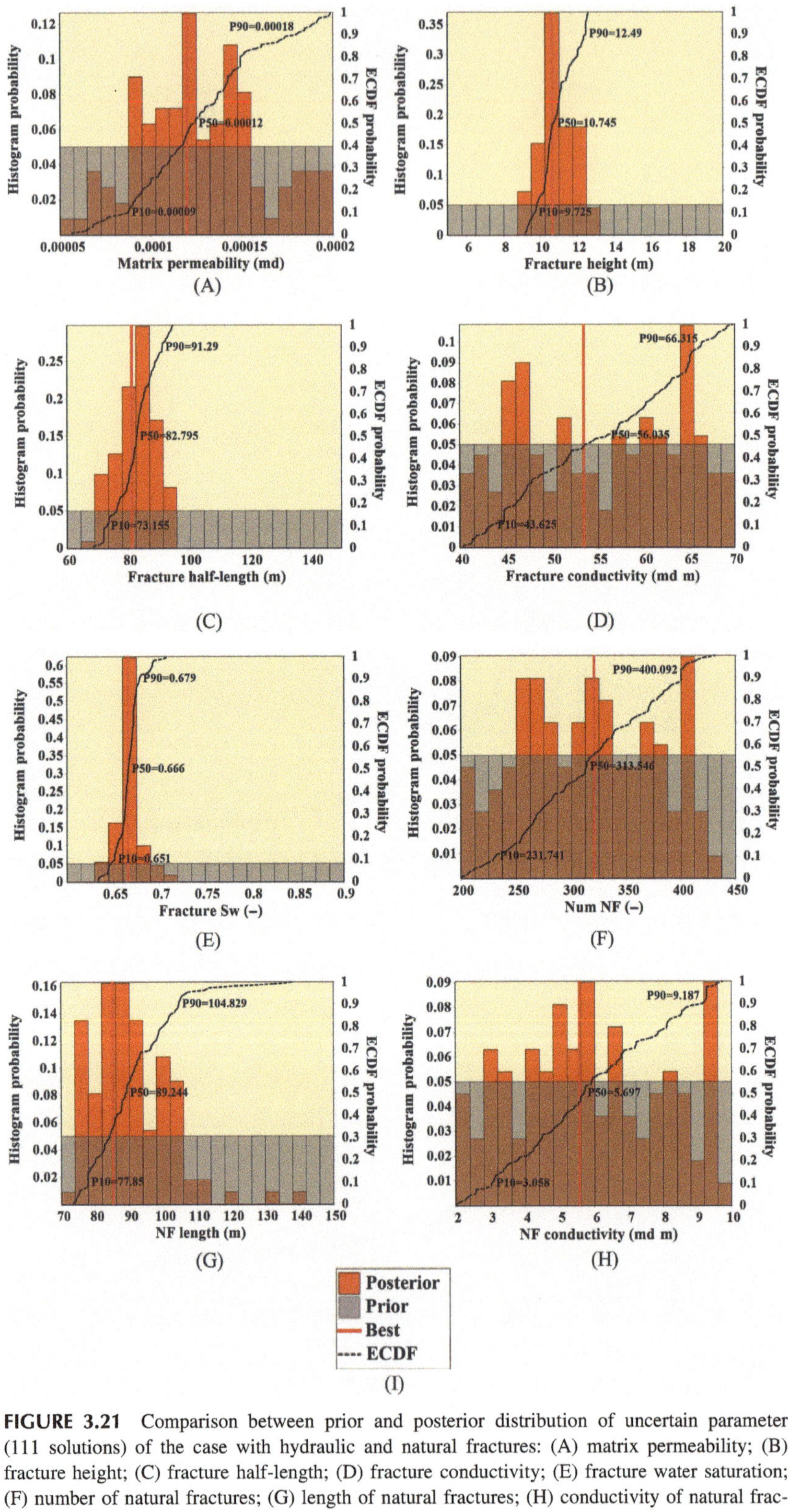

FIGURE 3.21 Comparison between prior and posterior distribution of uncertain parameter (111 solutions) of the case with hydraulic and natural fractures: (A) matrix permeability; (B) fracture height; (C) fracture half-length; (D) fracture conductivity; (E) fracture water saturation; (F) number of natural fractures; (G) length of natural fractures; (H) conductivity of natural fractures; and (I) figure legend.

TABLE 3.6 Summary of the comparisons between posterior P50 values and the synthetic model's true values for the case with hydraulic and natural fractures.

Uncertain parameters	True value	Posterior P50	Unit	% Error
Matrix permeability	0.0001	0.00012	md	+20
Fracture height	10	10.75	m	+7.5
Fracture half-length	90	82.80	m	−8.0
Fracture conductivity	50	56.04	md m	+12.1
Fracture water saturation	70	66.6	%	−4.9
Number of natural fractures	300	314	–	+4.7
Length of natural fractures	100	89.2	m	−10.8
Conductivity of natural fractures	5	5.70	md m	+13.9

error for each objective function was used to screen history-matching solutions and 111 history-matching solutions were found from 325 simulation runs (34%). Fig. 3.18 shows the propagation of solution numbers versus the iteration number. The number of solutions gradually increases as the iteration progresses. Figs. 3.19 and 3.20 show the responses from all simulation results and the best match case, respectively. We found that bottomhole pressure matching quality is not as good as in the scenario of hydraulic fractures only. This is because the spatially random generations of natural fractures are challenging to match exactly to the synthetic model.

3.3.3 Posterior distribution

The posterior distributions of uncertain parameters from 111 history-matching solutions are shown in Fig. 3.21. One observation is that the workflow successfully found the posterior distribution of fracture half-length to be lower than the case with only hydraulic fractures due to the presence of natural fractures. In addition, the narrower ranges enhance the effectiveness of the AHM workflow, as the algorithm did not explore many realizations outside the history-matching solution ranges in case the specified uniform prior distributions were too wide.

For the number of natural fractures and natural fracture conductivity, posterior distributions do not change significantly from prior distribution, implying any values in their specified ranges with a combination of other parameters will still satisfy the history-matching solution criteria.

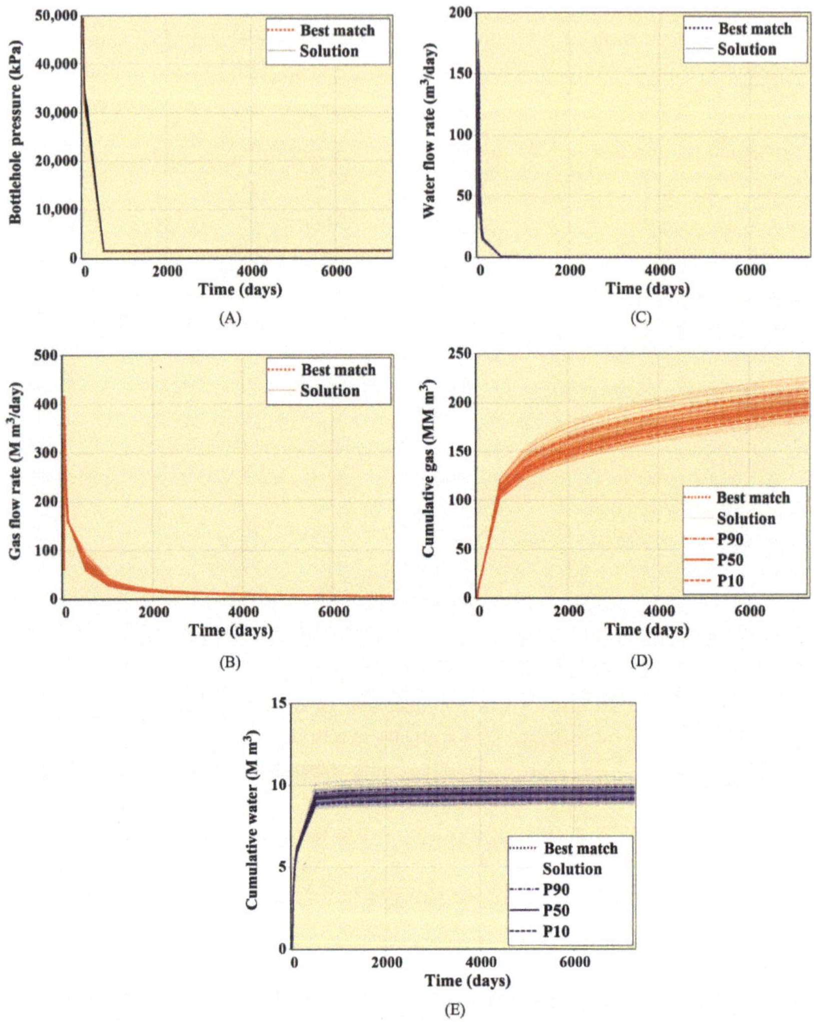

FIGURE 3.22 20-year production forecast of the 111 solutions for the case with hydraulic and natural fractures: (A) flowing BHP; (B) gas flow rate; (C) water flow rate; (D) cumulative gas production; and (E) cumulative water production. *BHP*, Bottomhole pressure.

The comparisons between the synthetic model's true values and the P50 obtained from posterior distributions are summarized in Table 3.6. While the relative errors of some parameters tend to be larger than case 1 due to the stochastic nature of natural fracture parameters, the overall results with true values show an acceptable discrepancy with an average absolute relative error of 10.2%.

TABLE 3.7 Summary of 20-year gas and water estimated ultimate recovery (EUR) information for history-matching solutions, best match model, and synthetic model (hydraulic and natural fractures).

EUR statistics	Cumulative gas production	Cumulative water production
Unit	Million cubic meters	Thousand cubic meters
P10	190	9.09
P50	199	9.49
P90	212	9.84
Best match	196	9.14
Synthetic model	204	9.95

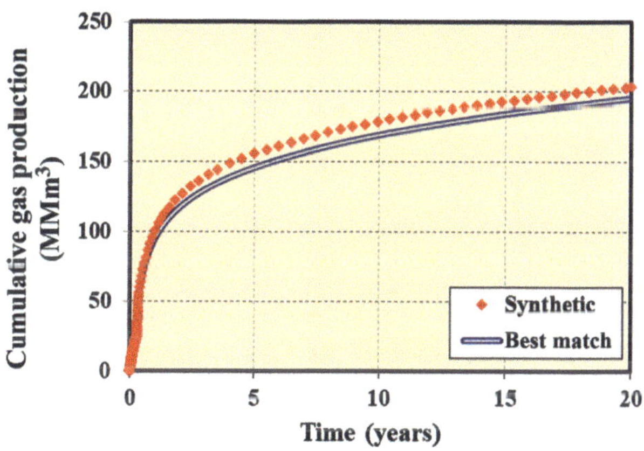

FIGURE 3.23 20-Year forecast of cumulative gas production profile of the synthetic model and best match model for the case with hydraulic and natural fractures.

Hence, even with challenges regarding natural fracture presence, the AHM workflow's robustness and accuracy are acceptable, seeing that the overall error is 10%.

3.3.4 Production forecast

Similarly, we performed the production forecast for this scenario. The predicted gas flow rate, water flow rate, cumulative gas production, and cumulative water production are shown in Fig. 3.22. The predicted 20-year gas EUR ranges approximately from 0.18 to 0.23 billion cubic meters and the

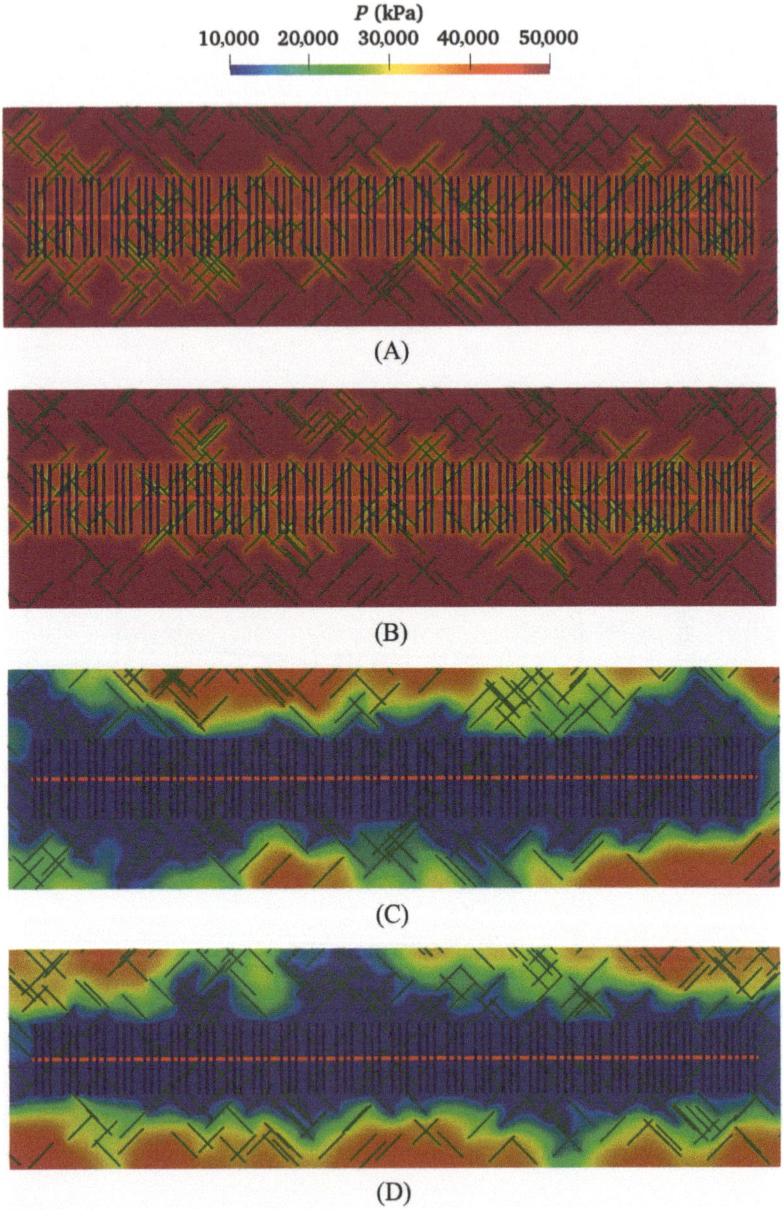

FIGURE 3.24 Matrix pressure visualization for the case with hydraulic and natural fractures: (A) synthetic model at 110 days; (B) best match model at 110 days; (C) synthetic model at 20 years; and (D) best match model at 20 years.

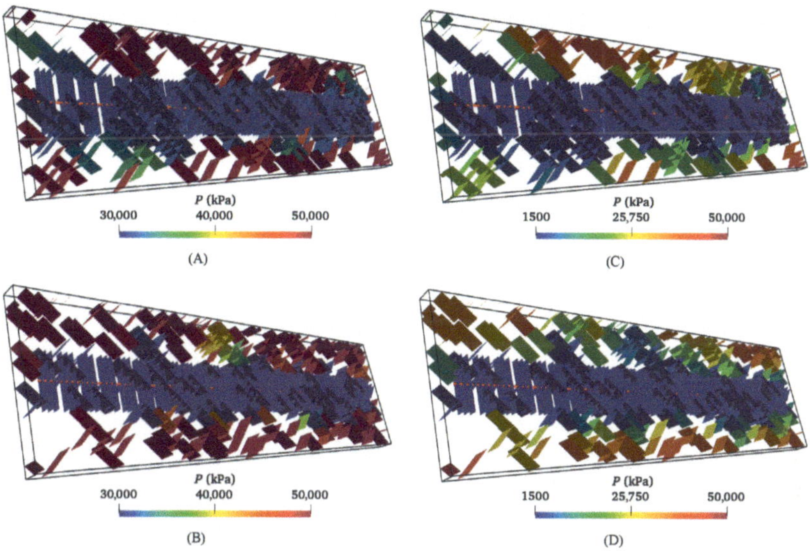

FIGURE 3.25 Fracture pressure visualization for the case hydraulic and natural fractures: (A) synthetic model at 110 days; (B) best match model at 110 days; (C) synthetic model at 20 years; and (D) best match model at 20 years.

20-year water EUR ranges roughly from 8.7 to 10.5 thousand cubic meters. The EUR information from the history-matching solutions, the best match case, and the synthetic model are summarized in Table 3.7, and the production forecast comparison between the best match model and synthetic model is plotted in Fig. 3.23.

Compared to the scenario without natural fractures, both gas and water EUR significantly improved, due to the contribution from the natural fractures and their connectivity with the hydraulic fractures. Again, we can see some minor discrepancies between P50 of posterior distributions, the true values, and the best match model values for both the gas EUR and water EUR. This is because natural fracture existence in each realization has some randomness associated with the spatial distributions and their connectivity with the wellbore/hydraulic fractures. However, the relative discrepancy is less than 5% and this confirms the robustness and accuracy of the AHM workflow.

3.3.5 Pressure visualization

We concluded the validation test with the pressure visualizations of the matrix blocks and the fractures (both hydraulic and natural fractures) at the end of the history and year 20. The visualizations for both the synthetic model and the best match model are compared in Fig. 3.24 for matrix pressure distribution and in Fig. 3.25 for fracture pressure distribution.

For matrix pressure, negligible discrepancy between both models is observed. For fracture pressure, both at the end of history and year 20, the shape of depleted pressure is different due to the stochastic nature of natural fractures generated during the history-matching workflow; however, the drainage area or depleted pressure area is approximately the same between the synthetic model and the best match model.

3.4 Remarks

Overall, this chapter has demonstrated the robustness and accuracy of the AHM workflow. Based on the information from the generated synthetic model for both scenarios (case 1 with hydraulic fractures only and case 2 with natural and hydraulic fractures), the workflow found multiple history-matching solutions giving P50 values that are close to the true solution model values, approximately 5% and 10% for case 1 and case 2, respectively. In addition, we obtained a similar production forecast and achieved indistinguishable pressure distribution visualizations in both matrix and fractures. Although the inclusion of natural fractures caused some discrepancies due to spatial randomness and its stochastic nature, the general deviation from the synthetic model is minimal and still acceptable. To conclude, the finding above confirms the robustness, reliability, and accuracy of the proposed AHM workflow to be applied in actual field studies.

Chapter 4

Shale-gas well in Longmaxi Shale with bi-wing hydraulic fractures

4.1 Introduction

We had 724 days of historical production data for a shale-gas well in the Longmaxi Shale of the Sichuan Basin in China. This included direct measurements of production data such as gas and water flow rates. Bottomhole pressure (BHP) data was estimated from the tubing head pressure with pressure-loss correlation.

4.2 Reservoir model

The Longmaxi Shale well that was studied is a horizontal well with a length of 5800 ft. The well was completed with 20 stages of hydraulic fracturing and, in each stage, three perforation clusters were shot (except for the first and last stage with 2 perforation clusters). The total number of hydraulic fracturing perforation clusters is 58.

Normally, a reservoir engineer receives a static reservoir model built from the integrated information obtained from geologists, geophysicists, and petrophysicists to represent the complexity of the shale reservoir—typically high heterogeneity, especially in the vertical direction, in terms of porosity, permeability, total organic carbon content, and water saturation (Aderibigbe et al., 2016; Amin et al., 2016). This information was not available for this study, so we built a homogeneous reservoir model with constant thickness for the purpose of history matching (HM). The model dimension is 6000 ft long × 2950 ft wide × fracture height. We assumed the fracture height as a model dimension in *Z* axis, meaning that fractures penetrate the whole reservoir thickness. The number of grid blocks is 120, 59, and 1 in *x*, *y*, and *z* directions, respectively. Basic reservoir and fracture properties are summarized in Table 4.1. The relative permeability curves for modeling gas and water flow are shown in Fig. 4.1. For gas desorption, we used the Langmuir isotherm, as shown

Assisted History Matching for Unconventional Reservoirs. DOI: https://doi.org/10.1016/B978-0-12-822242-3.00008-0
© 2021 Elsevier Inc. All rights reserved.

TABLE 4.1 Summary of basic reservoir and fracture parameters used in this study (Tripoppoom et al., 2019).

Reservoir description	Value	Unit
Model dimension ($x \times y$)	6000×2950	ft
Number of grid blocks ($x \times y \times z$)	$120 \times 59 \times 1$	–
Initial reservoir pressure	7397	psi
Reservoir temperature	200	°F
Residual water saturation	20%	–
Total compressibility	3×10^{-6}	psi^{-1}
Reservoir depth	8228	ft
Matrix porosity	12%	–
Well length	5800	ft
Number of stages	20	–
Clusters per stage (18 stages)	3	–
Clusters per stage (two stages)	2	–
Cluster spacing	100	ft

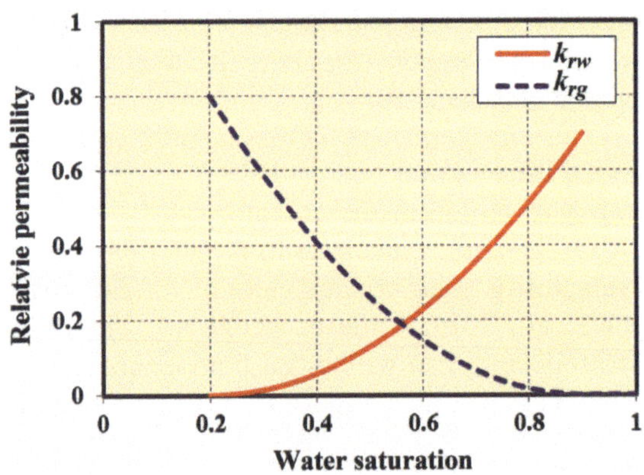

FIGURE 4.1 Relative permeability curves for modeling gas and water flow in shale-gas reservoirs (Tripoppoom et al., 2019).

in Table 4.2 (Yu et al., 2016; Yu and Sepehrnoori, 2018). The reservoir model that includes 58 hydraulic fractures in a horizontal well is demonstrated in Fig. 4.2.

TABLE 4.2 Langmuir isotherm model parameters for the gas desorption effect (Tripoppoom et al., 2019).

Langmuir isotherm parameters	Value	Unit
Inverse-pressure parameter for the Langmuir isotherm model, CH_4	0.000534	1/psi
Maximum moles of adsorbed component per unit mass of rock, CH_4	0.060148	gmole of component/ lb of rock
Rock density	162.32	lb/ft^3

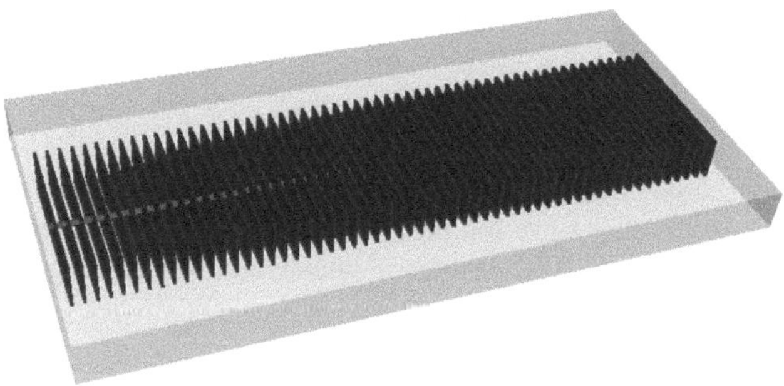

FIGURE 4.2 Reservoir model including 58 hydraulic fractures in a horizontal well (Tripoppoom et al., 2019).

4.3 Comparison between EDFM and LGR

We verified the embedded discrete fracture model (EDFM) with a traditional local grid refinement (LGR) to model fractures in this field case study. We used the same reservoir and fracture properties for both methods, as summarized in Tables 4.1 and 4.2, and other fracture and matrix parameters, summarized in Table 4.3. The only difference is fracture modeling technique. For LGR, we refined the parental cells containing fractures into seven cells in *I*-direction. For EDFM, we used the EDFM preprocessor to calculate nonneighboring connections and modified the reservoir simulator data file. Fig. 4.3 shows the production profile results from both methods. The cumulative gas and water production for LGR are 5.35 Bscf (billion standard cubic feet) and 15.74 MSTB (thousand stock tank barrel), while the cumulative gas and water production for EDFM are 5.47 Bscf and 15.71 MSTB, respectively. The relative differences are 2% and 0.2%, respectively. We can see that there is no significant difference between EDFM and LGR. Table 4.4 shows the CPU (central processing unit) computational time of LGR and EDFM. EDFM can complete the run twice as fast as LGR for this field case

TABLE 4.3 Summary of matrix and fracture parameters used in embedded discrete fracture model and local grid refinement comparison (Tripoppoom et al., 2019).

Matrix and fractures parameters	Value	Unit
Matrix permeability	0.02005	md
Fracture width	0.01	ft
Fracture conductivity	5.5	md ft
Fracture half-length	800	ft
Fracture height	65.5	ft
Fracture dip angle	90	degree
Initial water saturation	40%	–

study. To conclude, EDFM can mimic the fracture flow behavior of LGR with less computational time for this well application. That means we can perform assisted HM (AHM) faster for unconventional reservoirs by using EDFM.

4.4 Parameters identification and screening

For the proposed workflow, we can identify an unlimited number of uncertain parameters for the HM purpose, depending on how much available data we already had and the uncertainty of each data. In this study, we identified six uncertain parameters and their ranges were summarized in Table 4.5 with uniform distribution based on prior knowledge about this well. We modeled a fracture width of 1 ft and adjusted fracture permeability until the product of fracture width and fracture permeability were equal to fracture conductivity (Rubin, 2010). We then defined the objective function as the global error of both gas rate and water rate, shown in the following equation:

$$\text{Objective function} = \text{global error}(\varepsilon) = \frac{\sum_{j=1}^{m}\sum_{i=1}^{n}\left[\frac{\left(x_{ij,\text{model}} - x_{ij,\text{history}}\right)}{\text{Normalized Factor}_j} \times 100\right]^2 \times w_{ij}}{\sum_{j=1}^{m}\sum_{i=1}^{n} w_{ij}}, \tag{4.1}$$

where i is an index of data point, j an index of response parameter, n the number of data points, m the number of response parameters, $x_{ij,\text{model}}$ the value from the model (simulation run or proxy model) at index i of response parameter j, $x_{ij,\text{history}}$ the historical data at index i of response parameter j,

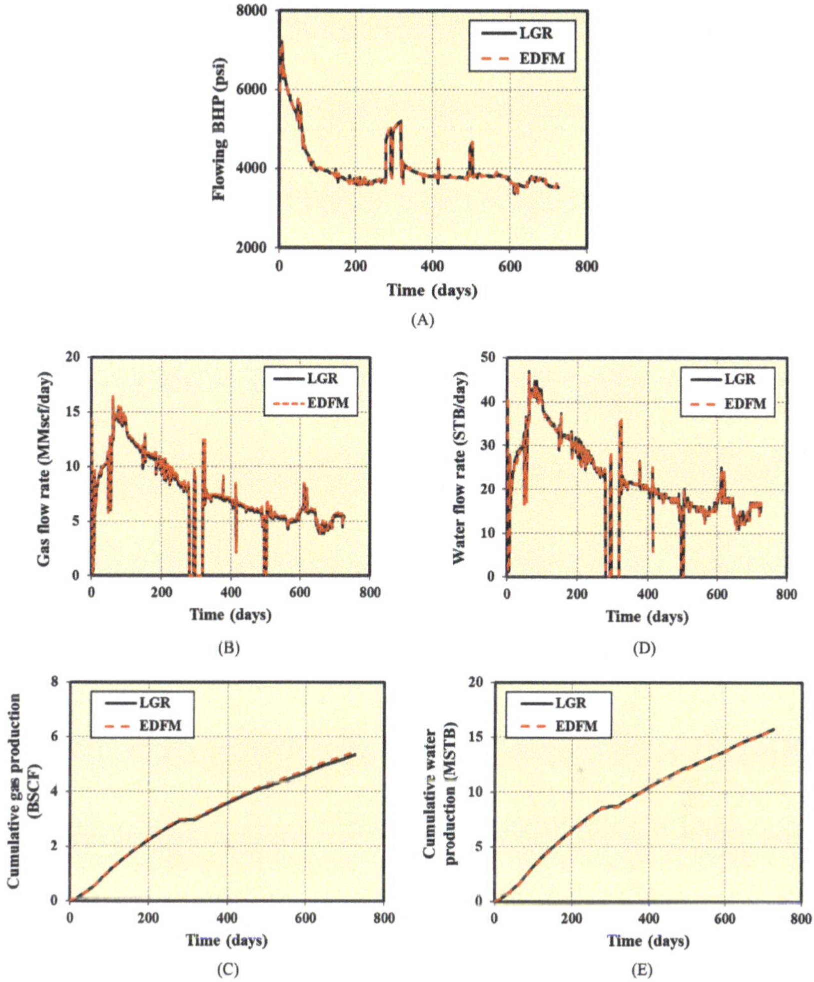

FIGURE 4.3 Comparison between simulation results from LGR and EDFM methods: (A) flowing BHP as constraint; (B) gas flow rate; (C) cumulative gas production; (D) water flow rate; and (E) cumulative water production (Tripoppoom et al., 2019). *BHP*, Bottomhole pressure; *EDFM*, embedded discrete fracture model; *LGR*, local grid refinement.

Normalized Factor_j defined as the maximum of data value minus the minimum of data value of response parameter j, and w_{ij} the weight of data points i of response parameter j.

Then, we screened uncertain parameters for HM by performing a two-level factorial design. Normally, 2^6 (64) cases should be generated for six parameters by using two-level full factorial design. The full factorial design can analyze all parameter interactions, but it requires too many runs for the screening purpose. 2^4–2^5 cases are sufficient for the screening step and the

TABLE 4.4 Comparison of CPU time and elapsed time between embedded discrete fracture model (EDFM) and local grid refinement (LGR) (Tripoppoom et al., 2019).

Method	CPU time (s)	Elapsed time (s)	Ratio
LGR	2318	340	1:0.5
EDFM	1159	176	

TABLE 4.5 Summary of six uncertain parameters and their prior distributions (Tripoppoom et al., 2019).

Code	Uncertain parameters	Unit	Distribution	Min value	Max value
A	Matrix permeability	md	Uniform	0.0001	0.04
B	Fracture half-length	ft	Uniform	300	1300
C	Fracture height	ft	Uniform	16	115
D	Fracture conductivity	md ft	Uniform	1	10
E	Matrix water saturation	–	Uniform	0.30	0.50
F	Fracture water saturation	–	Uniform	0.40	0.90

analysis of partial parameter interaction (Stat-Ease Inc, 2017). Therefore we used 2^5 (32) cases to screen significant uncertain parameters. We constrained the production by BHP and monitored the global error of gas rate and water rate. We used the EDFM preprocessor coupled with a compositional reservoir simulator to create simulation files automatically. The results of the 32 cases are shown in Fig. 4.4. The simulated data covering the historical data confirms that the specified ranges of six uncertain parameters are acceptable and the uncertain-parameter range does not need revision at this step.

To select significant uncertain parameters, first, we evaluated whether any transformation of response values (global error) is needed by looking at a Box–Cox plot (Box and Cox, 1964). The lowest lambda value indicates what transformation is needed ($\lambda = 1$ for no transformation, $\lambda = 0.5$ for square root, $\lambda = 0$ for natural log, $\lambda = -0.5$ for inverse square root, $\lambda = -1$ for inverse). In this case, log transformation was most appropriate for both gas and water global error.

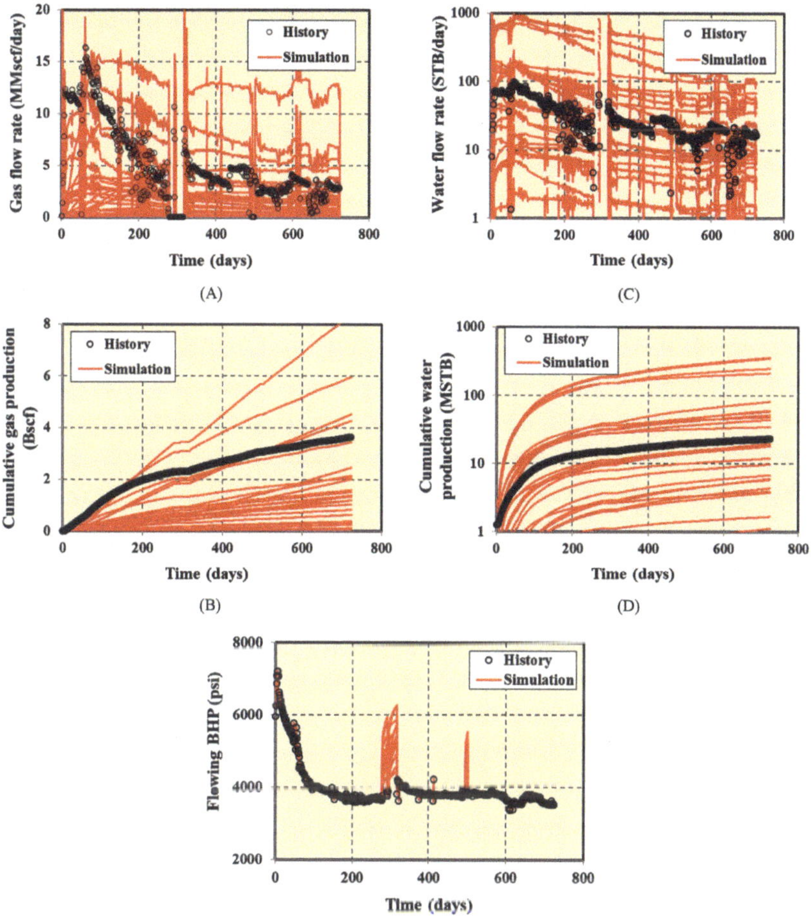

FIGURE 4.4 Comparison between simulation results and historical data of cases from the two-level factorial design: (A) gas flow rate; (B) cumulative gas production; (C) water flow rate; (D) cumulative water production; and (E) flowing BHP (Tripoppoom et al., 2019). *BHP*, Bottomhole pressure.

Then, we analyzed the Pareto chart of *t*-value, as shown in Figs. 4.5 and 4.6. The Pareto chart indicated that all parameters are lower than the Bonferroni limit but only AC and ACD are above the *t*-value limit, which are matrix permeability, fracture height, and fracture conductivity. This means fracture parameters other than fracture half-length have more impact on gas flow rate. For water global error, in the Pareto chart, CF, DF, F, C, and D are above the Bonferroni limit. C, D, and F are fracture height, fracture conductivity, and fracture water saturation, respectively. Even the next one ranking after D is above the *t*-value limit, but we did not select it as a

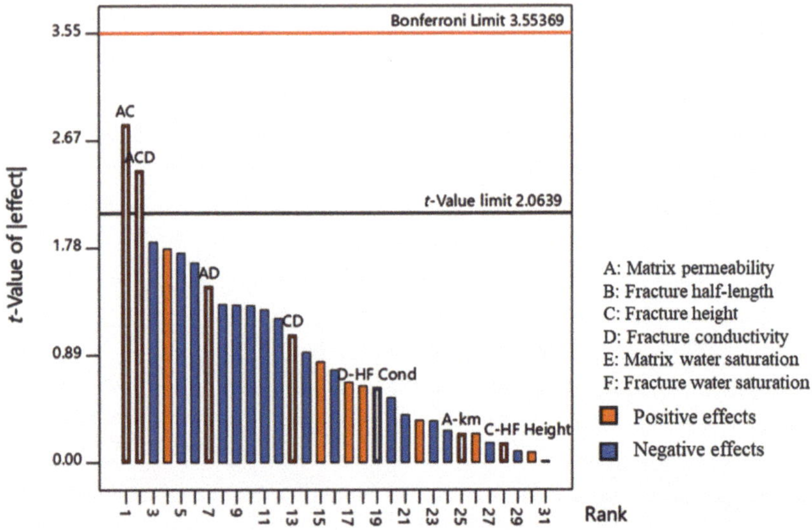

FIGURE 4.5 *t*-Value ranking of two-level factorial design for global error of gas flow rate (Tripoppoom et al., 2019).

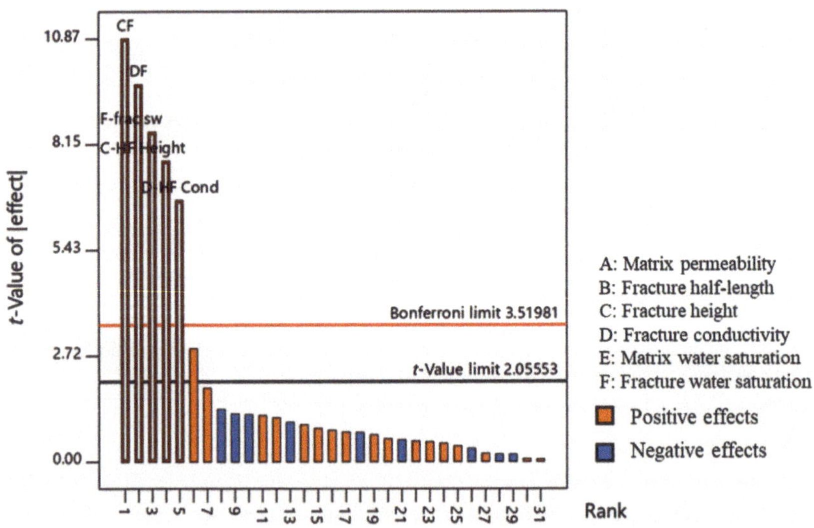

FIGURE 4.6 *t*-Value ranking of two-level factorial design for global error of water flow rate (Tripoppoom et al., 2019).

significant parameter because there is a large gap between D and the next one. In sum, we reduced six uncertain parameters to four significant uncertain parameters including matrix permeability, fracture height, fracture conductivity, and fracture water saturation. However, for the later demonstration, we included matrix water saturation for HM.

4.5 History matching

After screening for uncertain parameters, we used Latin hypercube design to sample the initial 25 cases from 5 uncertain parameters, summarized in Table 4.6, scattering over the uncertain-parameter domain. The fracture half-length is a constant value of 700 ft for all cases of HM. Fig. 4.7 shows the results of the initial 25 cases, which also cover the historical production data. Then, we created multiple response-parameter proxies from these initial 25 cases by using *k*-nearest neighbors. We created a total of 20 proxies including 10 proxies each for gas flow rate and water flow rate at days 20, 67, 199, 258, 325, 400, 508, 602, 680, and 723, respectively. Each proxy represents a different trend of historical production data. These multiple proxies will be later used in the proxy-based Markov chain Monte Carlo (MCMC) algorithm to propose new possible realizations as HM solutions in the next iteration.

The iterative HM workflow was continued automatically by our developed code. Using the proxy-based MCMC algorithm and filtering scheme, the next iteration cases were selected. We used one chain of MCMC and started the chain at the lowest global error point at each iteration. The iteration progressed until it reached the stopping criteria, which is the percentage of relative discrepancies between current and previous proxy with a threshold of 5% or the maximum iteration number (12 iterations or 300 simulation

TABLE 4.6 Summary of the five uncertain parameters after a screening step with a given fracture half-length of 700 ft (Tripoppoom et al., 2019).

Code	Uncertain parameters	Unit	Distribution	Min value	Max value
A	Matrix permeability	md	Uniform	0.0001	0.04
B	Fracture half-length	ft	Deterministic value	700	
C	Fracture height	ft	Uniform	16	115
D	Fracture conductivity	md ft	Uniform	1	10
E	Matrix water saturation	–	Uniform	0.30	0.50
F	Fracture water saturation	–	Uniform	0.40	0.90

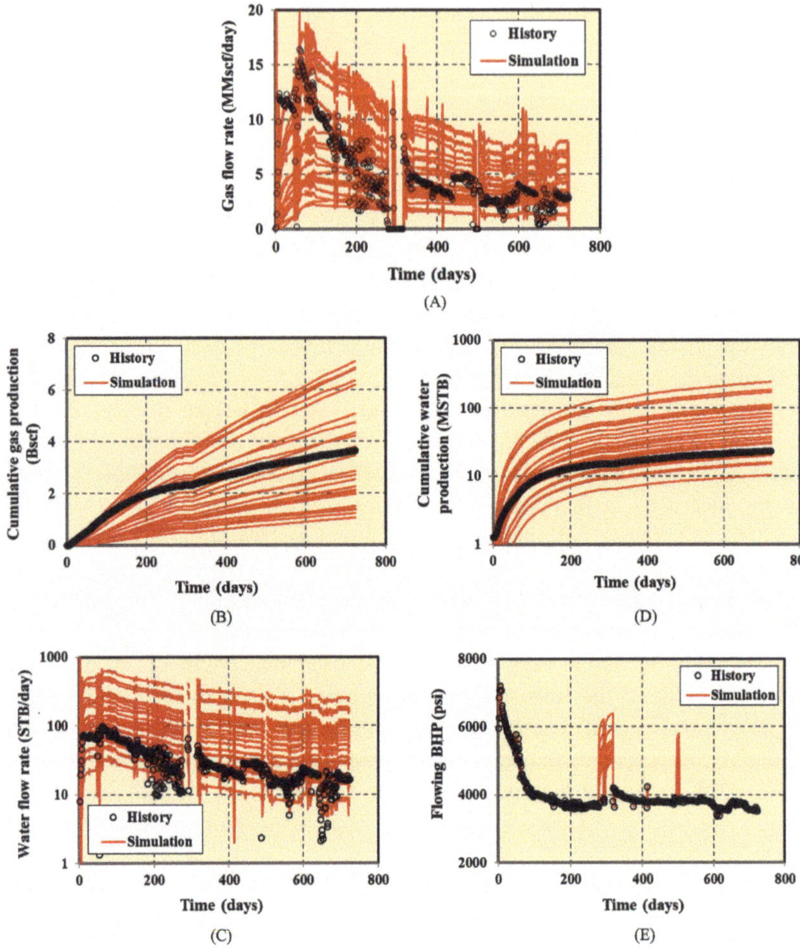

FIGURE 4.7 Comparison between simulation results and historical data of cases from Latin hypercube design (iteration 1): (A) gas flow rate; (B) cumulative gas production; (C) water flow rate; (D) cumulative water production; and (E) flowing BHP (Tripoppoom et al., 2019). *BHP*, Bottomhole pressure.

runs). For this study, the workflow stopped at iteration 11 because the percentage of relative discrepancies between the current and previous proxy was below the threshold of 5%, as shown in Fig. 4.8. The total computational time for each iteration—including the proxy-based MCMC algorithm simulation-running time—is around 2 hours on 4 processors, 3.0 GHz Dell computers. The total computational time for the HM section is around 1 day.

After that, the proxy-based MCMC algorithm generated the last iteration cases with a different filtering scheme using the lowest global error value for all 25 cases. Then, we evaluated all 300 cases from iteration 1 to the last

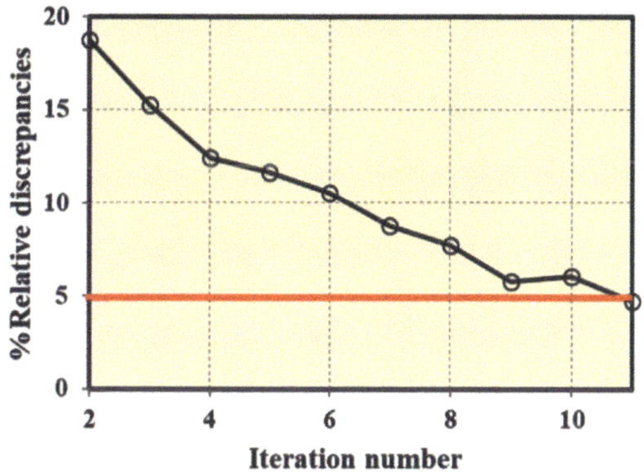

FIGURE 4.8 The percentage of relative discrepancies between the current and previous proxy versus iteration number (Tripoppoom et al., 2019).

iteration and screened them for HM solutions. First, the global error used as the threshold value for screening solutions was unknown. The global error is specified after simulation results are available so that we can compare to production history and identify the case and maximum global error that are least acceptable as HM solutions. We used the gas global error of 220 and water global error of 220 as the HM solutions threshold. Sixty-four cases passed the threshold out of 300 simulation runs, which is approximately 21%. With a more relaxed threshold, the number of HM solutions can be as high as 98 cases, which is approximately 33% out of 300 simulation runs. For this study, we used 64 HM solutions for the production forecast.

The simulation results of 64 HM solutions are shown in Fig. 4.9. The best match case is also illustrated in Fig. 4.10. The best match realization of 64 HM solutions has properties as follows: matrix permeability of 0.027 md, fracture height of 40 ft, fracture conductivity of 9.9 md ft, matrix water saturation of 0.38, and fracture water saturation of 0.48. In addition, a parallel plot of 64 HM solutions is also shown together with non-HM solution cases in Fig. 4.11. One can observe that only some combinations of each uncertain parameter (i.e., low fracture height values and high fracture conductivity value) are HM solutions shown as green lines. The prior and posterior distributions of each set of uncertain parameters from 64 HM solutions are shown in Fig. 4.12. To get a smoother posterior distribution, we also performed MCMC sampling to obtain 10,000 HM solutions using multiple response-parameter proxies. The prior and posterior distributions of each uncertain parameter from 10,000 proxy-generated HM solutions are shown in Fig. 4.13. Both Figs. 4.12 and 4.13 show a similar distribution shape. We observed an obvious change from prior uniform distribution to posterior distribution for all five parameters. For fracture height, fracture conductivity, and fracture water saturation the posterior distribution changed significantly

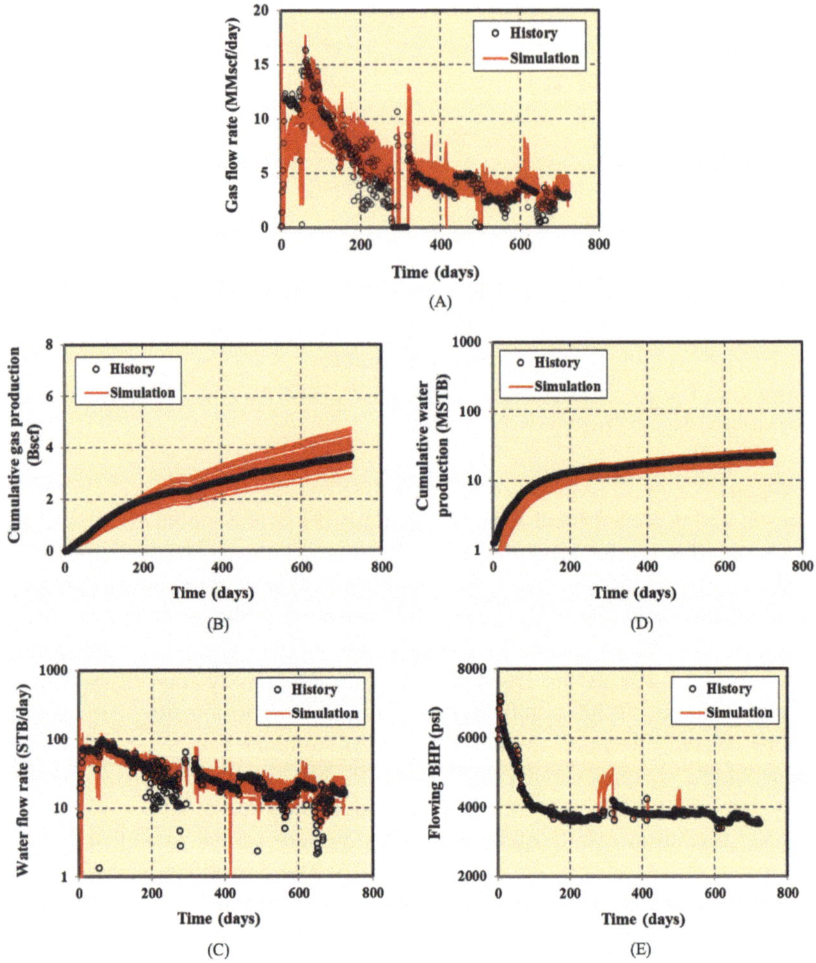

FIGURE 4.9 Comparison between simulation results and historical data of 64 HM solutions: (A) gas flow rate; (B) cumulative gas production; (C) water flow rate; (D) cumulative water production; and (E) flowing BHP (Tripoppoom et al., 2019). *BHP*, Bottomhole pressure; *HM*, history matching.

and had narrower ranges. This implied that these three parameters are very sensitive to the objective function. On the contrary, matrix permeability and matrix water saturation had relatively wider ranges. This indicated that matrix permeability and matrix water saturation are not very sensitive to the objective function (Yang et al., 2015). The finding here is aligned with the parameter screening step which emphasized that matrix water saturation was not a significant uncertain parameter. The values with the highest frequency of each set of parameters are as follows: matrix permeability of 0.031 md,

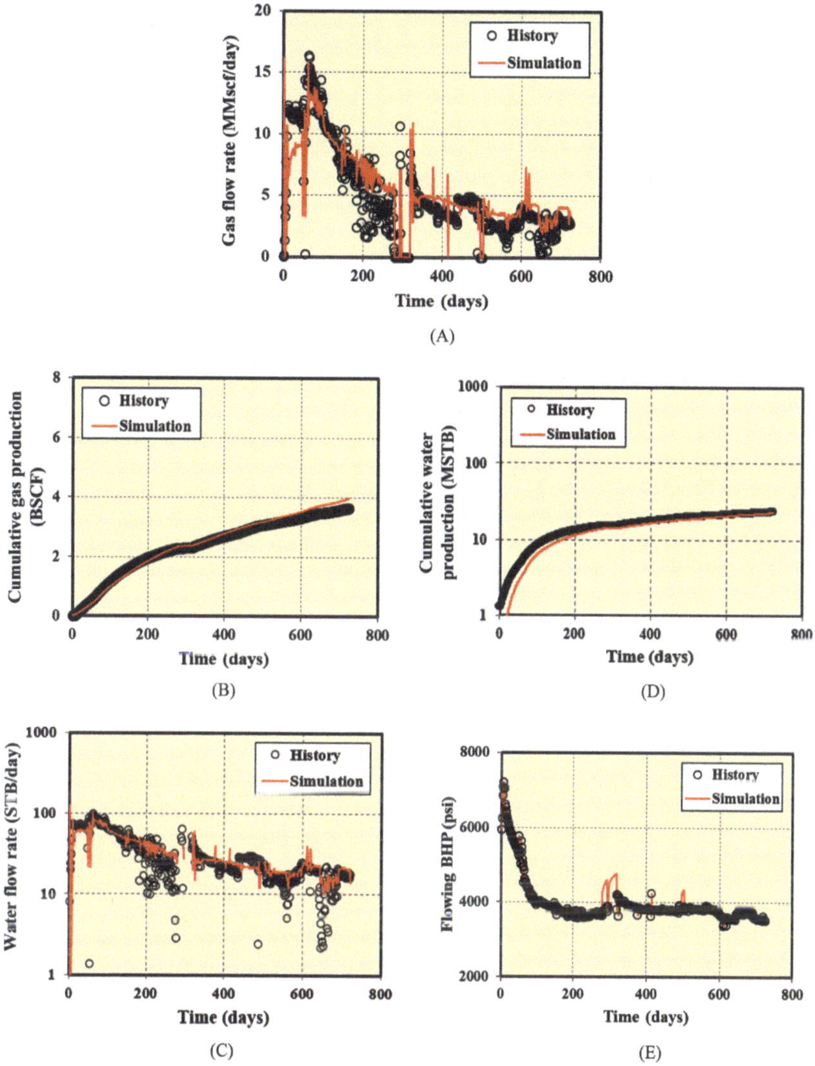

FIGURE 4.10 Comparison between simulation results and historical data of the best match case of 64 HM solutions: (A) gas flow rate; (B) cumulative gas production; (C) water flow rate; (D) cumulative water production; and (E) flowing BHP (Tripoppoom et al., 2019). *BHP*, Bottomhole pressure; *HM*, history matching.

fracture height of 39 ft, fracture conductivity of 9.5 md ft, matrix water saturation of 0.37, and fracture water saturation of 0.46.

Fig. 4.14A and B illustrates the improvement of proxy predictability in gas and water flow rates at day 723 (2 of 20 proxies). The final proxy before the stopping criteria triggered (iteration 11) is not perfect, but we can still

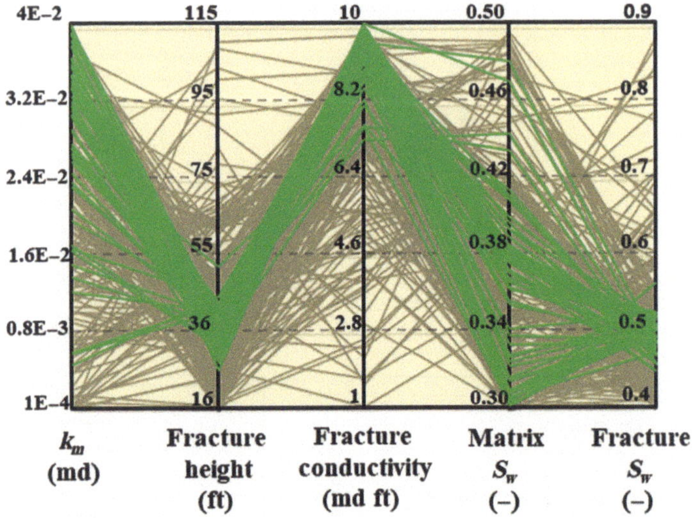

FIGURE 4.11 Parallel coordinate plot of 64 HM solutions. Each green line represents one realization of the solution. The gray lines represent non-HM solution realizations (Tripoppoom et al., 2019). *HM*, History matching.

see the gradual improvement of the proxy compared to the simulation results. The points get closer to a unit slope line as more iterations continue. Fig. 4.14C also shows the objective function estimated from multiple response-parameter proxies versus the actual one from simulation runs. The improving trend is clearly seen when the iteration progresses and there are more measured points for proxy modeling. One thing to be aware of is that the objective in using a proxy for HM is not to create the most accurate proxy to represent the high nonlinearity in the actual simulation results. To achieve that, many more simulation cases have to be run as measured points. In statistics, there is a saying that most proxies are wrong; only some of them can be used. As a result, the actual objective for using a proxy is to adopt the proxy as a guide for where the Metropolis–Hastings MCMC algorithm should propose possible HM solutions in the iterative HM workflow while being aware of the inaccuracy of the proxy.

The workflow gradually improved proxy quality in the interested area in an iterative manner and explored HM solutions at the same time. First, it added more points in the possible HM solutions domain, and second, by using the farthest distance from measured points criteria, it added measured points to explore new possible HM solutions in other areas. Fig. 4.15 shows the number of solutions versus the iteration number. The number of solutions increases in the later iterations. This is because the proxy accuracy in the later iteration is better due to having more measured points.

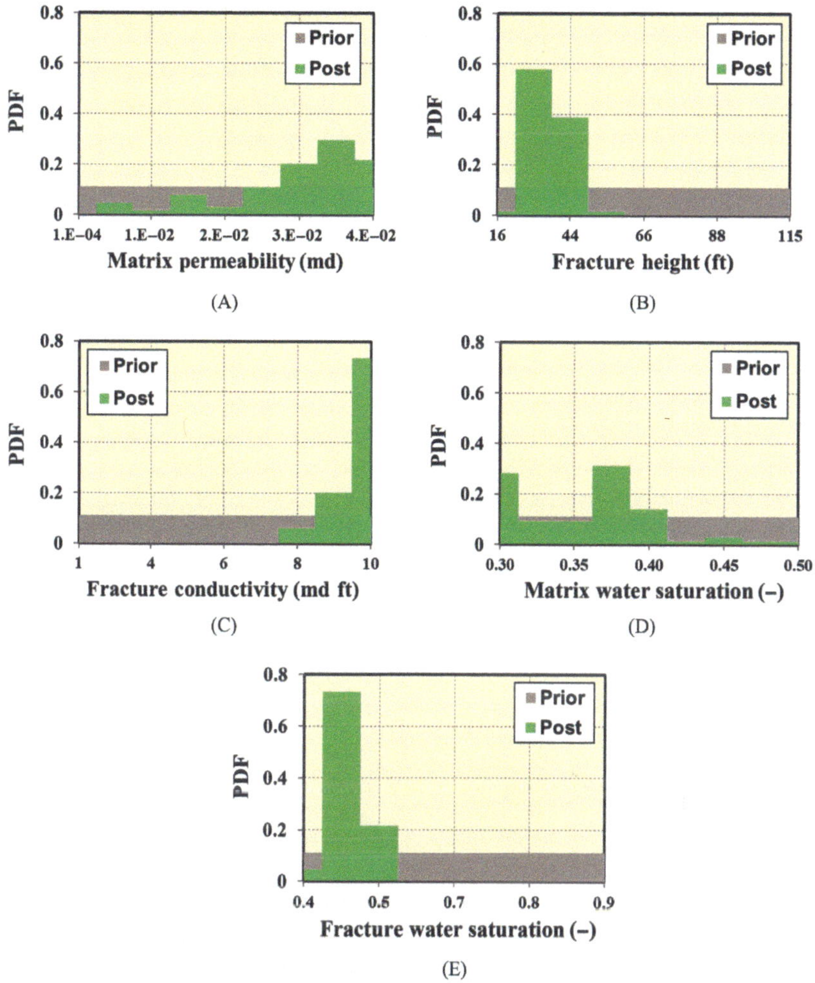

FIGURE 4.12 Comparison between prior and posterior probability density function of five parameters from 64 HM solutions: (A) matrix permeability; (B) fracture height; (C) fracture conductivity; (D) matrix water saturation; and (E) fracture water saturation (Tripoppoom et al., 2019). *HM*, History matching.

Fig. 4.16 shows the objective function and each uncertain parameter of the proposed history realization cases versus simulation run index. As shown, the objective function is improving when we have more simulation runs. This is to confirm the improvement of proxy modeling accuracy when we have more measured points. In the initial iteration where we used Latin hypercube design, we can see that the simulation run cases are scattered over the parameters domain. As the iteration progresses and more simulation runs are available, the ranges of the proposed cases in each iteration are narrowed

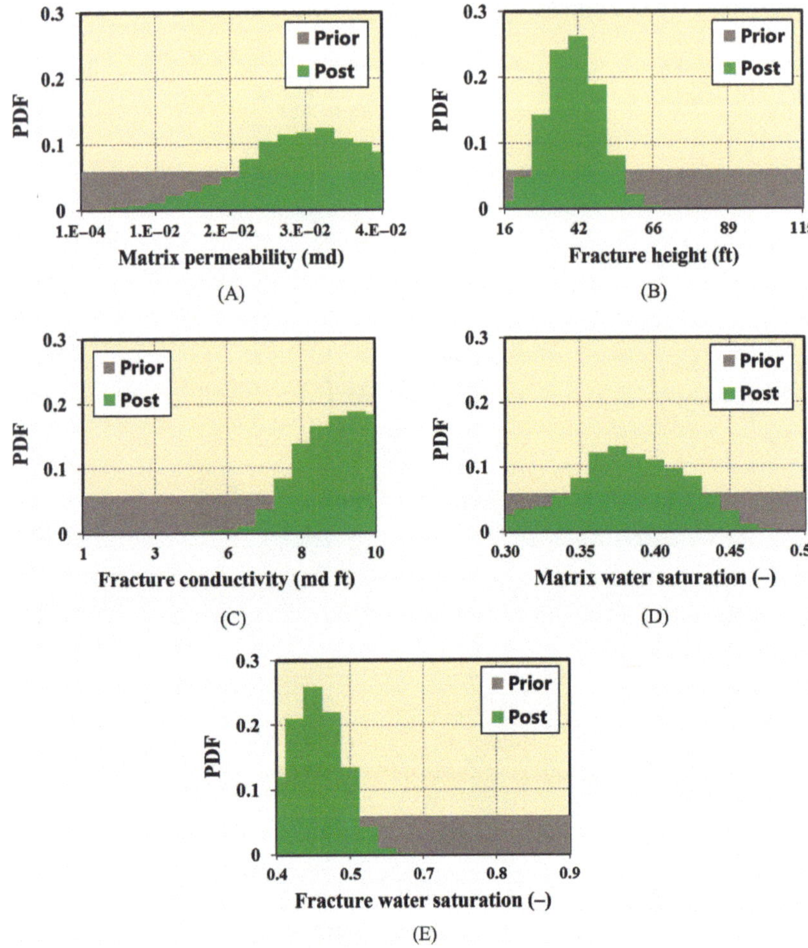

FIGURE 4.13 Comparison between prior and posterior probability density function of five parameters from 10,000 HM solutions generated by the proxy-based MCMC algorithm: (A) matrix permeability; (B) fracture height; (C) fracture conductivity; (D) matrix water saturation; and (E) fracture water saturation (Tripoppoom et al., 2019). *HM*, History matching; *MCMC*, Markov chain Monte Carlo.

down to a possible HM solutions domain for fracture height, fracture conductivity, and fracture water saturation, while there is a weaker narrowing trend for matrix permeability and matrix water saturation. As mentioned, the proxy-based MCMC algorithm samples and proposes more points around HM solutions to improve proxy accuracy in the areas of interest, while not wasting too many simulation runs in other areas. These other areas are still available to find new possible HM solutions domains.

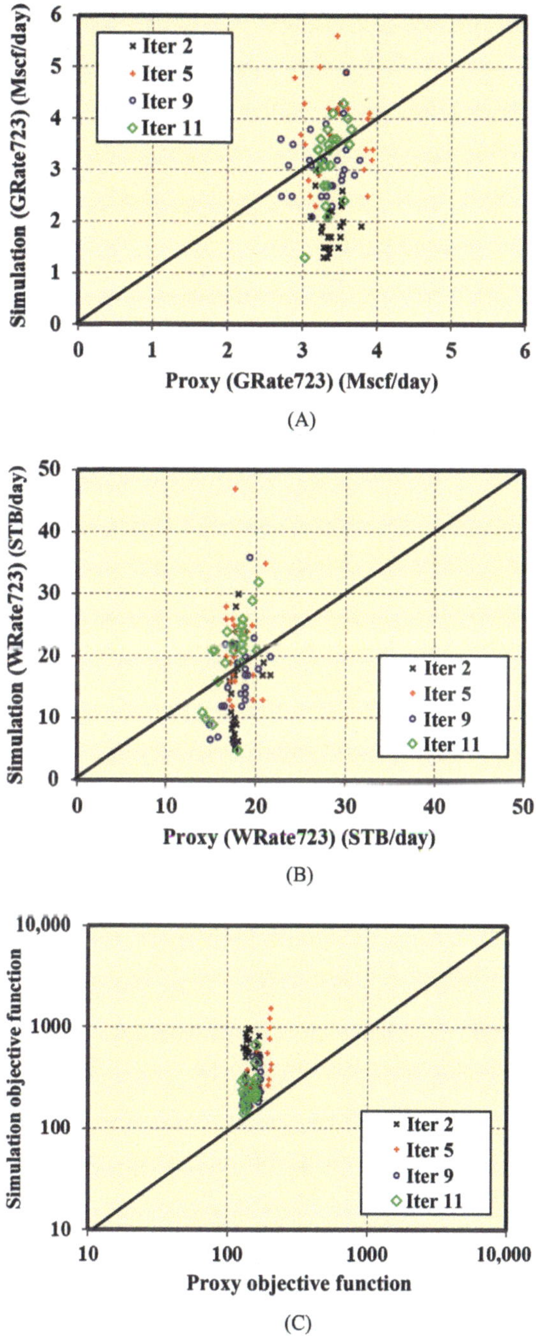

FIGURE 4.14 Cross plot of simulation values and proxy-estimated values showing the improvement of proxy predictability as the iteration progresses: (A) gas flow rate at day 723 (GRate723); (B) water flow rate at day 723 (WRate723); and (C) objective function (global error) (Tripoppoom et al., 2019).

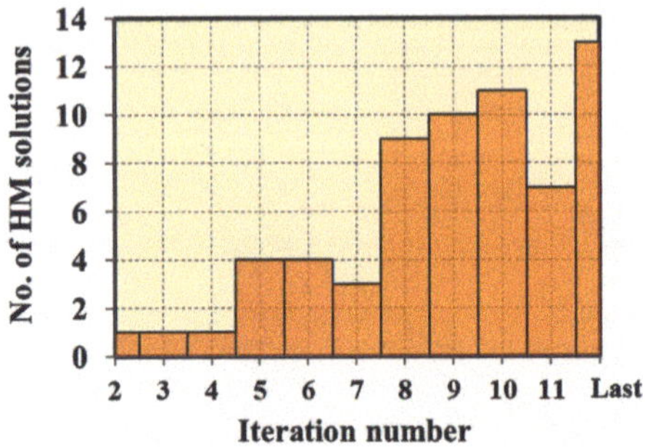

FIGURE 4.15 Number of HM solutions versus iteration number (Tripoppoom et al., 2019). *HM*, History matching.

Fig. 4.16 also shows that the proxy-based MCMC algorithm discovers more HM solutions (*red dot*) in the later iteration where proxies have more measured points. In addition, the algorithm can still find new solutions in each new iteration and is not trapped in any local HM solutions as the iteration progresses, especially for matrix permeability and matrix water saturation. In addition, we can see that the HM solutions for fracture height, fracture conductivity, and fracture water saturation domain quickly converge to specific ranges because these three parameters have more impact on the objective function. This confirms that the proxy-based MCMC algorithm is an efficient and practical way to explore HM solutions and capture the uncertainty of HM realizations as posterior distribution.

4.6 Probabilistic production forecasting

After obtaining HM solutions from the iterative HM workflow, we performed the production forecast. The code we developed will automatically generate the simulation files for production forecasting of all 64 HM solutions. We used a BHP of 500 psi as a production constraint after the HM period. The production forecast period is 30 years. The results of gas flow rate, water flow rate, cumulative gas production, and cumulative water production, including the best match case and P10–P50–P90 cases, are shown in Fig. 4.17.

The empirical cumulative distribution function (ECDF) of gas estimated ultimate recovery (EUR) from all 64 HM solutions is shown in Fig. 4.18A. P10, P50, and P90 of gas EUR are 15.1, 18.8, and 20.3 Bscf, respectively, while the gas EUR at the end of the history is 3.65 Bscf. The P90/P10 ratio of gas EUR is 1.34. The realization corresponding to P50 gas EUR has values for each parameter as follows: matrix permeability of 0.027 md, fracture height of 38 ft, fracture conductivity of 8.9 md ft, matrix water saturation of 0.37, and fracture water saturation of 0.51. For the best match case, the gas EUR is 19.3 Bscf.

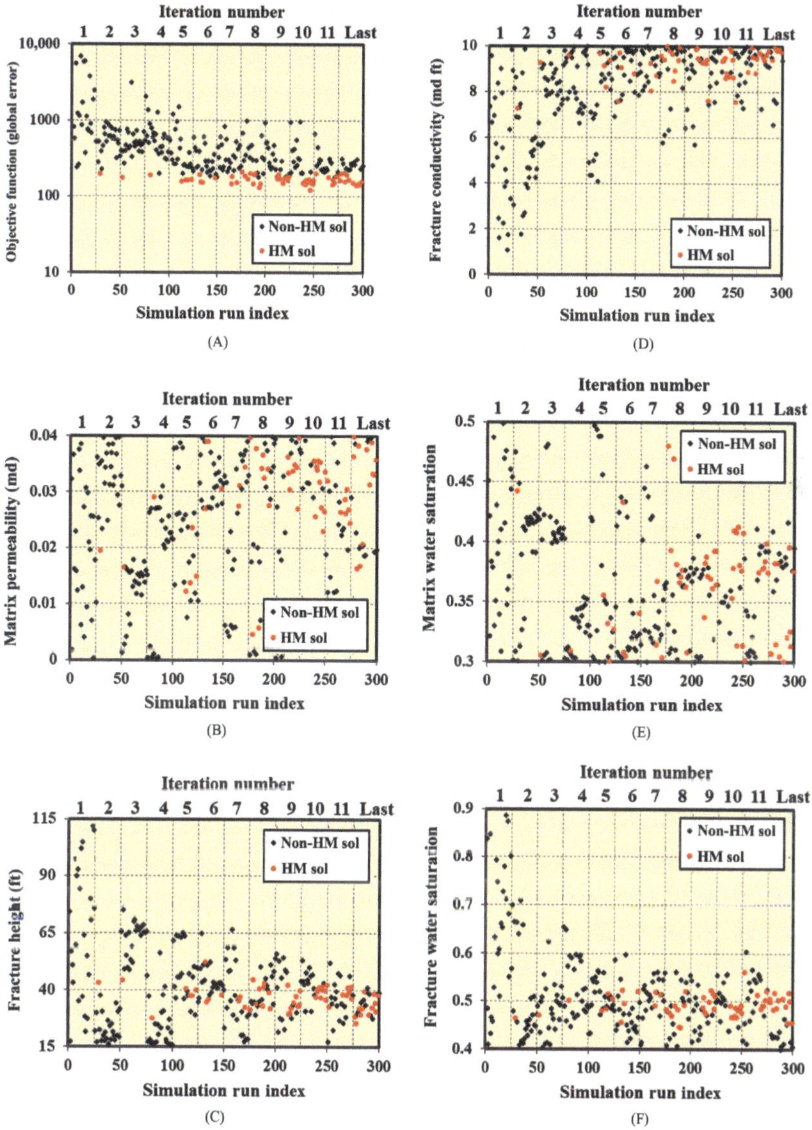

FIGURE 4.16 Proposed points (HM solutions in red, non-HM solutions in black) versus simulation run index and iteration number: (A) objective function (global error); (B) matrix permeability; (C) fracture height; (D) fracture conductivity; (E) matrix water saturation; and (F) fracture water saturation (Tripoppoom et al., 2019). *HM*, History matching.

To illustrate, we also generated 10,000 HM solutions via the proxy-based MCMC algorithm and constructed a gas EUR proxy to estimate gas EUR for all 10,000 HM solutions. The comparison of the ECDF for gas EUR of all

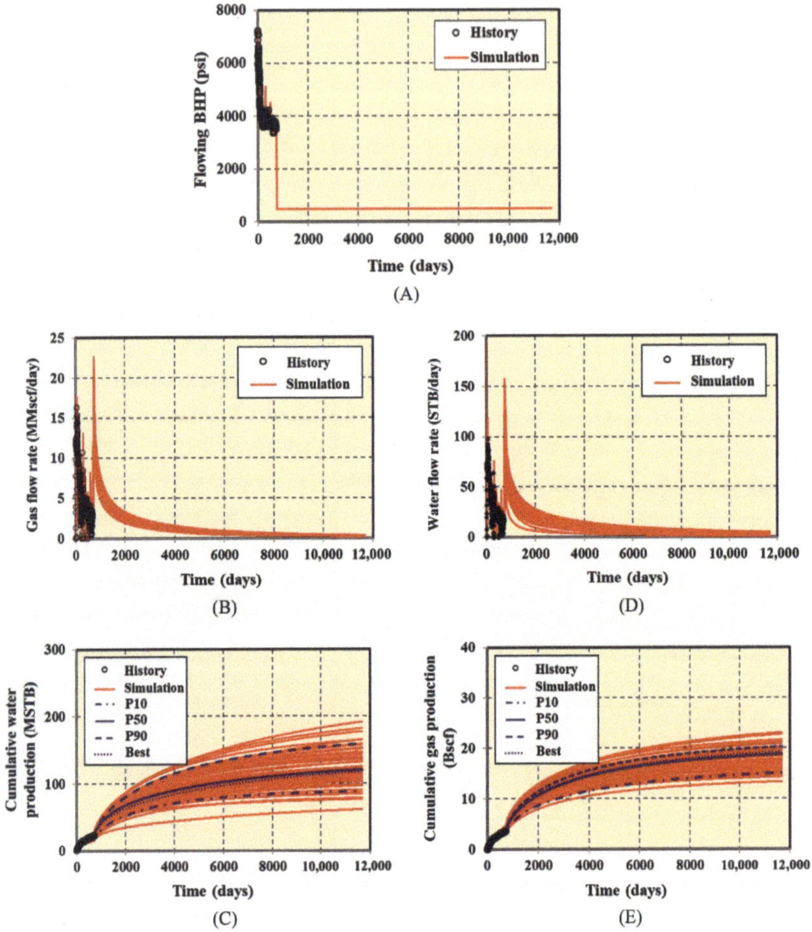

FIGURE 4.17 Posterior production forecast of 64 HM solutions: (A) flowing BHP as constraint; (B) gas flow rate; (C) cumulative gas production with P10–P50–P90 and the best match case; (D) water flow rate; and (E) cumulative water production with P10–P50–P90 and the best match case (Tripoppoom et al., 2019). *BHP*, Bottomhole pressure; *HM*, history matching.

10,000 and the 64 HM solutions is shown in Fig. 4.18B. P10, P50, and P90 of gas EUR from 10,000 HM solutions are 16.8, 19.4, and 20.7 Bscf, respectively. The P50 value from both 10,000 and 64 HM solutions is similar, while the P10 and P90 from the 10,000 HM solutions are narrower. This difference is from the accuracy of the gas EUR estimated from the gas EUR proxy. The gas EUR proxy predictability depends on the number of measured points, in this case 64 actual HM solutions, and how they are distributed over the uncertain-parameter space. The computational time for probabilistic forecasting is around 1 hour.

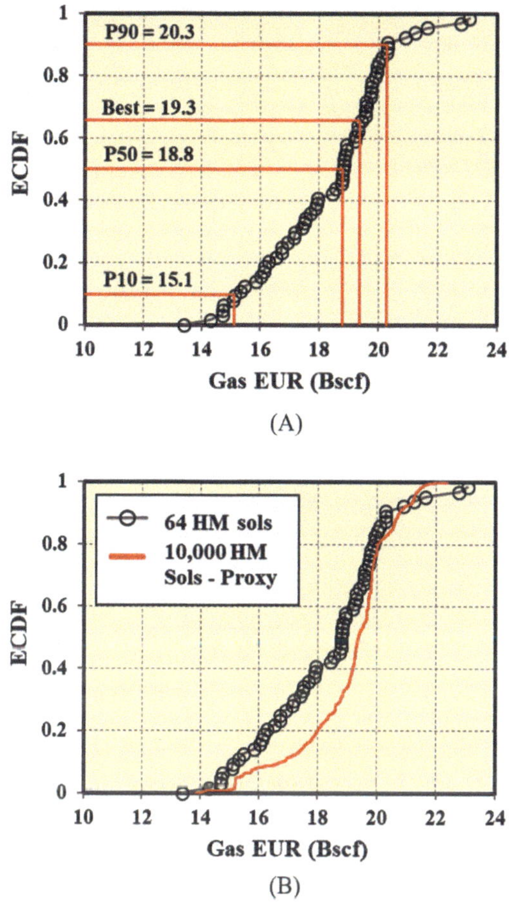

FIGURE 4.18 ECDF of gas EUR: (A) 64 HM solutions; and (B) comparison between 64 HM solutions and 10,000 HM solutions from the proxy-based MCMC algorithm (Tripoppoom et al., 2019). *ECDF*, Empirical cumulative distribution function; *EUR*, estimated ultimate recovery; *HM*, history matching; *MCMC*, Markov chain Monte Carlo.

Finally, pressure distributions at different times are illustrated in Fig. 4.19 to show reservoir depletion and the drainage area from 1 year of production until the end of production forecast—30 years after historical period.

4.7 Remarks

The AHM workflow was applied to history match and perform probabilistic production forecasting for the real case of a shale-gas well in the Longmaxi Shale, Sichuan Basin in China. In this case study, we concluded as follows:

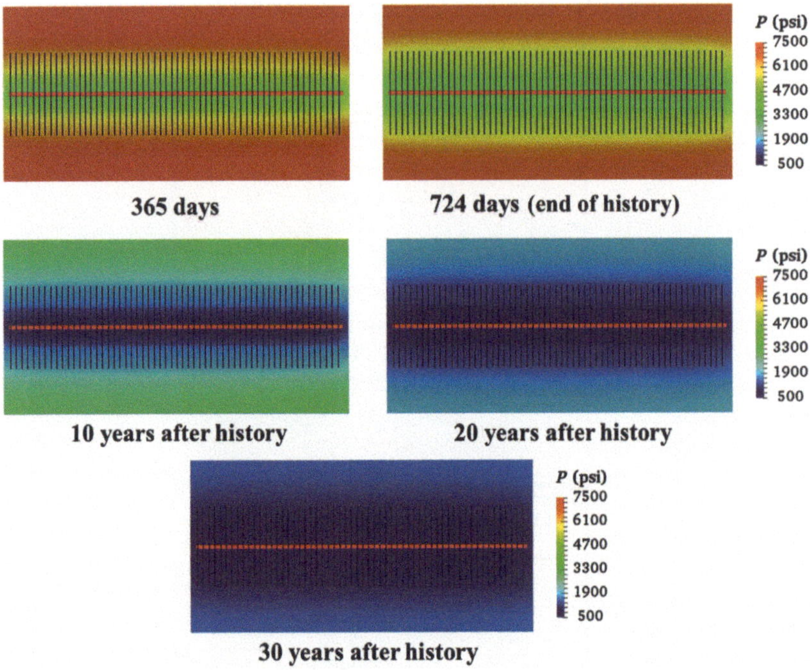

FIGURE 4.19 Pressure distributions of the best match case (Tripoppoom, 2019).

1. The AHM workflow efficiently performed HM of gas flow rate and water flow rate for a shale-gas well. The workflow found 64 HM solutions with 5 uncertain parameters from 11 iterations or 300 total simulation runs, which is around 21%. The computational time from parameter screening, HM, and probabilistic forecasting is around 1 day.
2. The posterior distributions of five uncertain parameters were obtained from the AHM workflow. The reservoir and fracture property uncertainties were effectively characterized and narrowed down.
3. The values with the highest frequency of each parameter are as follows: matrix permeability of 0.031 md, fracture height of 39 ft, fracture conductivity of 9.5 md ft, matrix water saturation of 0.37, and fracture water saturation of 0.46.
4. The P50 gas EUR realization has properties as follows: matrix permeability of 0.027 md, fracture height of 38 ft, fracture conductivity of 8.9 md ft, matrix water saturation of 0.37, and fracture water saturation of 0.51.
5. With a flowing BHP of 500 psi after the HM period, the probabilistic production forecast of gas EUR in this shale-gas well is 15.1, 18.8, and 20.3 Bscf for P10, P50, and P90, respectively.

References

Aderibigbe, A., Valdes, C.C., Heidari, Z., 2016. Integrated rock classification in the Wolfcamp Shale based on reservoir quality and anisotropic stress profile estimated from well logs. Interpretation 4 (2), 1–18.

Amin, S., Wehner, M., Heidari, Z., Tice, M., 2016. Rock classification in the Eagle Ford Shale through integration of petrophysical, geological, and geochemical characterization. In: Paper SPWLA-2016-KKK, Presented at SPWLA 57th Annual Logging Symposium, Reykjavik, Iceland, June 25–29, 2016.

Box, G.E.P., Cox, D.R., 1964. An analysis of transformations. Journal of the Royal Statistical Society, Series B 26 (2), 211–243.

Rubin, B., 2010. Accurate simulation of non Darcy flow in stimulated fractured shale reservoirs. In: Paper SPE 132093, Presented at SPE Western Regional Meeting, Anaheim, CA, May 27–29, 2010.

Stat-Ease Inc, 2017. Design-Expert Software (DX11). Stat-Ease Inc, Minneapolis, MN.

Tripoppoom, S., 2019. Assisted History Matching Workflow for Unconventional Reservoirs (M. S. thesis). The University of Texas at Austin.

Tripoppoom, S., Yu, W., Huang, H., Sepehrnoori, K., Song, W., Dachanuwattana, S., 2019. A practical and efficient iterative history matching workflow for shale gas well coupling multiple objective functions, multiple proxy-based MCMC and EDFM. Journal of Petroleum Science and Engineering 176, 594–611.

Yang, C., Nghiem, L., Erdle, J., Moinfar, A., Fedutenko, E., Li, H., et al., 2015. An efficient and practical workflow for probabilistic forecasting of brown fields constrained by historical data. In: Paper SPE 175122, Presented at SPE Annual Technical Conference and Exhibition, Houston, TX, September 28–30, 2015.

Yu, W., Sepehrnoori, K., 2018. Shale Gas and Tight Oil Reservoir Simulation. Gulf Professional Publishing. Elsevier.

Yu, W., Sepehrnoori, K., Patzek, T.W., 2016. Modeling gas adsorption in Marcellus Shale with Langmuir and BET isotherms. SPE Journal 21 (2), 589–600.

Chapter 5

Shale-gas well in Marcellus Shale with bi-wing hydraulic fractures

5.1 Introduction

We applied the assisted history-matching (AHM) workflow in an actual shale-gas well with available production data [gas and water flow rates and bottomhole pressure (BHP)] of 486 days from Marcellus Shale. The well was stimulated with 10 perforation stages that used 9 clusters per stage and 1 perforation stage using 7 clusters per stage. Cluster spacing was 50 ft. A total of 105,478 bbl of water and up to 6,384,490 pounds of proppant were injected.

5.2 Reservoir model

We built a reservoir model with dimensions of 5200 ft × 2040 ft, which corresponds to reservoir length and width, respectively. The nonintrusive embedded discrete fracture model in conjunction with a compositional numerical reservoir simulator was used to model 97 hydraulic fractures, as shown in Fig. 5.1. Each fracture is assumed to fully penetrate the whole

FIGURE 5.1 A reservoir model including a horizontal well and 97 hydraulic fractures for Marcellus Shale (Yu et al., 2018).

Assisted History Matching for Unconventional Reservoirs. DOI: https://doi.org/10.1016/B978-0-12-822242-3.00005-5
© 2021 Elsevier Inc. All rights reserved.

TABLE 5.1 Basic reservoir and fracture parameters used in this study (Yu et al., 2018).

Parameter	Value	Unit
Model dimension ($x \times y$)	5200 × 2040	ft
Initial reservoir pressure	3980	psi
Reservoir temperature	130	°F
Residual water saturation	20%	–
Total compressibility	3×10^{-6}	psi^{-1}
Reservoir depth	6179	ft
Well length	4534	ft
Number of stage	11	–
Clusters per stage (10 stages)	9	–
Clusters per stage (one stage)	7	–
Cluster spacing	50	ft

TABLE 5.2 Six uncertain reservoir and fracture parameters considered in this study (Yu et al., 2018).

Uncertain parameters	Coded symbol	Minimum	Maximum	Unit
Porosity	A	5%	13%	–
Permeability	B	0.0001	0.001	md
Fracture half-length	C	300	600	ft
Fracture conductivity	D	1	10	md ft
Fracture height	E	40	135	ft
Fracture water saturation	F	30%	90%	–

reservoir thickness. A grid block dimension 20 ft × 20 ft in x and y directions is applied. There is one block in the z direction and fracture height is uncertain. The other basic reservoir and fracture parameters used in the following simulations are listed in Table 5.1. Table 5.2 summarizes six uncertain

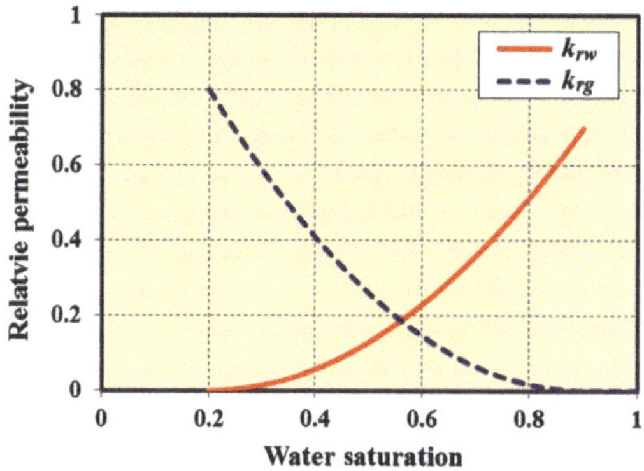

FIGURE 5.2 Relative permeability curve for modeling gas and water flow in shale-gas reservoirs (Yu et al., 2018).

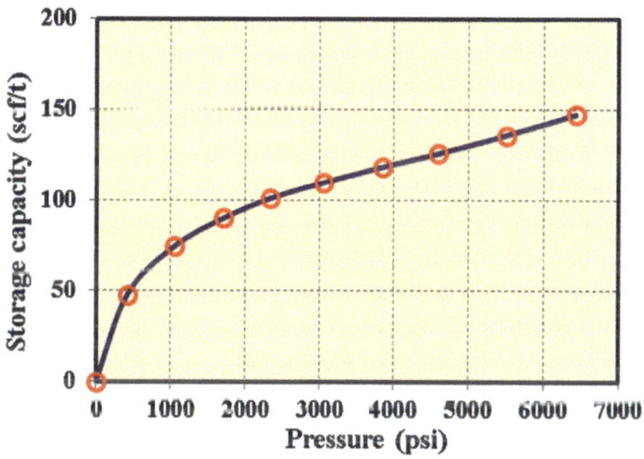

FIGURE 5.3 Measured gas content versus pressure from Marcellus Shale used in this study (Yu et al., 2018).

parameters including porosity, permeability, fracture half-length, fracture conductivity, fracture height, and fracture water saturation. Each uncertain parameter is given a reasonable range. The relative permeability curve for modeling gas and water flow in Fig. 5.2 and the gas desorption effect in Fig. 5.3 were used in the simulations. We used the measured flowing BHP in Fig. 5.4 as the simulation constraint and aimed to match gas flow rate and water flow rate.

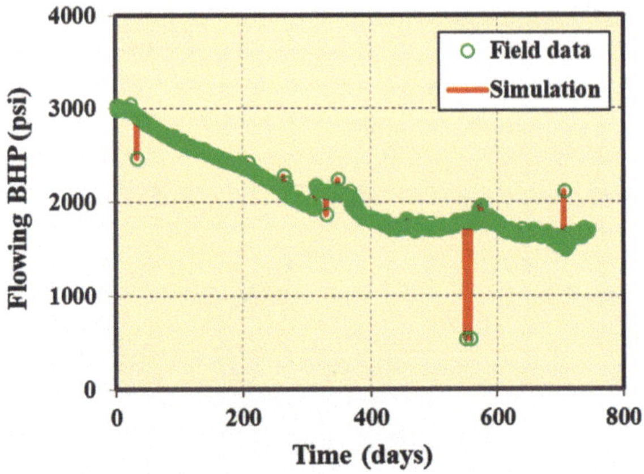

FIGURE 5.4 Measured flowing BHP as history-matching input for the shale-gas well (Yu et al., 2018). *BHP*, Bottomhole pressure.

5.3 Sensitivity analysis

Utilizing six variables, 32 simulation cases are generated based on two-level fractional factorial design. After numerical simulation of each case, comparisons of simulated gas and water production profiles of 32 cases against actual field data are shown in Figs. 5.5 and 5.6, respectively. It can be clearly seen that there is a wide range for gas and water production rates and that actual field data are effectively covered by the simulation results. It confirms that history-matching solutions can be guaranteed to be found within the given design space of six uncertain parameters. The rank of the significant parameters is determined by the *t*-value of their absolute effect, as shown in the Pareto plot of Fig. 5.7. It indicates that the interaction of D (fracture conductivity) and E (fracture height) is the most important parameter, followed by F (fracture water saturation), D, the interaction term of EF, ACF (A is the porosity and C is the fracture half-length), ABF (B is the permeability), and ADE. Therefore, all six uncertain parameters should be considered in the following history-matching process.

5.4 History matching

The first iteration with 25 initial design points for history matching is randomly generated based on the Latin hypercube design method from the design space of all six uncertain parameters. After performing production simulations of all cases, the *k*-nearest neighbors proxy model will be built. After that, another 25 design points per each iteration are sampled through

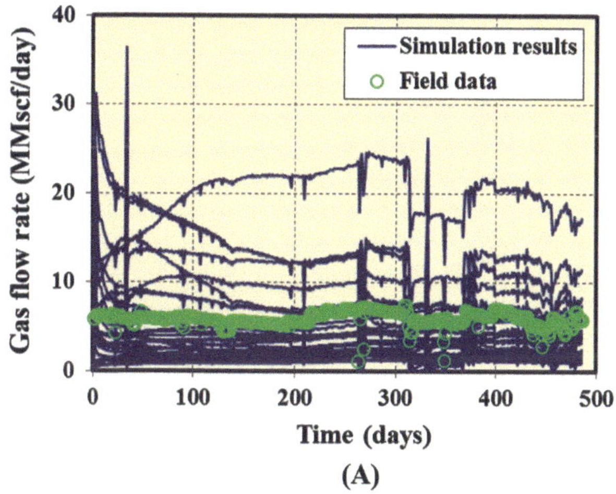

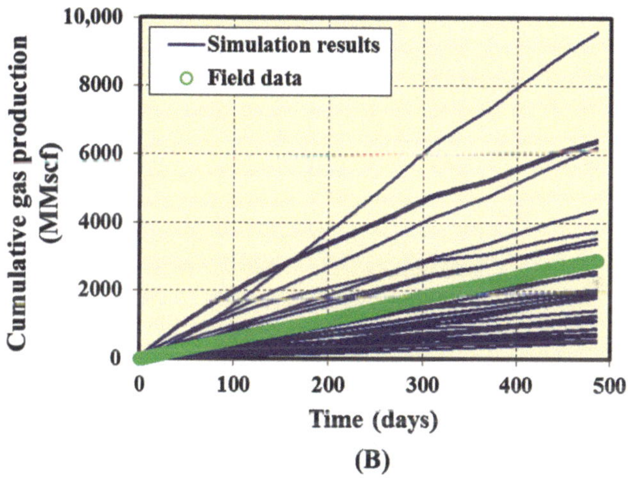

FIGURE 5.5 Comparison of simulated gas production profiles of 32 cases from the two-level fractional factorial experiment against actual field data: (A) gas flow rate and (B) cumulative gas production (Yu et al., 2018).

the Markov chain Monte Carlo (MCMC) method. In addition, the quality of the proxy models will be improved during the iterations. For proxy convergence, this is to evaluate whether the current proxy does not significantly change from the previous proxy. One hundred thousand random cases will be generated and a proxy-estimated global error from the current and previous proxy is calculated. Then, global error discrepancies of all 100,000 random cases are averaged and evaluated as to whether the average reaches the

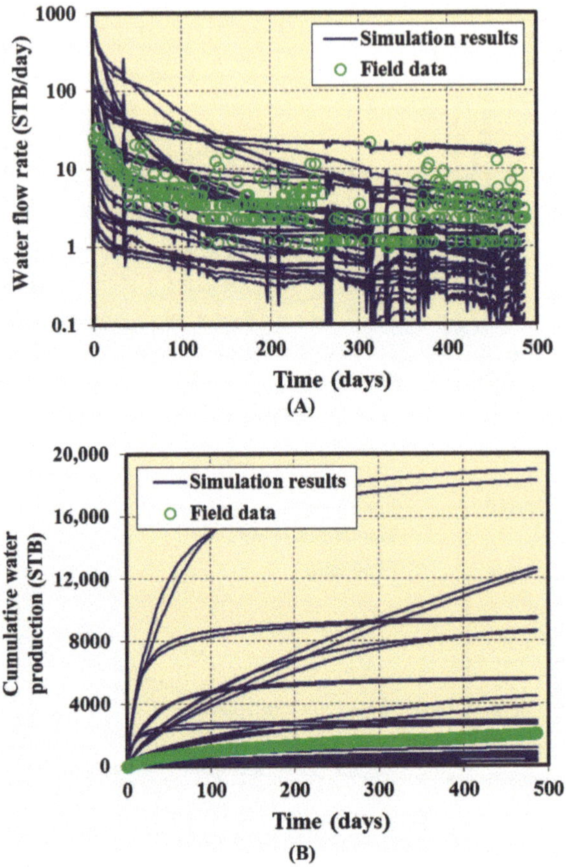

FIGURE 5.6 Comparison of simulated water production profiles of 32 cases from the two-level fractional factorial experiment against actual field data: (A) water flow rate and (B) cumulative water production (Yu et al., 2018).

specified proxy convergence threshold. In this study, a threshold discrepancy of 5% is used. The trend of relative discrepancy with increasing iterations is presented in Fig. 5.8. In this field case study, the history-matching workflow terminates at the iteration number of 10. After the termination of the iterative history-matching workflow, all simulation runs are screened by the tolerance to retrieve all history-matching solutions. There are, in total, 250 simulations and, of those, 52 history-matching solutions are determined based on the tolerance of a global error of 210 for gas rate response and 100 for water rate response.

Figs. 5.9 and 5.10 show the comparisons of 52 history-matching solutions for gas and water production profiles against actual data, respectively. As

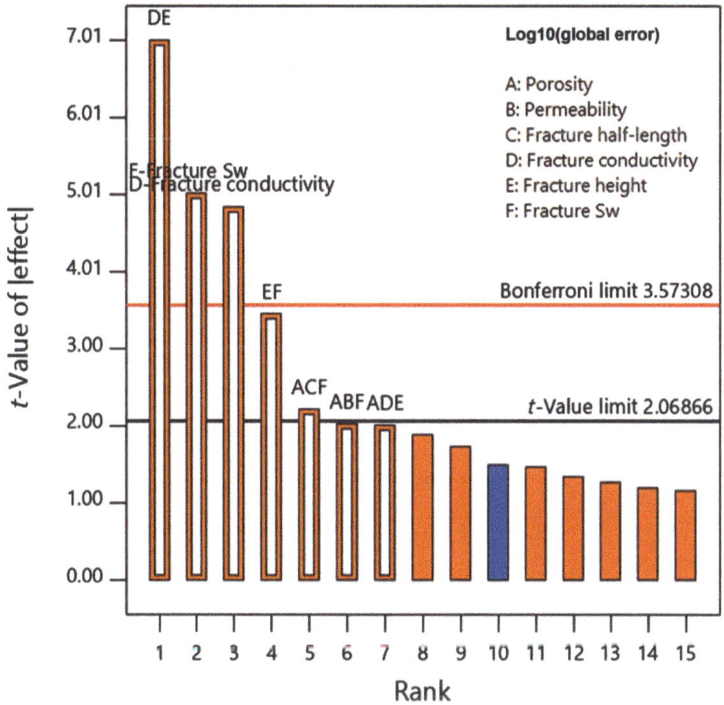

FIGURE 5.7 Pareto plot of global history-matching error based on the two-level fractional factorial design (Yu et al., 2018).

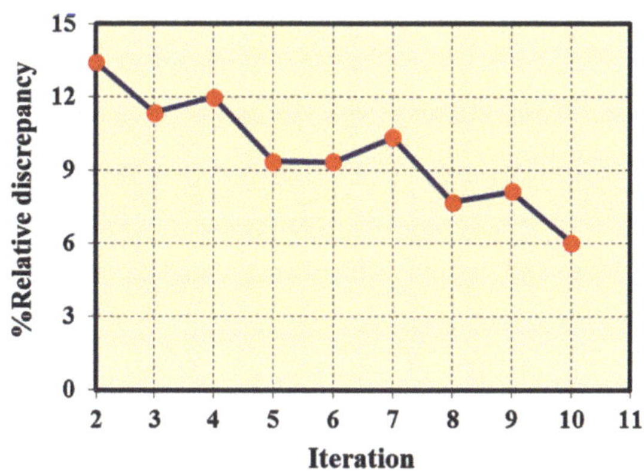

FIGURE 5.8 Trend of relative discrepancy with the increasing iterations during the automatic history-matching process (Yu et al., 2018).

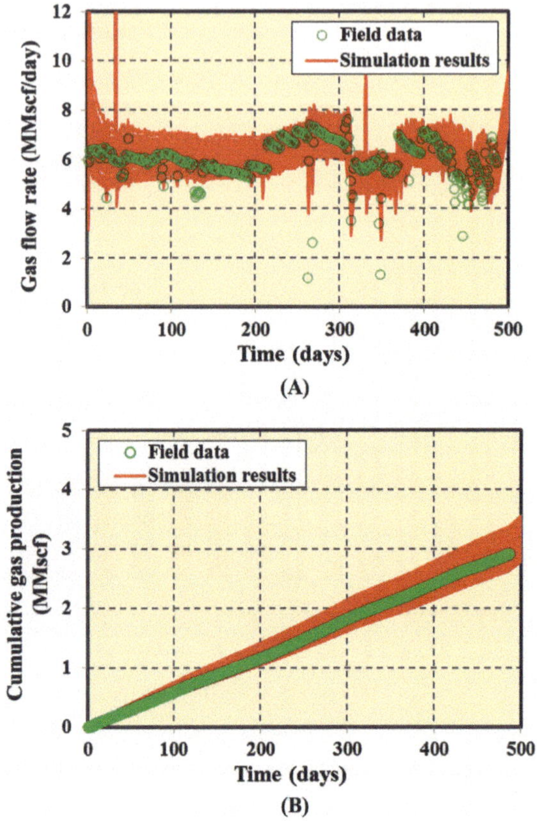

FIGURE 5.9 Comparison of 52 history-matching solutions for gas production profiles from the workflow against actual data: (A) gas flow rate and (B) cumulative gas production (Yu et al., 2018).

can be seen, these solutions have reasonable match results with the actual production data.

In addition, the best match case for both gas and water flow behaviors is shown in Figs. 5.11 and 5.12, respectively. As shown, great matches between simulation results and field data were achieved. The best match realization has properties as follows: matrix porosity of 12.4%, matrix permeability of 0.000525 md, fracture half-length of 417 ft, fracture conductivity of 2.4 md ft, fracture height of 90.4 ft, and fracture water saturation of 0.40.

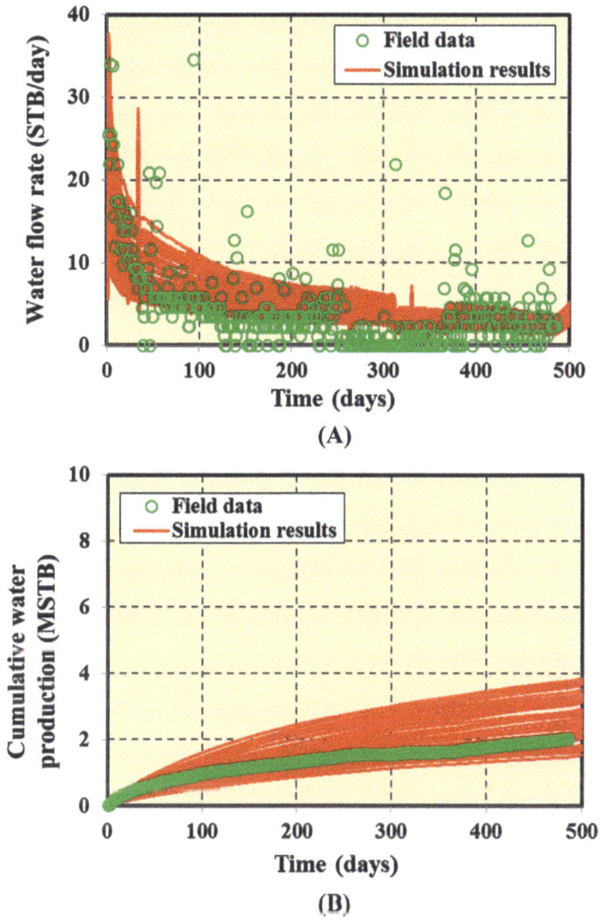

FIGURE 5.10 Comparison of 52 history-matching solutions for water production profiles from the workflow against actual data: (A) water flow rate and (B) cumulative water production (Yu et al., 2018).

5.5 Posterior distribution of matrix and fracture parameters

Fig. 5.13 displays the posterior distribution of matrix porosity, matrix permeability, fracture half-length, fracture conductivity, fracture height, and fracture water saturation from 10,000 history-matching solutions generated by the proxy-based MCMC algorithm. As shown, the posterior distribution for each uncertain parameter becomes narrower than the original range with a

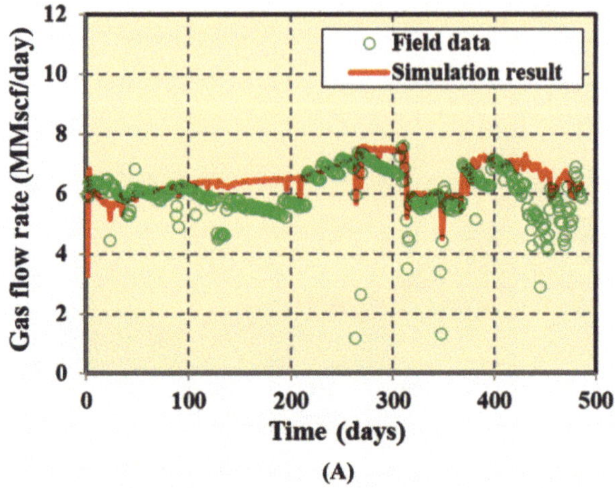

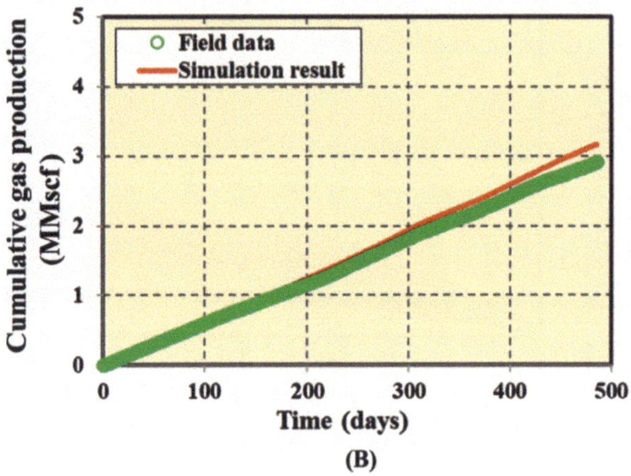

FIGURE 5.11 Comparison of the best history-matching solution for gas production profiles from the workflow against actual data: (A) gas flow rate and (B) cumulative gas production.

uniform distribution. Hence, the uncertainty is significantly reduced based on the assisted history-matching workflow. In addition, it provides critical insights for understanding effective fracture properties for this actual well stimulation. Finally, the values with the highest frequency for each uncertain parameter are captured: porosity is 10.4%, permeability is 0.00034 md, fracture half-length is 450 ft, fracture conductivity is 2.85 md ft, fracture height is 87.5 ft, and fracture water saturation is 38.8%.

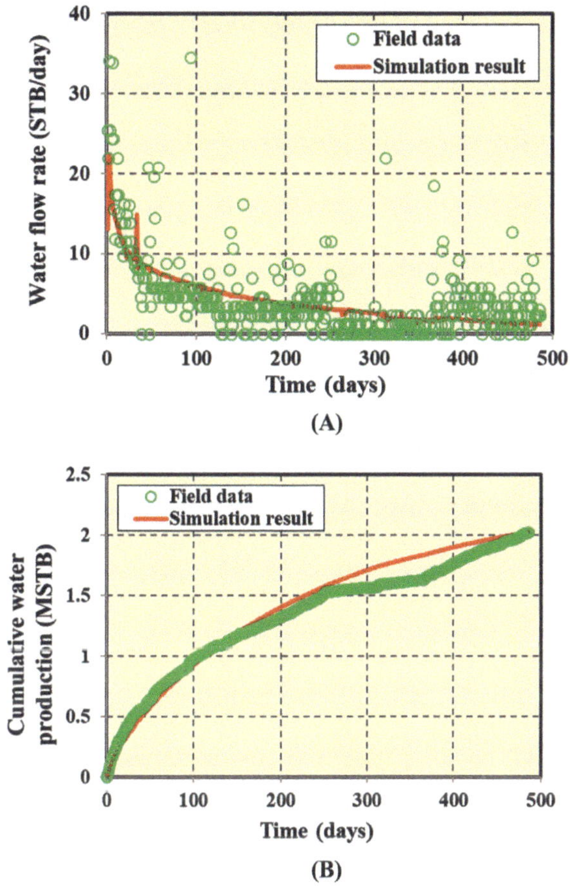

FIGURE 5.12 Comparison of the best history-matching solution for water production profiles from the workflow against actual data: (A) water flow rate and (B) cumulative water production.

5.6 Probabilistic production forecasting

After the history-matching period, we performed production forecasting for 30 years using 52 history-matching solutions at a given BHP of 500 psi, as shown in Fig. 5.14. Predicted gas and water production profiles are presented in Figs. 5.15 and 5.16, respectively. As shown, the P10, P50, and P90 of the estimated ultimate recovery (EUR) of gas production at the end of 30 years are 11.9 Bcf (billion cubic feet), 13.1 Bscf, and 16.4 Bscf, respectively.

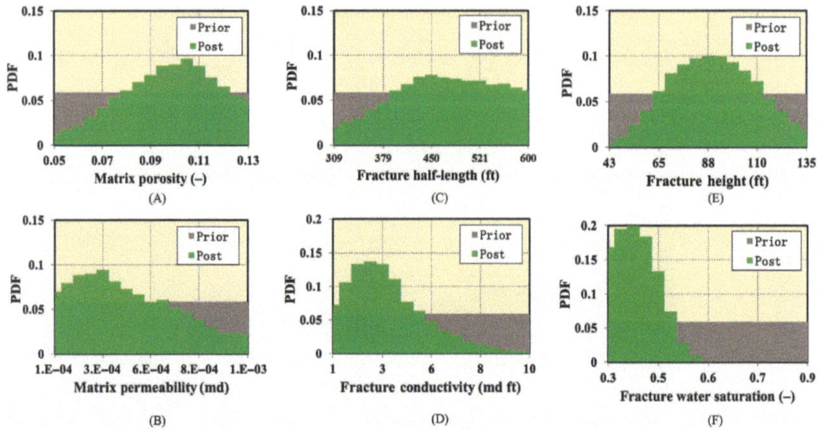

FIGURE 5.13 Posterior distributions of six uncertain parameters based on the history-matching workflow: (A) matrix porosity; (B) matrix permeability; (C) fracture half-length; (D) fracture conductivity; (E) fracture height; and (F) fracture water saturation (Yu et al., 2018).

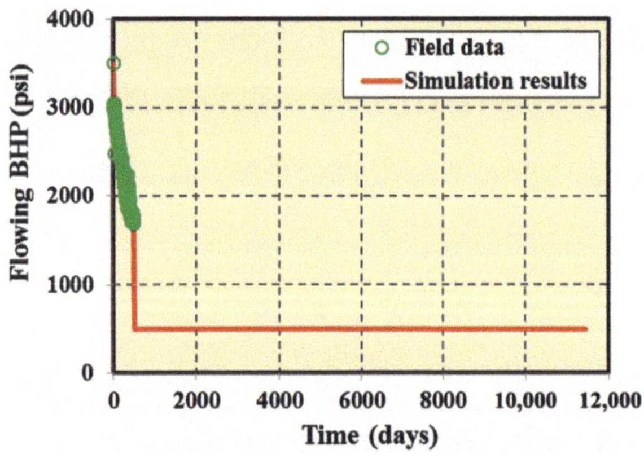

FIGURE 5.14 Flowing BHP used for long-term production forecasting (Yu et al., 2018). *BHP*, Bottomhole pressure.

Finally, pressure distributions at different production times are shown in Fig. 5.17, clearly illustrating reservoir depletion and the drainage area from 1 year of production until the end of production of 30 years.

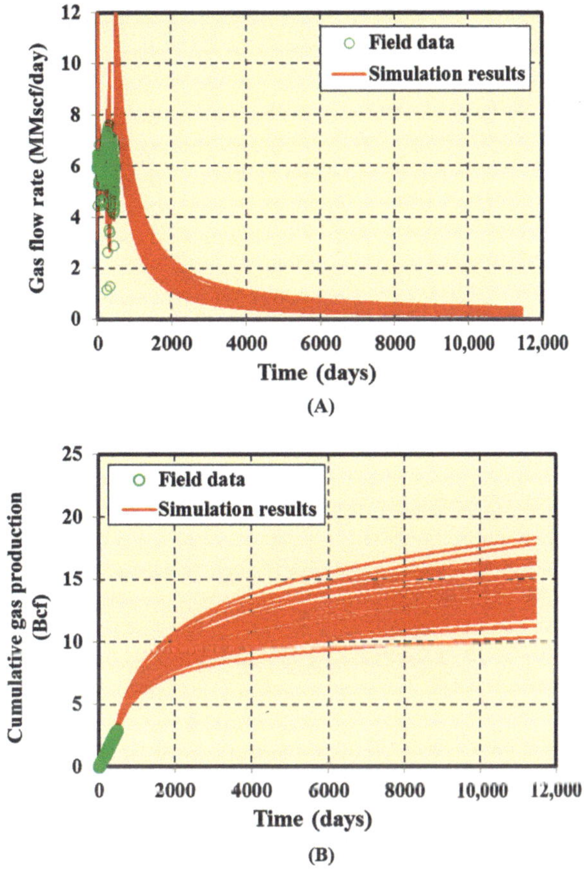

FIGURE 5.15 Predicted gas production profiles of 52 history-matching solutions from the workflow: (A) gas flow rate and (B) cumulative gas production (Yu et al., 2018).

5.7 Remarks

We applied the AHM workflow to analyze an actual shale-gas well with gas and water flow rates from the Marcellus Shale. The following conclusions can be drawn from this study:

1. The AHM workflow automatically determined 52 history-matching solutions with reasonable match results against actual gas and water flow rates using 10 iterations.

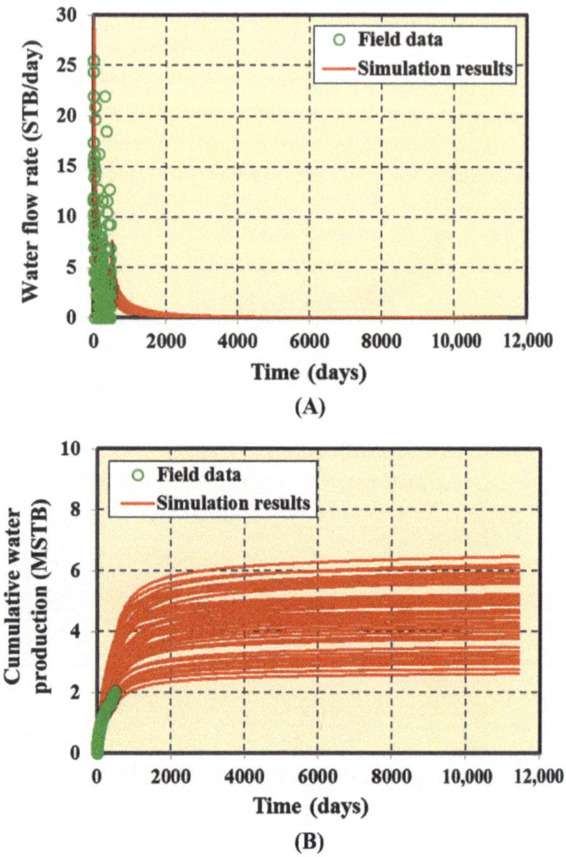

FIGURE 5.16 Predicted water production profiles of 52 history-matching solutions from the workflow: (A) water flow rate and (B) cumulative water production (Yu et al., 2018).

2. The AHM workflow generated more reliable P10, P50, and P90 of EUR predictions of gas recovery after 30 years for this well, which are 11.9, 13.1, and 16.4 Bcf, respectively.
3. The workflow identified narrower posterior distributions for matrix porosity and permeability and fracture properties such as fracture half-length, fracture conductivity, fracture height, and fracture water saturation, which are very useful for understanding the stimulation effectiveness of the well.
4. The values with the highest frequency for each uncertain parameter are determined: porosity is 10.4%, permeability is 0.00034 md, fracture half-length is 450 ft, fracture conductivity is 2.85 md ft, fracture height is 87.5 ft, and fracture water saturation is 38.8%.

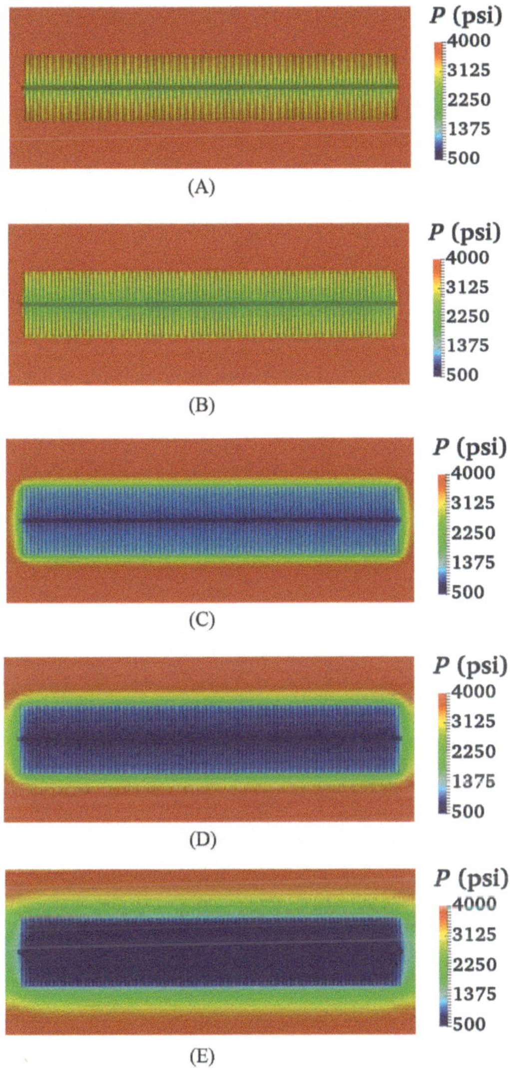

FIGURE 5.17 Pressure distributions of the best match case at different production times: (A) 1 year; (B) 486 days (end of history); (C) 5 years after history; (D) 10 years after history; and (E) 30 years after history.

Reference

Yu, W., Tripoppoom, S., Sepehrnoori, K., Miao, J., 2018. An automatic history-matching workflow for unconventional reservoirs coupling MCMC and non-intrusive EDFM methods. In: Paper SPE 191473, Presented at SPE Annual Technical Conference and Exhibition, Dallas, TX, September 24–26, 2018.

Chapter 6

Proxy comparison between neural network and *k*-nearest neighbors

6.1 Introduction

In this study, we used an actual shale gas well with 74-day historical production data. We had direct measurements of production data including gas and water flow rates, whereas the bottom-hole pressure (BHP) data were estimated from the wellhead pressure (WHP) with pressure-loss correlation. The pressure loss was estimated from the well model, which is a function of reservoir depth, water–gas ratio (WGR), gas flow rate, and WHP. We performed history matching (HM) by using two different proxies including *k*-nearest neighbors (KNN) and neural network (NN).

6.2 Reservoir model

The well is a horizontal well with a length of 4921 ft with 30 stages of hydraulic fractures and, in each stage, three perforation clusters were shot. The total number of hydraulic fractures is 90. In cases where the static model integrating geological and petrophysical data is available, this static model can be used as the starting point for HM to capture reservoir heterogeneity. However, in this study, the static model was not available at the time of the study. Therefore we constructed a homogeneous reservoir model with a constant thickness to characterize hydraulic fracture properties and key reservoir properties from HM instead. The model dimension is 5500-ft long, 2200-ft wide, and 28-ft thick. We also assumed that the hydraulic fractures penetrate the whole reservoir thickness, the model dimension in the z axis. The number of grid blocks is 275, 22, and 1 in the x, y, and z directions, respectively. Basic reservoir and fracture properties are summarized in Table 6.1. For shale gas reservoirs, we modeled gas desorption using the Langmuir isotherm as shown in Table 6.2. Also, the effect of geomechanics during production depletion was included. The pressure-dependent permeability for both matrix and fractures was modeled to capture the permeability reduction when reservoir pressure was depleted, as shown in Eq. (6.1). The permeability modulus

Assisted History Matching for Unconventional Reservoirs. DOI: https://doi.org/10.1016/B978-0-12-822242-3.00001-8
© 2021 Elsevier Inc. All rights reserved.

TABLE 6.1 Summary of basic reservoir and fracture parameters used in this study (Tripoppoom et al., 2020).

Reservoir description	Value	Unit
Model dimension ($x \times y \times z$)	5500 × 2200 × 28	ft
Number of grid blocks ($x \times y \times z$)	550 × 22 × 1	—
Grid block dimension ($x \times y \times z$)	10 × 100 × 28	ft
Initial reservoir pressure	8847	psi
Reservoir temperature	215	°F
Residual water saturation	20%	—
Matrix water saturation	39%	—
Matrix porosity	5.6%	—
Total compressibility	3×10^{-6}	psi^{-1}
Reservoir depth	10,499	ft
Well length	4921	ft
Number of stages	30	—
Cluster per stage	3	—
Cluster spacing	55	ft

TABLE 6.2 Langmuir isotherm model parameters for the gas desorption effect (Tripoppoom et al., 2020).

Langmuir isotherm parameters	Value	Unit
Inverse-pressure parameter for the Langmuir isotherm model, CH_4	0.000985	1/psi
Maximum moles of adsorbed component per unit mass of rock, CH_4	0.071	gmole of component/ lb of rock
Rock density	158.6	lb/ft^3

of 0.055 and 0.056 was used for matrix and fractures, respectively. The reservoir model including 90 hydraulic fractures in a horizontal well modeled by embedded discrete fracture model (EDFM) is demonstrated in Fig. 6.1. The planar fractures were used in the study to estimate the effective properties of hydraulic fractures from HM because more complicated fracture network and geometry information were not available at the time of the study. However,

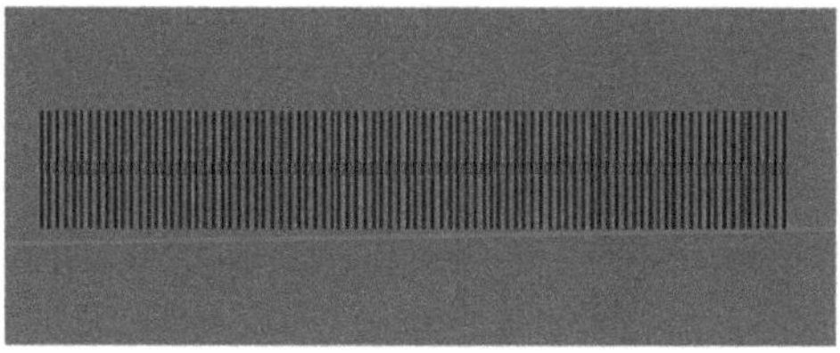

FIGURE 6.1 Reservoir model including 90 hydraulic fractures in a horizontal well (Tripoppoom et al., 2020).

with more information on fracture network and geometry, more complex fractures from the coupled three-dimensional (3D) hydraulic fracture propagation model can also be used to constrain the HM as an initial model.

$$k = k_i \exp^{-\gamma \cdot (P_i - P)} \tag{6.1}$$

where k is pressure-dependent permeability (md), k_i is permeability at initial reservoir pressure (md), γ is permeability modulus (1/psi), P_i is initial reservoir pressure (psi), and P is reservoir pressure (psi).

6.3 Parameters identification and screening

Initially, we identified eight uncertain parameters for HM. The ranges of each uncertain parameter with uniform distribution were determined based on the prior knowledge of the well, as summarized in Table 6.3. The uncertain parameters can be grouped into two categories, fracture and matrix properties. We also included relative permeability curve parameters in uncertain parameters.

Then, we screened uncertain parameters for HM by performing the two-level fractional factorial design. For the screening purpose, we generated the 2^{8-2} or 64 cases to investigate each parameter effect and partial interactions of parameters. We did not investigate all interactions as full factorial design, which requires 2^8 or 256 cases. We constrained the production by gas flow rate and monitored BHP and the WGR. The BHP is calculated from the available WHP and pressure loss estimation, as shown in Fig. 6.2A. The pressure loss is related directly to depth, as calculated by Eq. (6.2). However, to account for a high initial WGR, we also added an extra pressure loss of 4000 psi at day 1, estimated from the pressure loss at initial WGR and this well reservoir depth, and then linearly declining to 0 psi at day 30, as shown in Fig. 6.2B.

$$\text{BHP} = \text{WHP} + 0.1\text{psi/ft} \times \text{depth} \tag{6.2}$$

For WGR, the reason that we selected WGR instead of matching water flow rate is that the WGR can capture the relationship between gas and water flow rate. This can be related to the relative permeability curve parameters

TABLE 6.3 Summary of eight uncertain parameters and their prior distributions (Tripoppoom et al., 2020).

Code	Uncertain parameters	Unit	Distribution	Min value	Max value	Sources
A	Matrix permeability	md	Uniform	0.00001	0.0001	• Nearby well core • Diagnostic Fracture Injection Test (DFIT) analysis
B	Fracture half-length	ft	Uniform	200	1000	• Historical microseismic data • History-matched parameters from other wells
C	Fracture conductivity	md ft	Uniform	1	100	• DFIT analysis
D	Fracture water saturation	—	Uniform	0.6	0.9	• History-matching experience
E	Exponent of k_{ro}	—	Uniform	1	4	
F	Endpoint of k_{rw}	—	Uniform	0.5	1	
G	Exponent of k_{rw}	—	Uniform	1	4	
H	Fracture width	ft	Uniform	0.01	5	

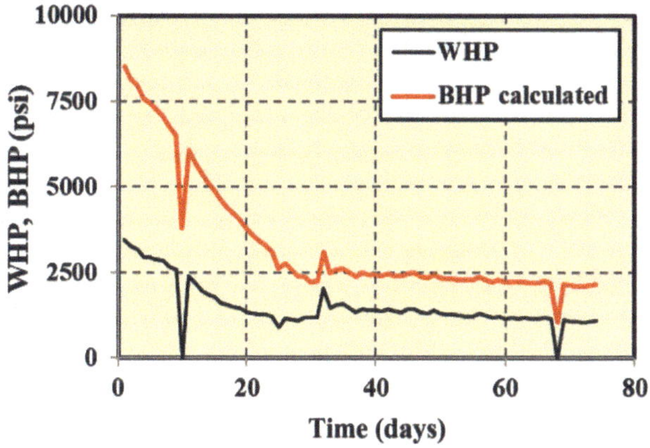

(A)

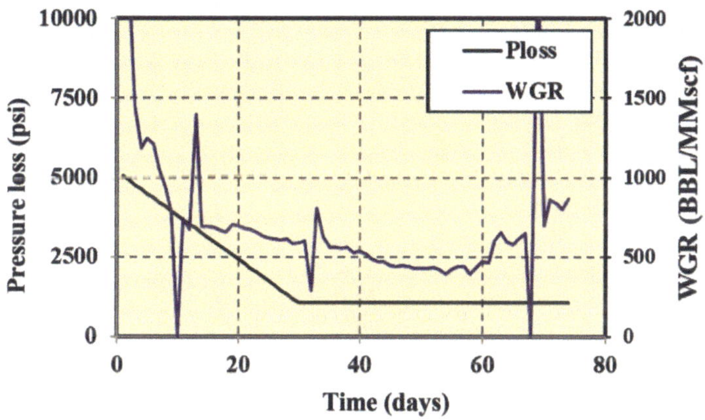

(B)

FIGURE 6.2 Calculate BHP from WHP by assuming pressure loss in the wellbore: (A) WHP and calculated BHP; and (B) pressure loss and WGR (Tripoppoom, 2019). *BHP*, Bottom-hole pressure; *WGR*, water–gas ratio; *WHP*, wellhead pressure.

and water saturation. Also, the WGR trend normally starts at high values and shifts to lower constant values so we can use a smaller number of WGR proxy points to match the same historical gas and water flow rate. If we used water flow rate, we may need more proxy points to capture the up and down of water flow rate values, causing more computational time in the workflow.

To model hydraulic fractures, we used EDFM preprocessor coupling with a compositional reservoir simulator to create simulation files automatically. The simulation results of 64 cases are shown in Fig. 6.3. It is important to check whether the simulated data covered the historical data or not. If they did not cover the historical data, the initial ranges needed to be revised.

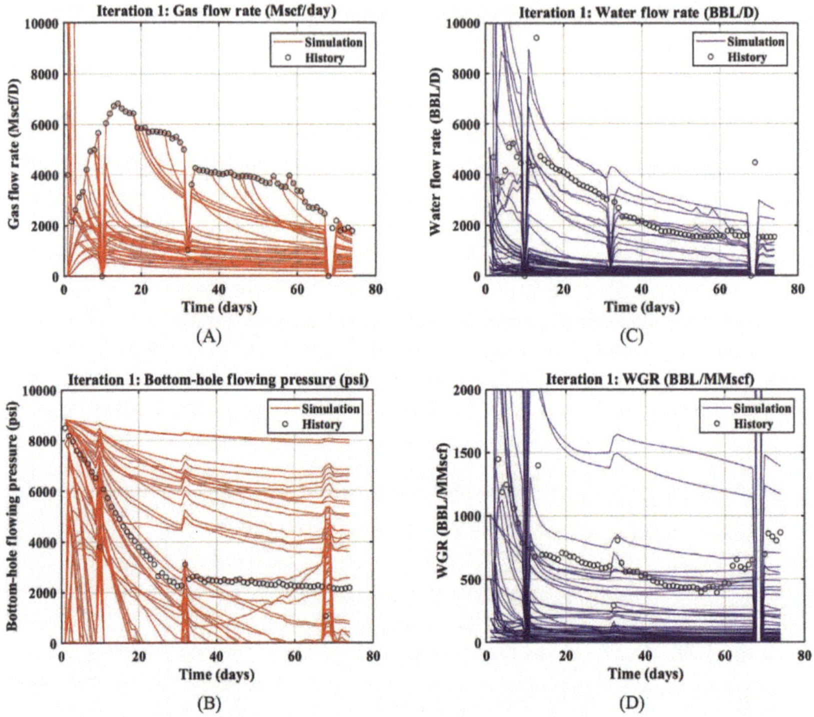

FIGURE 6.3 Comparison between simulation results and historical data of 64 cases from the two-level fractional factorial design: (A) gas flow rate; (B) flowing BHP; (C) water flow rate; and (D) water–gas ratio (Tripoppoom 2019). *BHP*, Bottom-hole pressure.

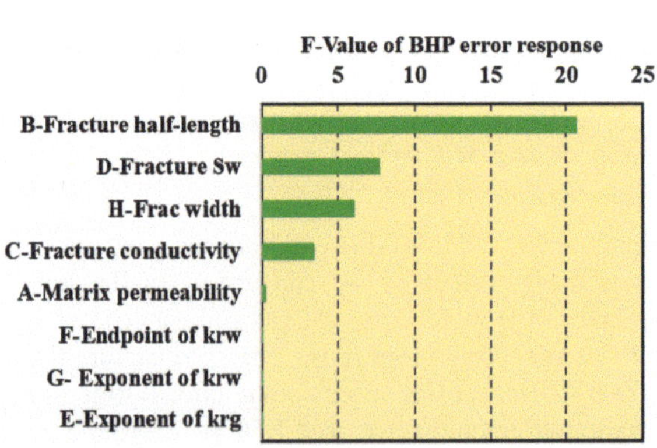

FIGURE 6.4 *F*-value ranking of the two-level fractional factorial design for the BHP error response (Tripoppoom 2019). *BHP*, Bottom-hole pressure.

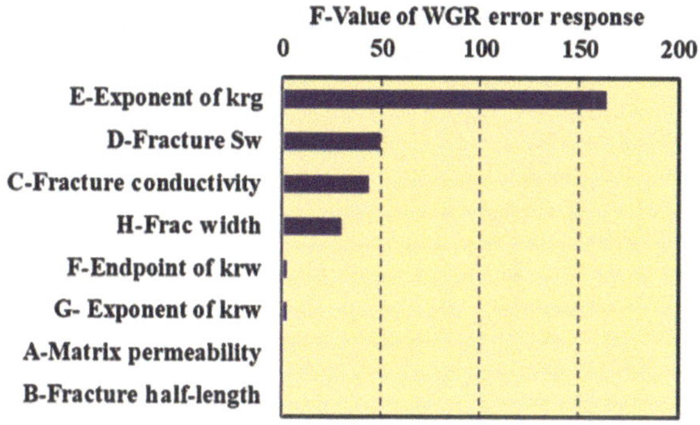

FIGURE 6.5 *F*-value ranking of the two-level fractional factorial design for the WGR error response (Tripoppoom 2019). *WGR*, Water–gas ratio.

Nevertheless, in this case, the results confirm that the specified ranges of eight uncertain parameters are acceptable and do not need the revision of the uncertain parameter range at this step.

Then, we perform a *t*-value test or *f*-value test to see what uncertain parameters have more impact on BHP and WGR. In this study, we used the *f*-value test and the plots for *f*-values of BHP error response and WGR error response of the individual parameters are shown in Figs. 6.4 and 6.5, respectively. The formula to calculate response error is shown in Eq. (6.3)

$$\text{Objective Function} = \text{global error}(\varepsilon) = \frac{\sum_{j=1}^{m}\sum_{i=1}^{n}\left[\frac{\left(x_{ij,model}-x_{ij,history}\right)}{x_{ij,history}} \times 100\right]^2 \times w_{ij}}{\sum_{j=1}^{m}\sum_{i=1}^{n} w_{ij}}, \tag{6.3}$$

where i is an index of data point, j is an index of response parameter, n is a number of data points, m is a number of response parameters, $x_{ij,model}$ is the value from model (simulation run or proxy model) at index i of response parameter j, $x_{ij,history}$ is the historical data at index i of response parameter j, and w_{ij} is the weight of data points i of response parameter j.

For the BHP error response, B, C, D, and H (fracture half-length, fracture conductivity, fracture water saturation, and fracture width) have more effect on the response. For the WGR error response, C, D, E, and H (fracture conductivity, fracture water saturation, exponent of k_{rg}, and fracture width) have more impact than the others. By considering both responses, we could screen from eight parameters to five parameters by not including matrix permeability, exponent of k_{rw}, and endpoint of k_{rw}. However, for a later demonstration, we will include all eight uncertain parameters in the HM part.

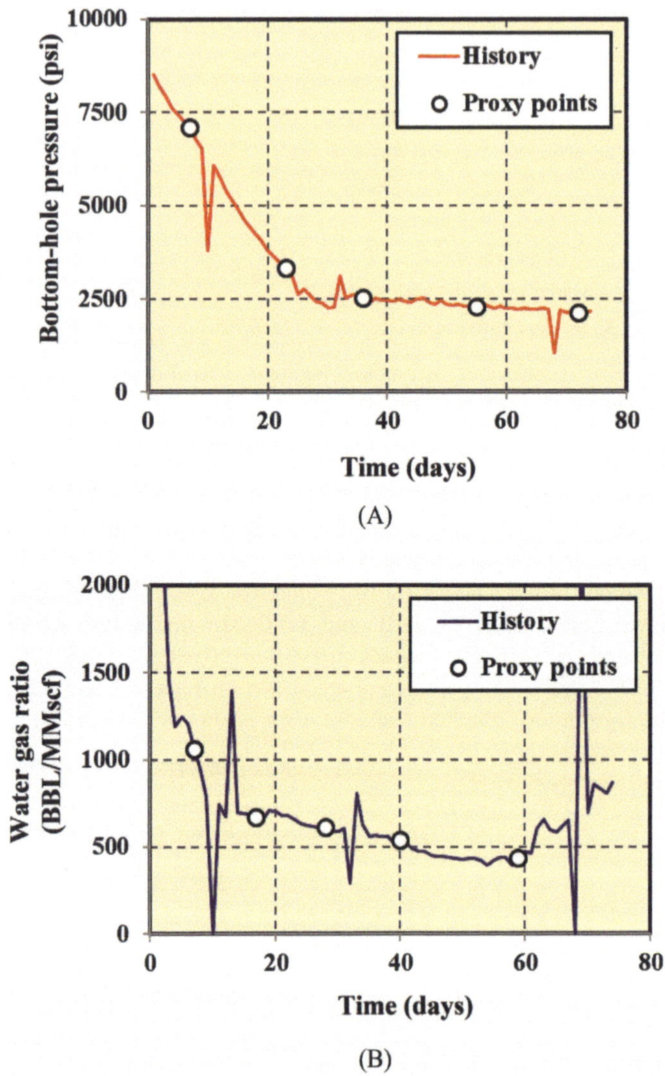

FIGURE 6.6 Ten multiple proxies including (A) five proxies of BHP; and (B) five proxies of WGR (Tripoppoom et al., 2020). *BHP*, Bottom-hole pressure; *WGR*, water–gas ratio.

Next, we had to identify the points needed to create multiple proxy models because we used multiple proxies instead of a single proxy for each error response to lessen the nonlinearity of the problem. We selected ten proxy points, five proxy points for BHP at day 7, 23, 36, 55, and 72, and another five proxy points for WGR at day 7, 17, 28, 40, and 59, as shown in Fig. 6.6. Adding more proxy points should yield more valuable information, such as different production trends. In this case, the selected 10 proxy points can capture the major trend of production data. In cases where more proxy points were selected, the workflow would need more computational time for proxy model training and proxy points value estimation in every Markov chain Monte Carlo (MCMC) step and every iteration of the HM workflow. Therefore, we should not add redundant points into the proxy point set. On the other hand, if we used a smaller number of proxy points, the quality of the final HM realization would be deteriorated because we had fewer anchor points to be matched.

In this study, the weight of data points, defined as w_{ij} in Eq. (6.3), was assigned as 2 for all points of BHP response and as 1 for all points of WGR response. This means the global error will have more effect from BHP than WGR. The reason is that BHP is a more important response parameter than WGR when matching the historical data.

6.4 History matching

After we obtained uncertain parameters to be used for HM, we generated initial iteration cases by using a Latin hypercube design to sample the initial 50 cases from eight uncertain parameters summarized as shown in Table 6.3, scattering over the uncertain parameter domain. As shown in Fig. 6.7, the concept of Latin hypercube sampling (LHS) is that the parameter domain is divided into different squares in the case of two parameters or different cubes for three parameters. Any higher dimension is also applicable with the same concept. The sampling algorithm progresses by sampling within the squares or cubes to determine that there is no sampling point already existing. Eventually, the LHS samples the points scattering the entire parameter domain. With this method, the LHS creates initial points that are good candidates for proxy modeling unlike Monte Carlo that samples randomly or the two-level factorial design that evaluates only at the maximum and minimum values (Tripoppoom et al., 2019).

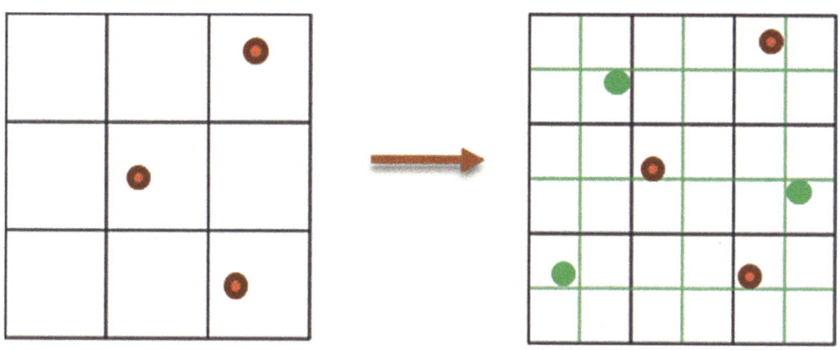

FIGURE 6.7 Latin hypercube sampling concept (Tripoppoom et al., 2020).

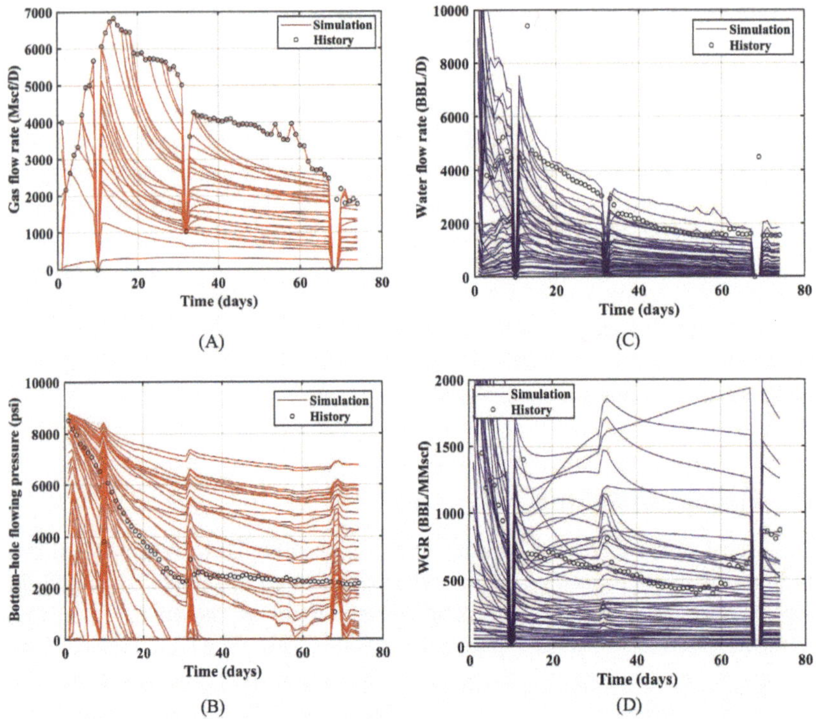

FIGURE 6.8 Comparison between simulation results and historical data of 50 cases from the Latin hypercube design: (A) gas flow rate; (B) flowing BHP; (C) water flow rate; and (D) WGR (Tripoppoom et al., 2020). *BHP*, Bottom-hole pressure; *WGR*, water–gas ratio.

Fig. 6.8 shows the results of the initial 50 cases from the Latin hypercube design. By using the Latin hypercube design, it creates simulation results that are more uniformly distributed. This will be beneficial when constructing multiple proxy models in the next step with either NN or KNN. These multiple proxies will be later used in the multiple-proxy-based MCMC algorithm to propose new possible realizations as HM solutions in the next iteration.

The iterative HM workflow continues automatically according to the developed code. We used one chain of MCMC and started the chain at the lowest global error points in each iteration. Then, we calculated the acceptance probability ratio or α^* of every MCMC step to determine whether to accept or reject each MCMC step. One parameter to be estimated is the initial variance values, σ^2. We estimated the initial variance from the variance of the global error values gleaned from the initial iteration cases (50 cases in Latin hypercube design). In cases where the acceptance ratio of MCMC, defined as the number of accepted steps to the total number of MCMC steps, was not between 15% and 50%, the variance must be adjusted. If the acceptance ratio was below 15%, the variance would be multiplied with a factor of two. On the other hand, if the acceptance ratio was above 50%, the variance would be divided by a factor of two. This continues until we can achieve an acceptance ratio within the range of 15%–50%. Finally, we obtained the MCMC ensemble after eliminating the 20% burn-in period.

Then, the proxy-based MCMC algorithm with filtering scheme proposed the next 25 cases for the next iteration to be validated with a reservoir simulator. Then, the reservoir simulation results in each iteration were included in the training data set to further train and improve the quality of the proxies. After that, the assisted history matching (AHM) workflow progressed until the stopping criteria were triggered, which was either that relative discrepancies between the current and previous proxy reached a threshold of 5% or that the maximum iteration number (12 iterations or 325 simulation runs) was reached. For this study, the workflow stopped at the maximum iteration of 12 for both NN and KNN.

Next, we screened all cases from all iterations with the HM solution criteria of 20% for both BHP and WGR errors. The algorithms found 111 HM solutions (34%) and 20 HM solutions (6%) from a total 325 simulation runs for NN and KNN, respectively, as shown in Fig. 6.9. We can observe that NN performs much better in terms of proposing a higher number of HM solutions. One thing to point out is that more solutions are found in the later iteration because the proxy model was improved by having more cases to train in the interested domain.

With the low number of HM solutions for KNN, we continued running until the number of HM solutions was approximately the same as the number of solutions from NN for the reason of comparison. In this case, we ran until

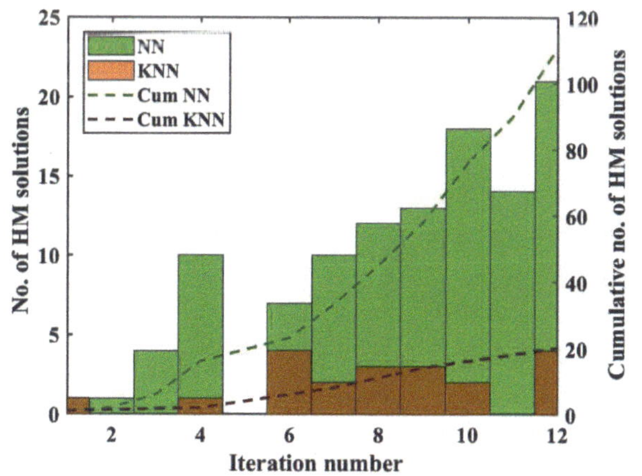

FIGURE 6.9 Comparison of the number of history matching solutions versus iteration number between NN and KNN until iteration 12 (Tripoppoom et al., 2020). *KNN*, *k*-nearest neighbors; *NN*, neural network.

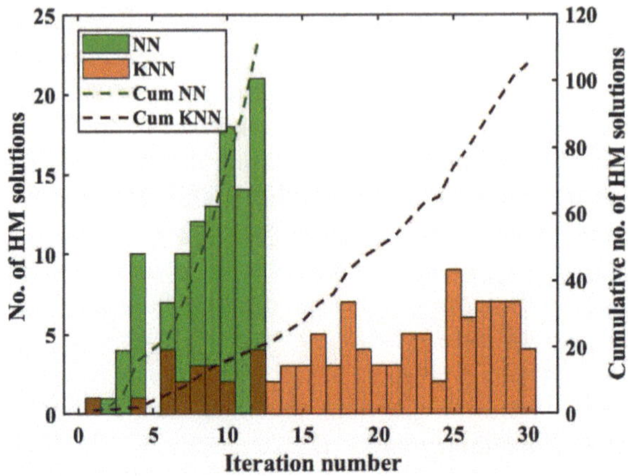

FIGURE 6.10 Comparison of the number of history matching solutions versus iteration number between NN (until iteration 12) and KNN (continued running until obtaining around 100 history matching solutions) (Tripoppoom et al., 2020). *KNN*, *k*-nearest neighbors; *NN*, neural network.

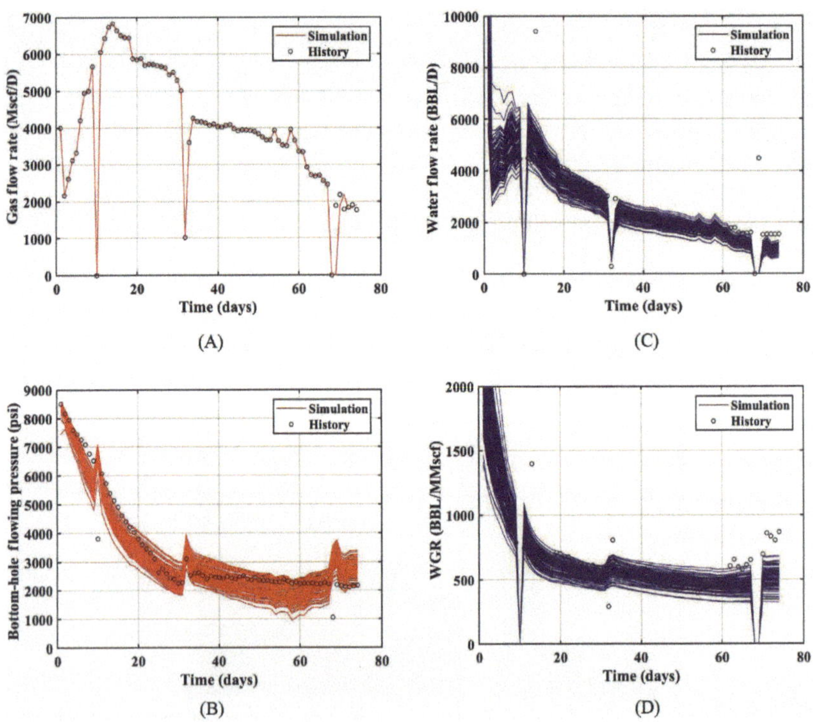

FIGURE 6.11 Comparison between simulation results and historical data of 111 history matching solution of using NN as a proxy model: (A) gas flow rate; (B) flowing BHP; (C) water flow rate; and (D) WGR (Tripoppoom et al., 2020). *BHP*, Bottom-hole pressure; *WGR*, water–gas ratio.

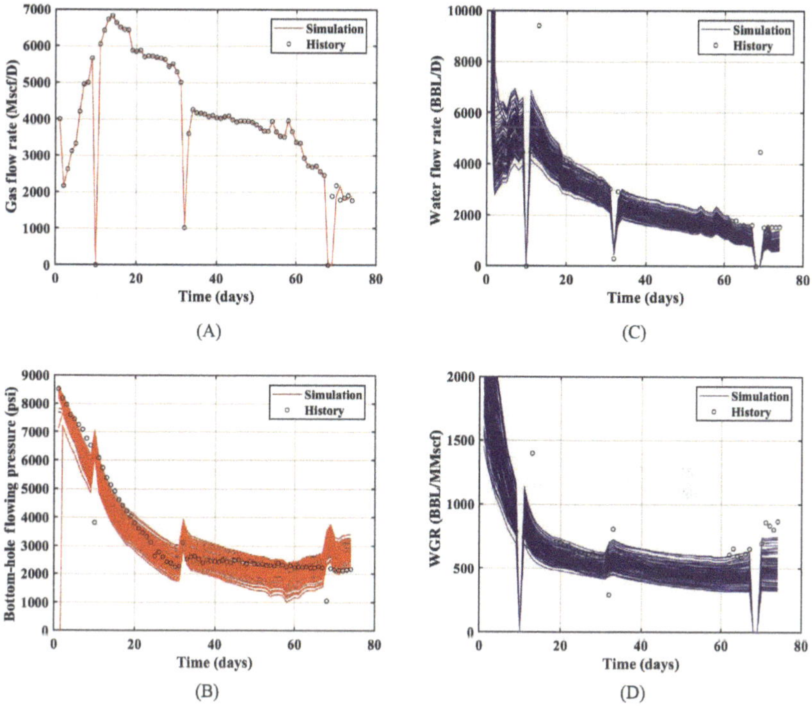

FIGURE 6.12 Comparison between simulation results and historical data of 105 history matching solution of using KNN as a proxy model: (A) gas flow rate; (B) flowing BHP; (C) water flow rate; and (D) WGR (Tripoppoom et al., 2020). *BHP*, Bottom-hole pressure; *WGR*, water–gas ratio.

30 iterations were completed and obtained 105 HM solutions from 775 simulation runs (14%), as shown in Fig. 6.10. The simulation results of HM solutions for NN and KNN are plotted in Figs. 6.11 and 6.12.

Next, we obtained the posterior distribution of HM solutions from the AHM workflow for both NN and KNN, as shown in Fig. 6.13. Overall, both NN and KNN provide similar posterior distributions of each uncertain parameter. Posterior distributions of fracture half-length, fracture conductivity, fracture water saturation, and fracture width have changed significantly from the prior distribution, for example, on the higher side of fracture conductivity, above 50 md ft, fracture half-length around 400–700 ft, fracture water saturation around 0.75–0.85, and the higher side of fracture width above 2 ft. On the other hand, the posterior distribution of uncertain parameters such as matrix permeability changed less significantly from the prior distribution and any values of matrix permeability can provide HM solutions. These minor changes in posterior distribution from prior distribution imply that parameters such as matrix permeability are less significant than others.

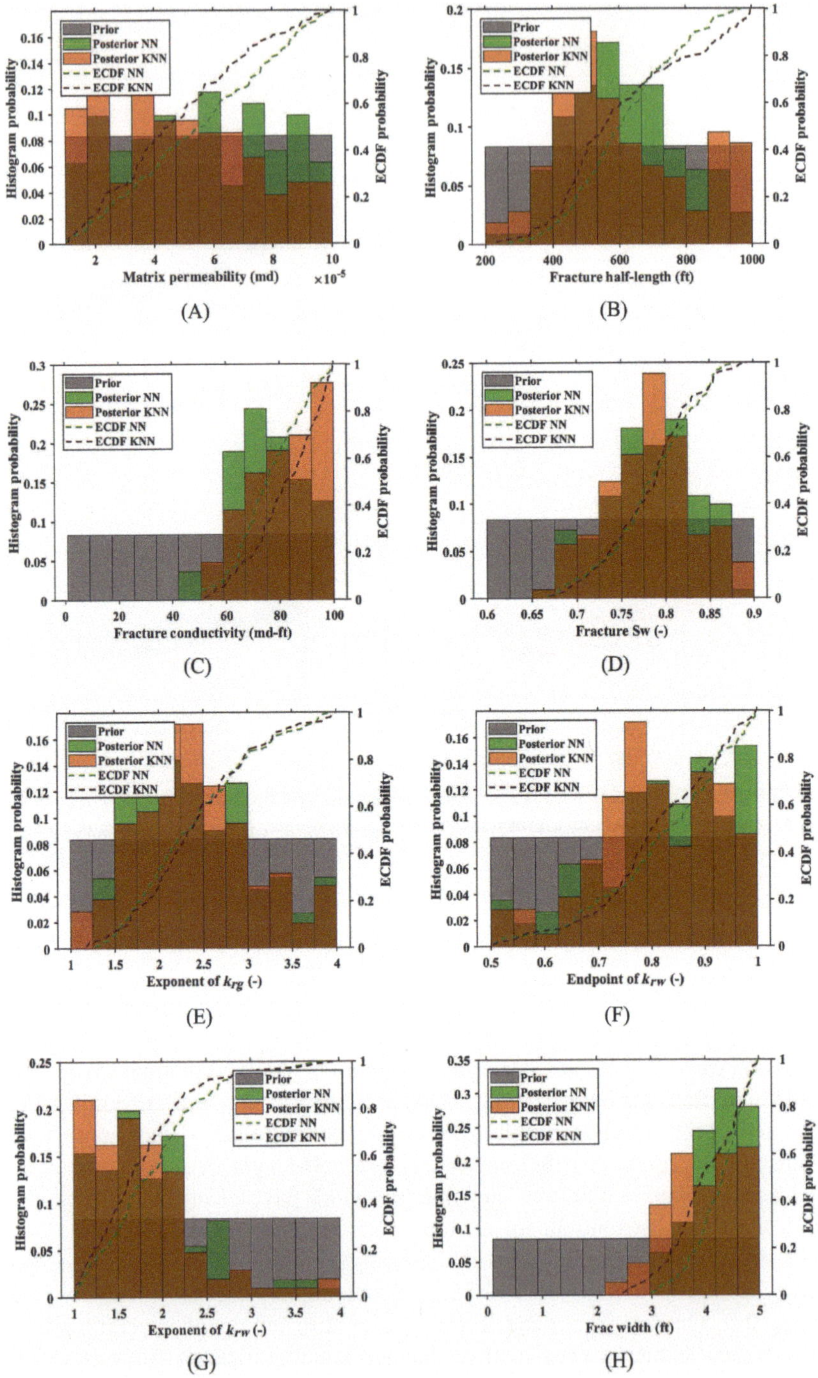

FIGURE 6.13 Comparison between prior and posterior distribution of uncertain parameters for NN (111 solutions) and KNN (105 solutions) proxy model: (A) matrix permeability; (B) fracture half-length; (C) fracture conductivity; (D) fracture water saturation; (E) exponent of k_{rg}; (F) endpoint of k_{rw}; (G) exponent of k_{rw}; and (H) fracture width (Tripoppoom et al., 2020). *KNN*, *k*-nearest neighbors; *NN*, neural network.

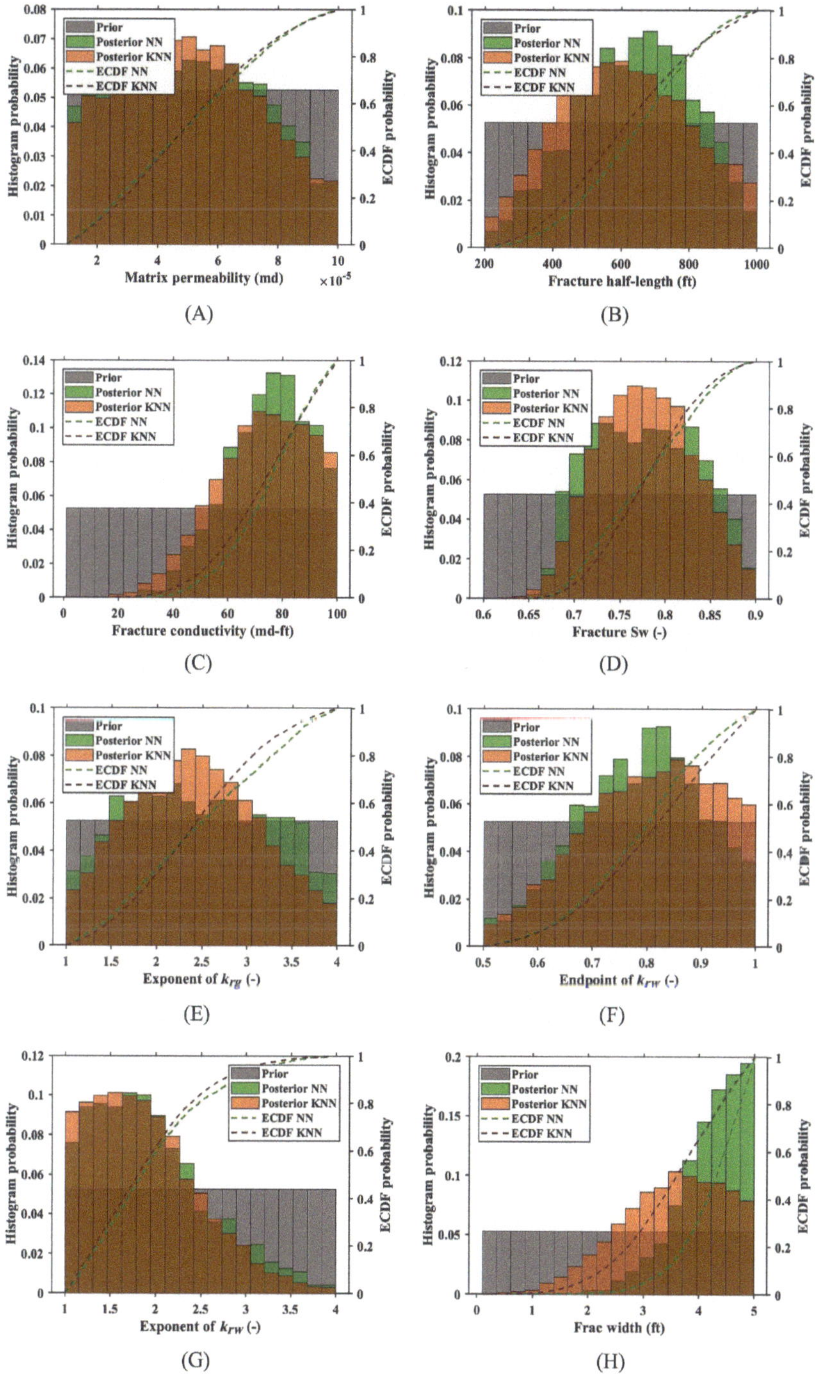

FIGURE 6.14 Comparison between prior and posterior distribution of uncertain parameters for NN and KNN proxy model (1000 solutions from proxy): (A) matrix permeability; (B) fracture half-length; (C) fracture conductivity; (D) fracture water saturation; (E) exponent of k_{rg}; (F) endpoint of k_{rw}; (G) exponent of k_{rw}; and (H) fracture width (Tripoppoom et al., 2020).

This is also in line with the previous parameter screening study that found that matrix parameters are less significant uncertain parameters. In sum, the results of in-lined posterior distribution between NN and KNN and changes in posterior distribution guarantee the accuracy and efficiency of the proxy-based MCMC algorithm that we used as a HM algorithm.

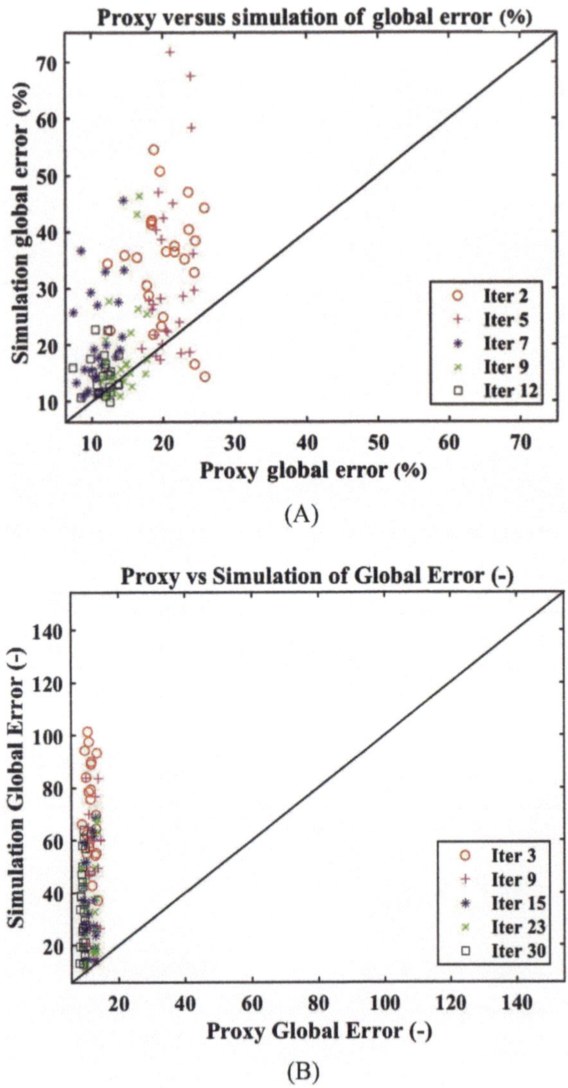

FIGURE 6.15 Proxy versus simulation values of global error: (A) neural network; and (B) *k*-nearest neighbors (Tripoppoom et al., 2020).

To obtain smoother posterior distributions, we generated 1000 solutions from proxy models for both NN and KNN, as shown in Fig. 6.14. However, one thing to be highlighted is that these 1000 solutions are not validated with the actual reservoir simulator, this means some cases will not be HM solutions when they validate with the simulator. This has been illustrated in Fig. 6.10 where KNN has a low accuracy rate at about 5 solutions out of 25

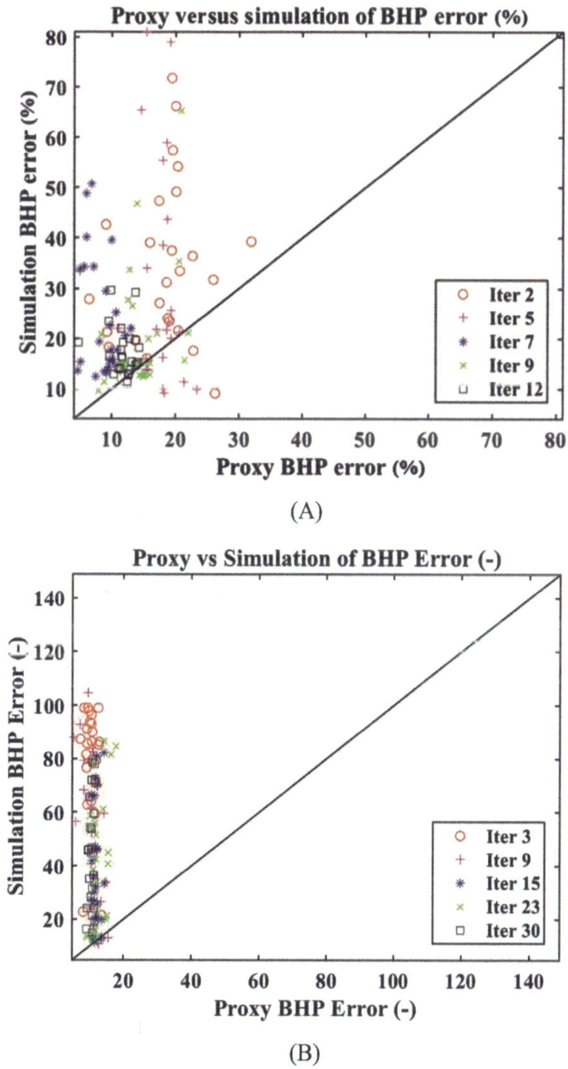

FIGURE 6.16 Proxy versus simulation values of BHP error: (A) neural network; and (B) *k*-nearest neighbors (Tripoppoom et al., 2020). *BHP*, Bottom-hole pressure.

proposed cases at iteration 30, whereas NN has a high accuracy rate of 21 solutions out of 25 proposed cases at iteration 12. Hence, as shown in Fig. 6.14H for fracture width, we can observe the posterior distribution difference of proxy solutions between NN and KNN. However, for other uncertain parameters, there is no significant difference between NN and KNN for proxy solutions.

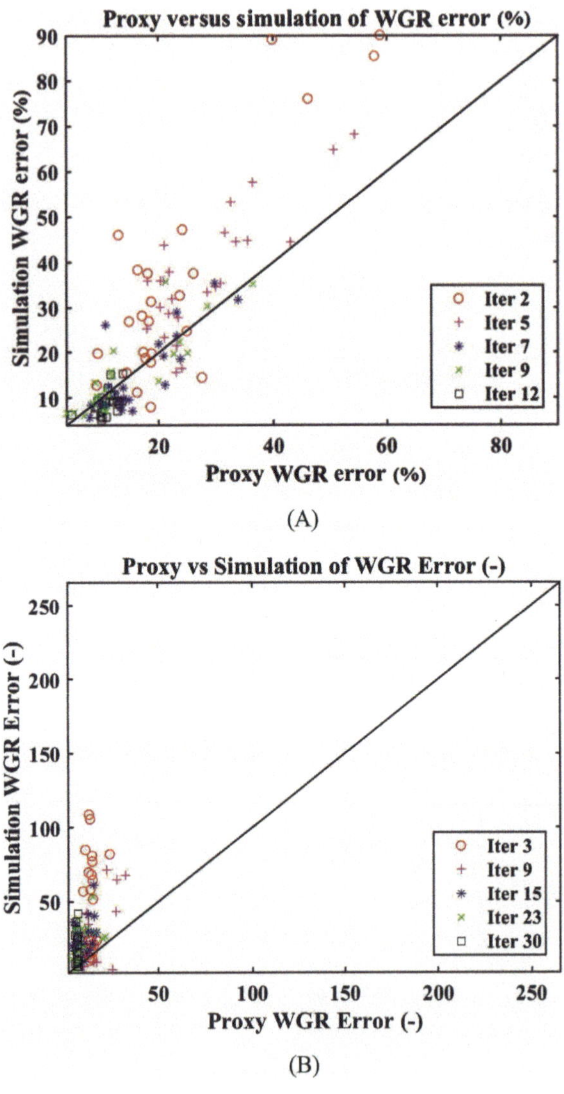

FIGURE 6.17 Proxy versus simulation values of WGR error: (A) neural network; and (B) *k*-nearest neighbors (Tripoppoom et al., 2020). *WGR*, Water–gas ratio.

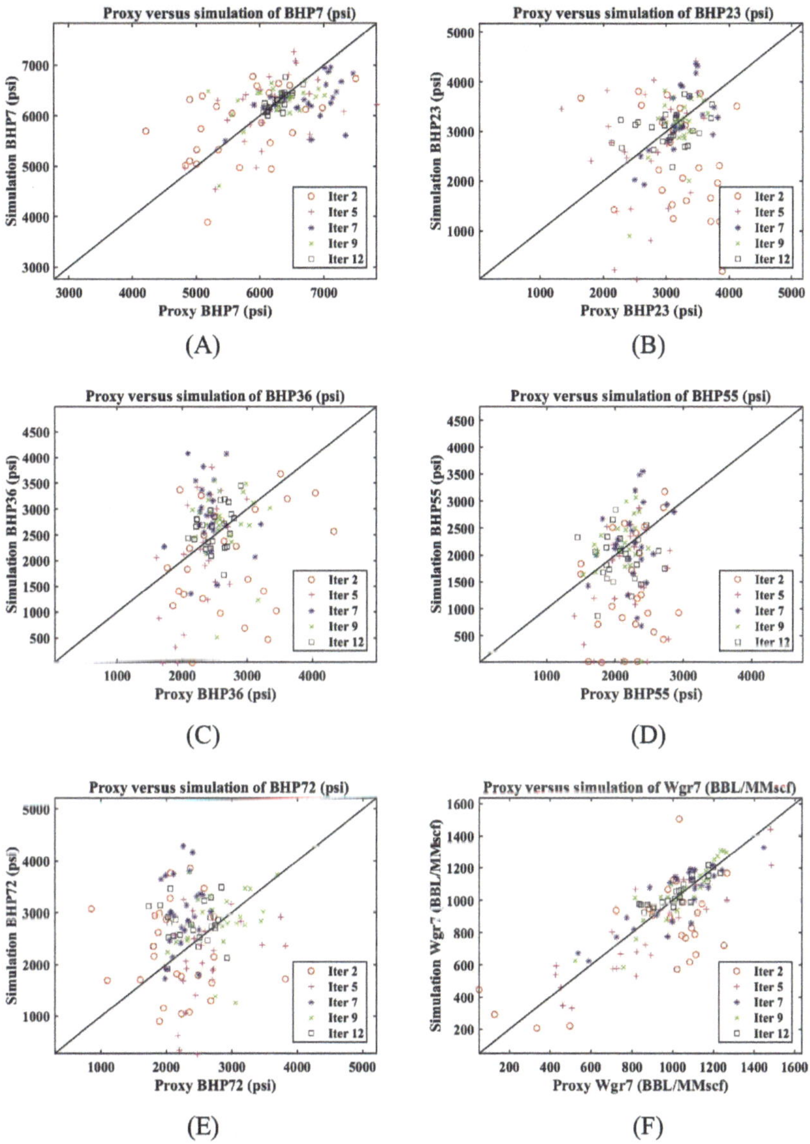

FIGURE 6.18 Cross plot between proxy and simulation values of 10 proxies for neural network as proxy model: (A) BHP at day 7; (B) BHP at day 23; (C) BHP at day 36; (D) BHP at day 55; (E) BHP at day 72; (F) WGR at day 7; (G) WGR at day 17; (H) WGR at day 28; (I) WGR at day 40; and (J) WGR at day 59 (Tripoppoom, 2019). *BHP*, Bottom-hole pressure; *WGR*, water−gas ratio.

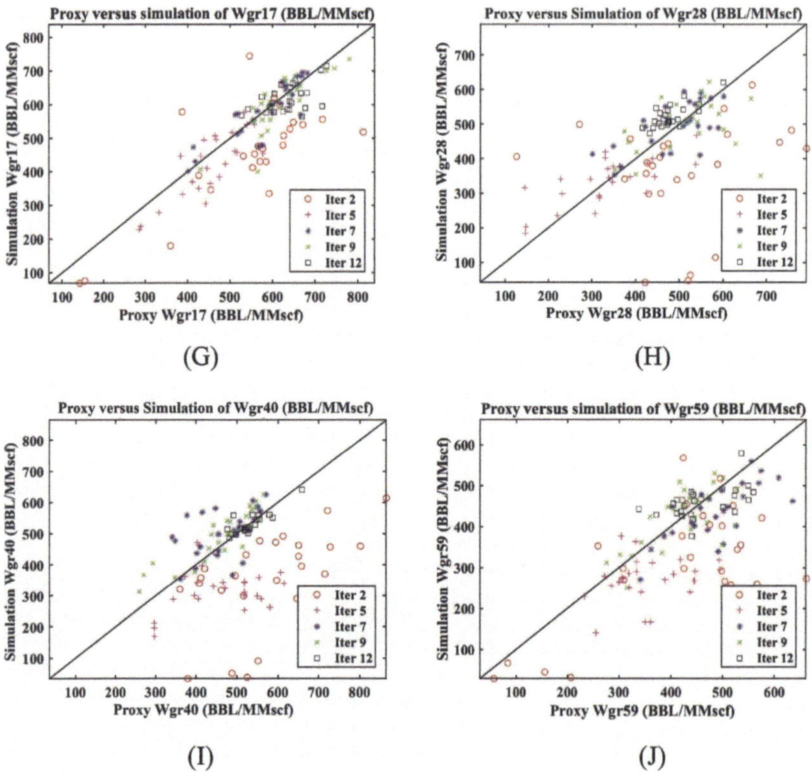

FIGURE 6.18 (Continued).

To compare the performance between NN and KNN, we plotted the proxy values versus simulation values cross plot for both NN and KNN, as shown in Figs. 6.15–6.17, for global error, BHP error, and WGR error. The unit slope line represents 100% alignment between proxy and simulation values. As you can see, as the iteration progressed, the proposed points came closer to the unit slope line for both NN and KNN. However, NN predictability is obviously superior to KNN as more points were located in the unit slope line in the later iteration while we can observe the vertical points lining up for KNN. The vertical lineup implies that the proxy suggests lower error values, but the simulation result values are still spreading from low to high error values. This reflects less accuracy and predictability of KNN when compared to NN. In addition, the plots developed for all 10 multiple proxies for both NN and KNN are also illustrated in Fig. 6.18 and Fig. 6.19, respectively. Similarly, the proxy accuracy is improved as the iteration progressed. NN provided more accuracy than KNN in all proxies, and we still observed vertical lining up for KNN.

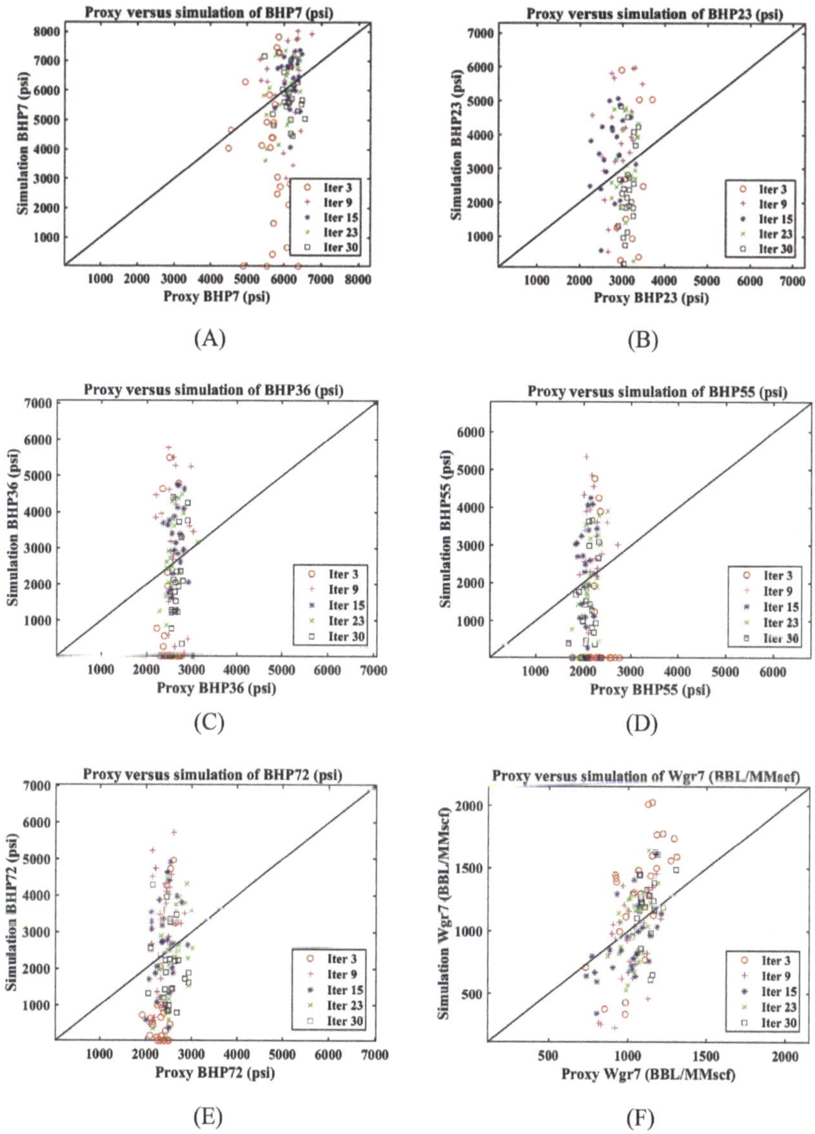

FIGURE 6.19 Cross plot between proxy and simulation values of 10 proxies for *k*-nearest neighbors as proxy model: (A) BHP at day 7; (B) BHP at day 23; (C) BHP at day 36; (D) BHP at day 55; (E) BHP at day 72; (F) WGR at day 7; (G) WGR at day 17; (H) WGR at day 28; (I) WGR at day 40; and (J) WGR at day 59 (Tripoppoom, 2019). *BHP*, Bottom-hole pressure; *WGR*, water–gas ratio.

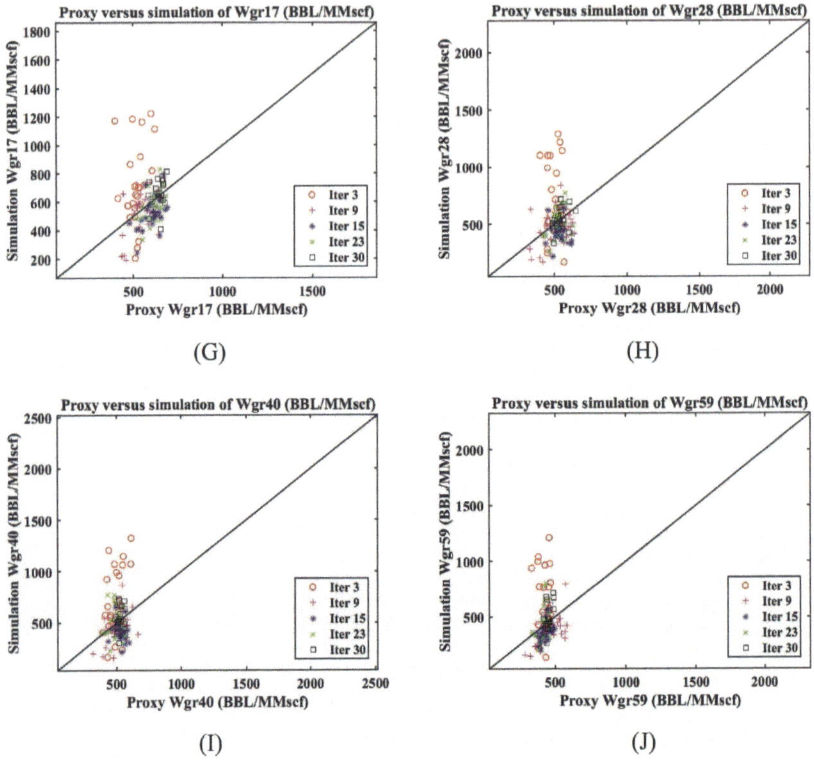

FIGURE 6.19 (Continued).

Fig. 6.20 shows the objective function vs. simulation number index for both NN and KNN. This plot also emphasizes the better accuracy of NN over KNN as we observed a faster improvement of the objective function value. Fig. 6.21 and Fig. 6.22 depict scatter plots of the proposed cases in each iteration versus simulation number index for each uncertain parameter for NN and KNN, respectively. First, the proxy-based MCMC algorithm proposed points over the uncertain parameter domain; however, only a certain range of values can produce HM solutions. Second, we can see that the proposed values were converged to the final range as the simulation number index was higher for some parameters such as fracture half-length, fracture conductivity, fracture water saturation, exponent of k_{rw}, and fracture width. This is because these parameters had more impact or were more sensitive to the objective function. In sum, for both NN and KNN, Figs. 6.20–6.22 reaffirm the accuracy and efficiency of the proxy-based MCMC algorithm for HM.

For a total running time, we plotted the cumulative number of HM solutions versus elapsed time, as shown in Fig. 6.23. The elapsed time included all steps that occurred in the AHM workflow such as automatic case

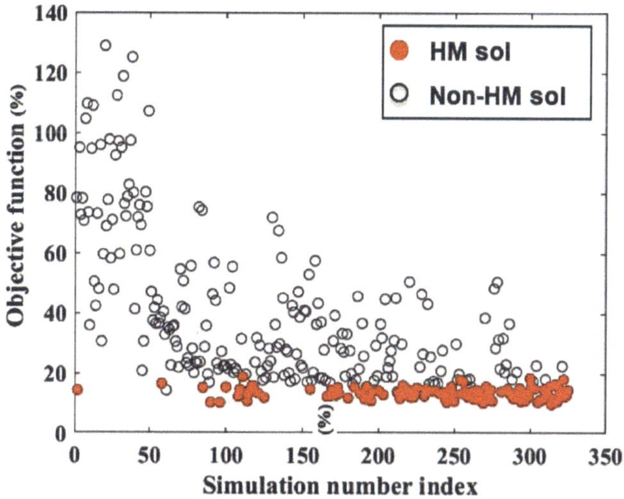

(A)

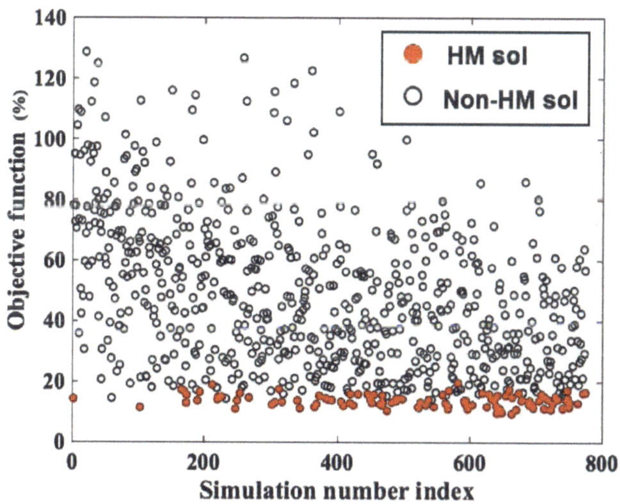

(B)

FIGURE 6.20 Scatter plot of objective function (global error) versus simulation number index: (A) neural network; and (B) *k*-nearest neighbors (Tripoppoom et al., 2020).

generation, proxy model training, proxy-based MCMC algorithm runs, EDFM preprocessor runs, simulation runs, and stopping criteria evaluation. The difference of using either NN or KNN would contribute to proxy model training, proxy-based MCMC algorithm, and stopping criteria evaluation. By considering the elapsed time required to obtain the same number of solutions, NN

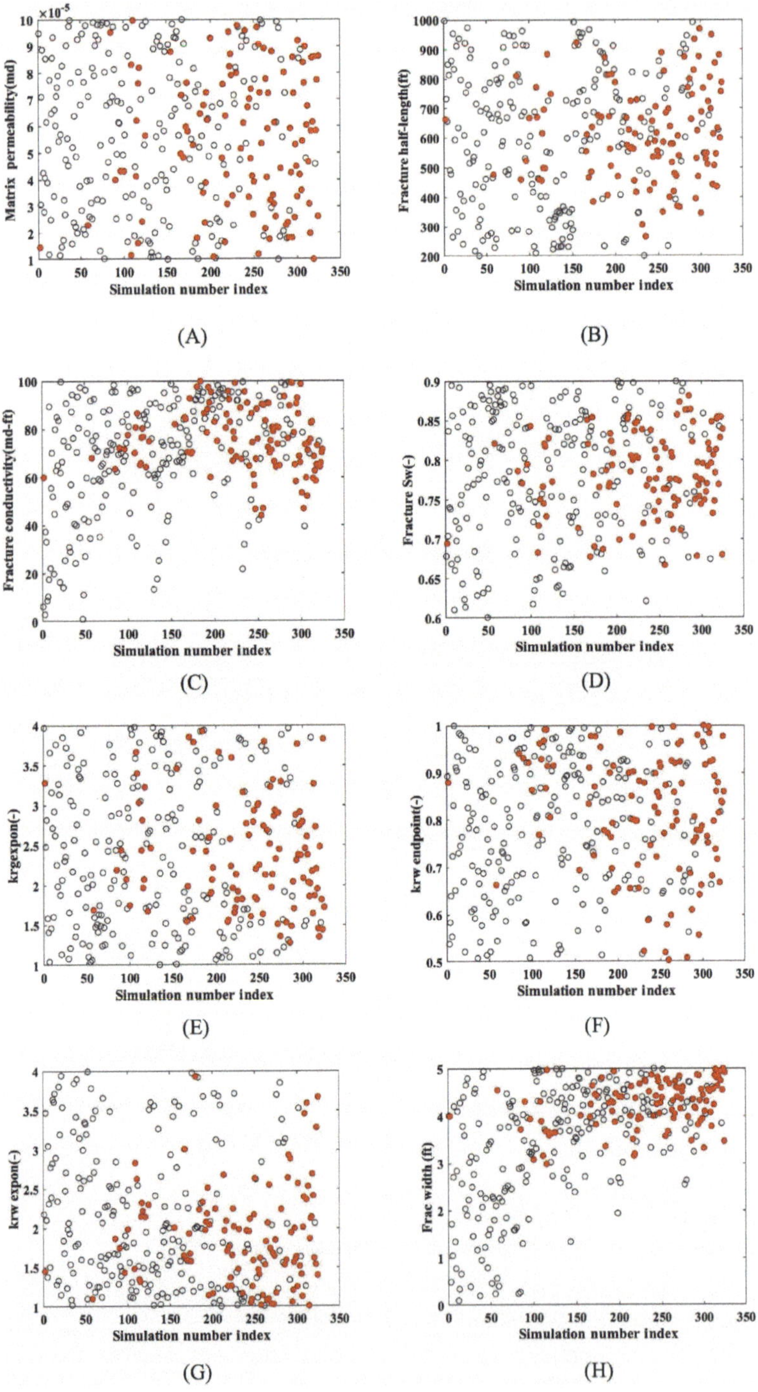

FIGURE 6.21 Uncertain parameter values versus simulation number index for neural network: (A) matrix permeability; (B) fracture half-length; (C) fracture conductivity; (D) fracture water saturation; (E) exponent of k_{rg}; (F) endpoint of k_{rw}; (G) exponent of k_{rw}; and (H) fracture width (Tripoppoom, 2019).

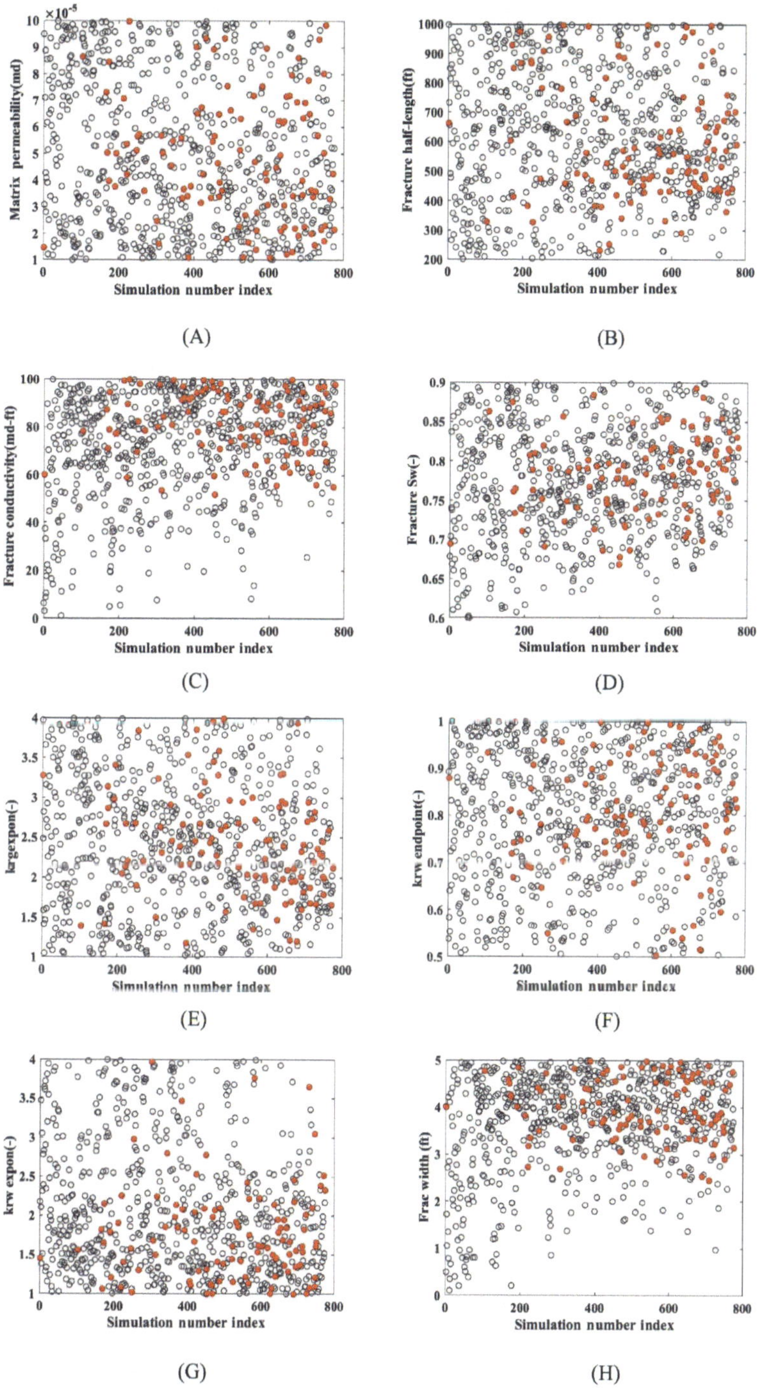

FIGURE 6.22 Uncertain parameter values versus simulation number index for *k*-nearest neighbors: (A) matrix permeability; (B) fracture half-length; (C) fracture conductivity; (D) fracture water saturation; (E) exponent of k_{rg}; (F) endpoint of k_{rw}; (G) exponent of k_{rw}; and (H) fracture width (Tripoppoom, 2019).

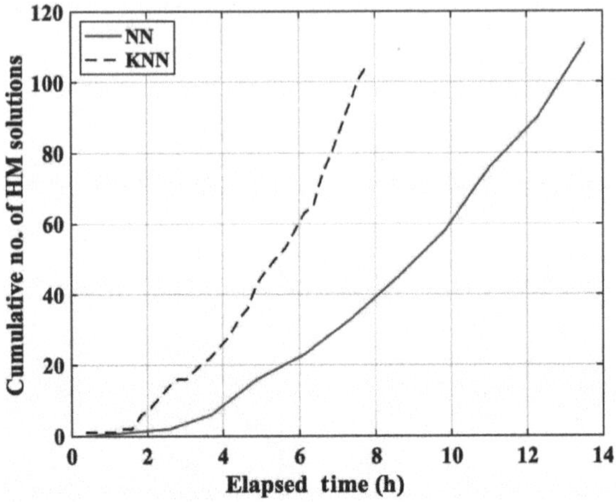

FIGURE 6.23 Comparison between NN and KNN for cumulative number of history matching solutions versus elapsed time (Tripoppoom et al., 2020). *KNN*, *k*-nearest neighbor; *NN*, neural network.

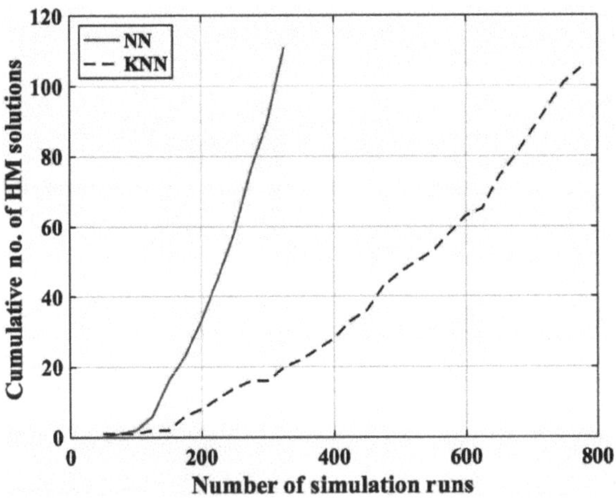

FIGURE 6.24 Comparison between NN and KNN for cumulative number of history matching solutions versus number of simulation runs (Tripoppoom et al., 2020). *KNN*, *k*-nearest neighbors; *NN*, neural network.

spent 1.6 times longer in elapsed time than KNN. However, if we consider the number of simulation runs or total resources of simulation license, as shown in Fig. 6.24, NN performed significantly better than KNN by obtaining 111 solutions from fewer total simulation runs than KNN by a factor of two. The main portions of elapsed time including the EDFM preprocessor, simulation runs, and multiple-proxy-based MCMC algorithm for NN and KNN are shown in

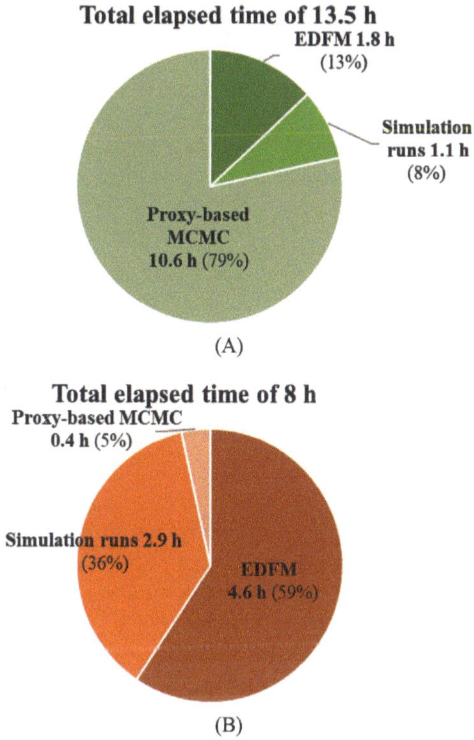

FIGURE 6.25 Pie chart of total elapsed time including EDFM preprocessor, simulation runs, and proxy-based MCMC algorithm: (A) neural network; and (B) *k*-nearest neighbors (Tripoppoom et al., 2020). *EDFM*, Embedded discrete fracture model; *MCMC*, Markov chain Monte Carlo.

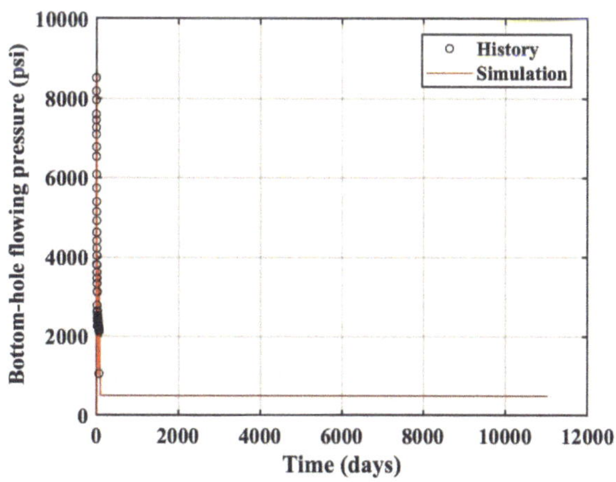

FIGURE 6.26 BHP of 500 psi was used as a production constraint for 30-year production forecast (Tripoppoom, 2019). *BHP*, Bottom-hole pressure.

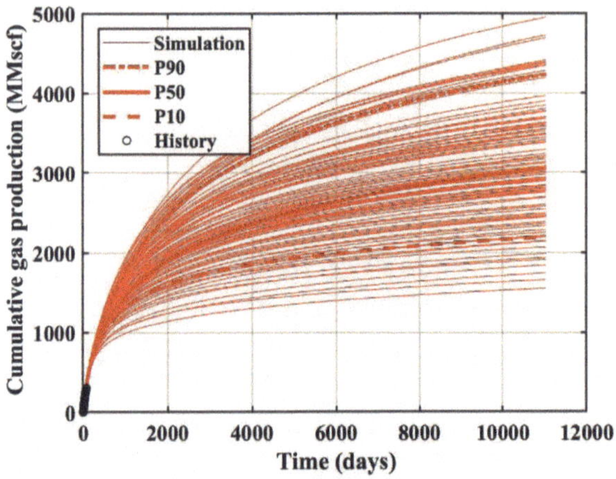

(A)

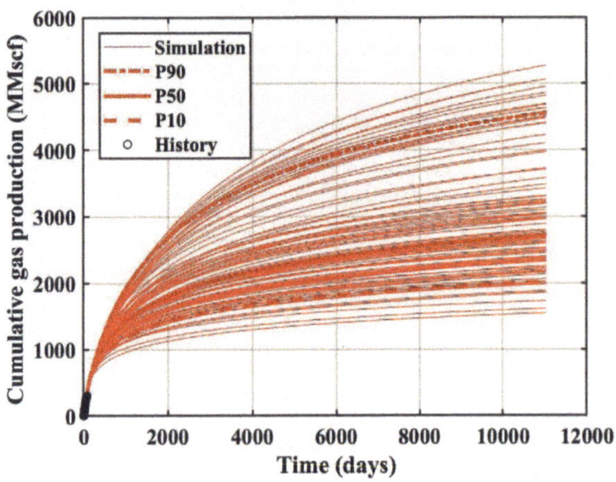

(B)

FIGURE 6.27 Cumulative gas production forecast of all history matching solutions from AHM workflow: (A) neural network; and (B) *k*-nearest neighbors (Tripoppoom et al., 2020).

Fig. 6.25. For NN, 79% of the elapsed time is contributed to the proxy-based MCMC algorithm, whereas for KNN, only 5% is contributed to the proxy-based MCMC algorithm. However, for KNN, a higher portion of 95% is contributed to the EDFM preprocessor and simulation runs due to requiring more simulation cases to obtain the same number of HM solutions as NN. Therefore, to select which proxy model should be used, consideration would depend on the project limitations and whether the running time or the available reservoir simulator resources were more important.

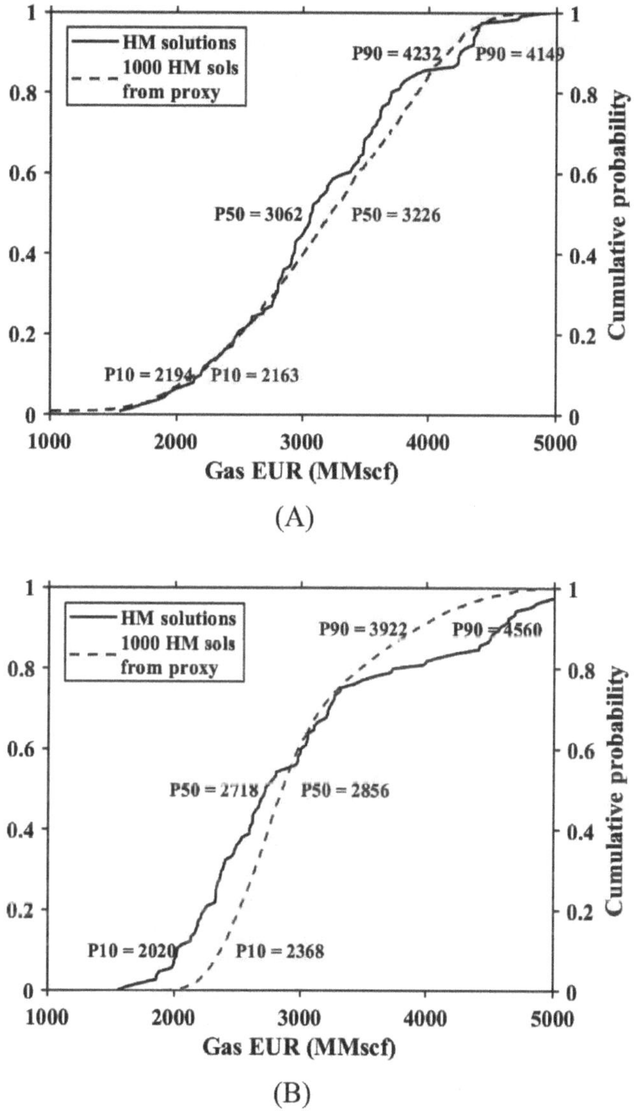

FIGURE 6.28 Empirical cumulative distribution function of gas estimated ultimate recovery for history matching solutions from AHM workflow and 1000 history matching solutions generated by proxy model: (A) neural network; and (B) *k*-nearest neighbors (Tripoppoom et al., 2020). *AHM*, Assisted history matching.

6.5 Probabilistic forecasting

Following the multiple HM solutions that we obtained from the AHM workflow, we performed a probabilistic production forecast for all solutions

TABLE 6.4 Summary of P50 cases of using neural network and *k*-nearest neighbors as a proxy model (Tripoppoom et al., 2020).

Uncertain parameters	Unit	Neural network (iteration 12 case14)	*k*-Nearest neighbors (iteration 26 case13)
Matrix permeability	md	0.000031	0.000013
Fracture half-length	ft	442	501
Fracture conductivity	md ft	59	73
Fracture water saturation	—	0.81	0.82
Exponent of k_{ro}	—	1.87	1.70
Endpoint of k_{rw}	—	0.89	0.93
Exponent of k_{rw}	—	1.01	1.94
Fracture width	ft	4.98	4.71

of both NN and KNN. In this study, we forecasted production for 30 years by constraining BHP at 500 psi, as illustrated in Fig. 6.26. The production forecast of cumulative gas production for both NN and KNN is shown in Fig. 6.27 with P10–50–90 cases. The definition of P10–50–90 means there are 10%, 50%, and 90% of the distribution that falls below those values, respectively. After that, we can estimate the empirical cumulative distribution function (ECDF) of gas estimated ultimate recovery (EUR) from all cases as shown in Fig. 6.28. The figures illustrate both the solutions from the AHM workflow and 1000 solutions generated by the proxy models. Both ECDF from the AHM workflow and proxy models are in line with each other, but more differences are seen in KNN due to its lower predictability as the proxy model. In comparing NN and KNN, the P10–50–90 of forecasted gas EUR are in the same range with a slight discrepancy of 10%.

The P50 realizations of both NN and KNN are summarized in Table 6.4. Most of parameters for both NN and KNN are similar while for some parameters, such as matrix permeability, the values are quite different because this parameter is less significant, as discussed in Fig. 6.13. The pressure distribution of both cases at the end of the HM period and 30 years of production forecast is also depicted in Fig. 6.29.

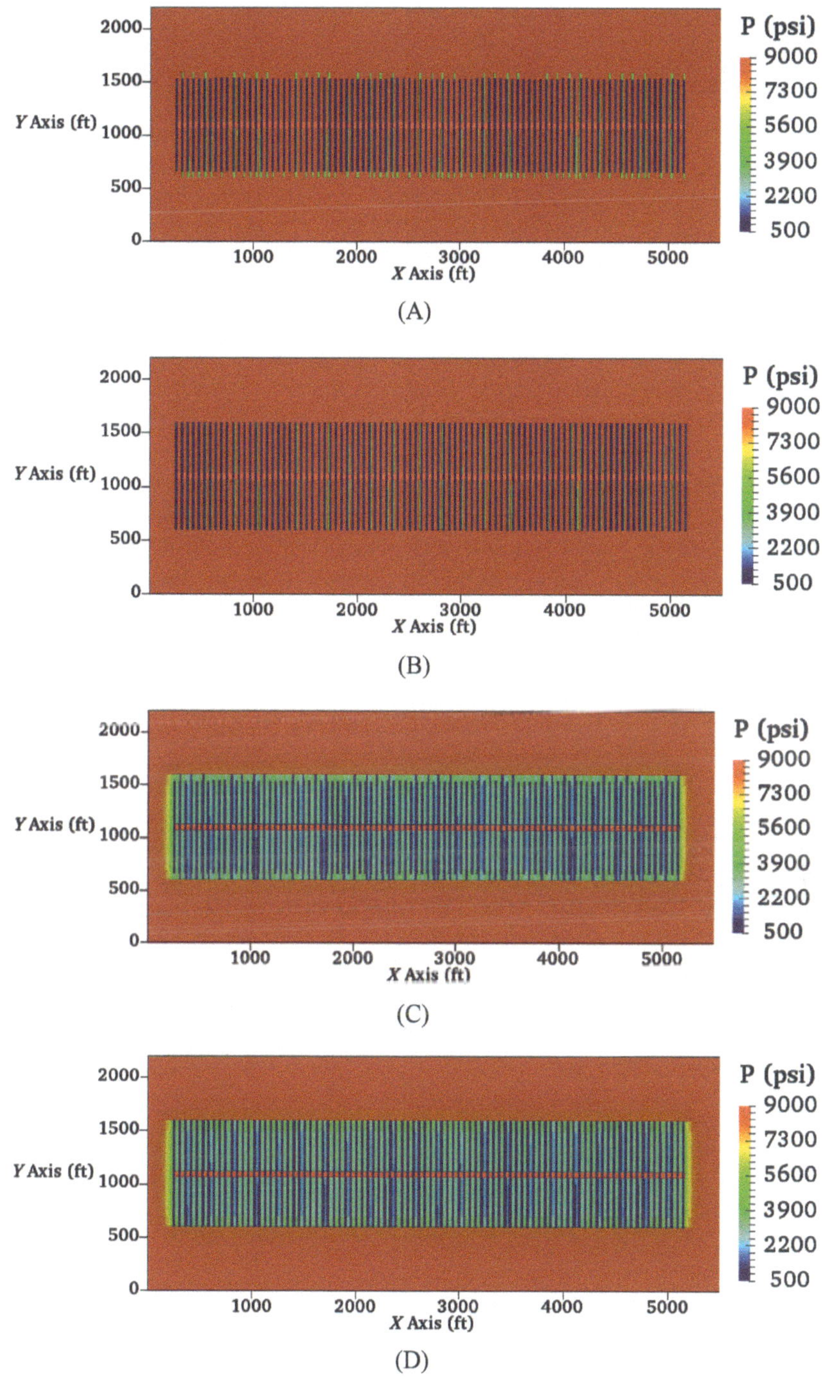

FIGURE 6.29 Pressure distribution of P50 cases: (A) after 74 days (end of history) using NN; (B) after 74 days (end of history) using KNN; (C) after 30 years using NN; and (D) after 30 years using KNN (Tripoppoom et al., 2020). *KNN*, *k*-nearest neighbors; *NN*, neural network.

6.6 Remarks

We applied the AHM workflow to an actual shale gas well and compared the performances of two different proxy models: NN and KNN. We found that NN performs better than KNN in terms of predictive accuracy when validating with the actual simulator. In other words, NN requires a smaller number of simulation runs than KNN to obtain the same number of solutions. On the other hand, KNN performs better in terms of running time because it required almost half of the run time that NN uses. The selection of either NN or KNN as proxy model depends on the project limitation of whether running time or simulation license is more restricted for the project. The in-lined HM results from the two different proxy models confirm the robustness of the proposed HM algorithm. From the AHM workflow, we can obtain multiple HM solutions and characterize hydraulic fractures and reservoir properties in a probabilistic manner. The probabilistic production forecast was also performed from multiple history-matched solutions. The information from AHM could be used further to improve hydraulic fracturing design and well placement optimization.

References

Tripoppoom S., 2019. Assisted history matching workflow for unconventional reservoirs. MS Thesis, The University of Texas at Austin.

Tripoppoom, S., Yu, W., Huang, H., Sepehrnoori, K., Song, W., Dachanuwattana, S., 2019. A practical and efficient iterative history matching workflow for shale gas well coupling multiple objective functions, multiple proxy-based MCMC and EDFM. Journal of Petroleum Science and Engineering 176, 594–611.

Tripoppoom, S., Ma, X., Yong, R., Wu, J., Yu, W., Sepehrnoori, K., et al., 2020. Assisted history matching in shale gas well using multiple-proxy-based Markov chain Monte Carlo algorithm: the comparison of K-nearest neighbors and neural networks as proxy model. Fuel 262, 116563.

Chapter 7

Shale-gas well with and without enhanced permeability area

7.1 Introduction

In conventional reservoirs, planar bi-wing hydraulic fractures may be acceptable for reservoir modeling purpose. However, in unconventional reservoirs, such as shale formations, the planar hydraulic fracture assumption alone cannot capture the actual complex fractures around main hydraulic fractures. This complex fracture geometry could be due to the interaction with preexisting natural fractures.

One method to generate complex fractures network is the application of a geomechanical model to simulate the fracture propagation phenomenon (Cipolla et al., 2012; Taleghani and Olson, 2013; Wu and Olson, 2015; Shrivastava et al., 2018). This model integrates reservoir properties, fracturing operation and completion, and the interaction with preexisting natural fractures, if they are known in the area. Nevertheless, one of the main challenges is the accuracy and reliability of the fracture network we obtain from the model, as there is high uncertainty from both input information and model assumptions. Moreover, the input information to construct the geomechanical model may be limited and cannot be performed promptly and routinely for history-matching purposes.

Therefore, many studies attempt to compromise the complexity of fracture networks and modeling assumption with a simpler concept of simulated rock volume (SRV) instead of the fully complex fractures network from the fracture propagation model. The SRV is defined as the stimulated reservoir volume generated by hydraulic fractures and their interaction with preexisting fractures (Mayerhofer et al., 2010). There are many techniques to mimic SRV including the dual-porosity model (Du et al., 2010), dual-permeability, dual-porosity model (Chang and Zhang, 2018), logarithmically spaced, locally refined, dual-permeability grid (LS-LR-DK) (Rubin, 2010), enhanced permeability area (EPA) with explicit hydraulic fractures (Yin et al., 2011), and embedded discrete fracture model (EDFM) (Shakiba and Sepehrnoori, 2015).

In this study, we used the concept of EPA together with EDFM for history matching. The EPA represents the complex fractures around main hydraulic

Assisted History Matching for Unconventional Reservoirs. DOI: https://doi.org/10.1016/B978-0-12-822242-3.00004-3
© 2021 Elsevier Inc. All rights reserved.

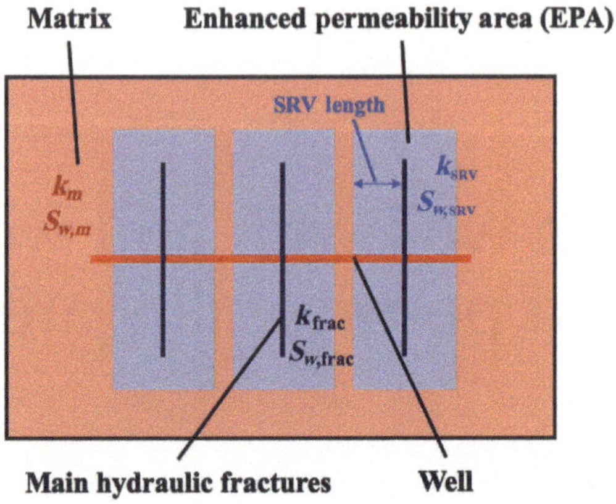

FIGURE 7.1 The concept of EPA with main hydraulic fractures (Tripoppoom, 2019). *EPA*, Enhanced permeability area.

fractures as a higher permeability value area. As illustrated in Fig. 7.1, there are three clusters of planar hydraulic fractures in EPA. The parameters of EPA consist of the SRV length from main hydraulic fractures, permeability of SRV region (k_{SRV}), and water saturation in SRV region ($S_{w,SRV}$). The production history we used is the same shale-gas well with 74 days of historical production data as in Chapter 6, Proxy Comparison Between Neural Network and *k*-Nearest Neighbors. Also, a general reservoir model description was discussed in that chapter.

7.2 Parameters identification and screening

For parameter identification and screening, we used the screening results (Figs. 6.4 and 6.5) from Chapter 6, Proxy Comparison Between Neural Network and *k*-Nearest Neighbors, where all determined uncertain parameters will be used in history matching. To model EPA, we identified two additional parameters: SRV permeability or the permeability in the EPA, and the length of SRV from hydraulic fractures, as illustrated in Fig. 7.1. In this study, we assumed the initial range of SRV permeabilities to be around 10% of hydraulic fracture conductivity range, and SRV length ranges from no EPA area to fully connected across fracture spacing. For SRV length, besides modifying permeability as SRV permeability, we also modified water saturation inside the SRV length to be the same as fracture water saturation. All parameters ranges for the EPA case are summarized in Table 7.1. The reservoir model using the EPA concept is illustrated in Fig. 7.2. The color represents permeability values in the $x-y$ plane. The matrix, EPA, and hydraulic fractures by EDFM are shown in dark blue, light blue, and red, respectively. In other words, the dark blue area is the value of matrix permeability, the

TABLE 7.1 Summary of 10 uncertain parameters and their prior distributions for the enhanced permeability area case (Tripoppoom, 2019).

Code	Uncertain parameters	Unit	Distribution	Min value	Max value
A	Matrix permeability	md	Uniform	0.00001	0.0001
B	Fracture half-length	ft	Uniform	200	1000
C	Fracture conductivity	md ft	Uniform	1	100
D	Fracture water saturation	–	Uniform	0.6	0.9
E	Exponent of k_{ro}	–	Uniform	1	4
F	Endpoint of k_{rw}	–	Uniform	0.5	1
G	Exponent of k_{rw}	–	Uniform	1	4
H	Fracture width	ft	Uniform	0.01	5
I	SRV permeability	md	Uniform	0.1	1
J	SRV length	ft	Uniform	0	27.5

SRV, Simulated rock volume.

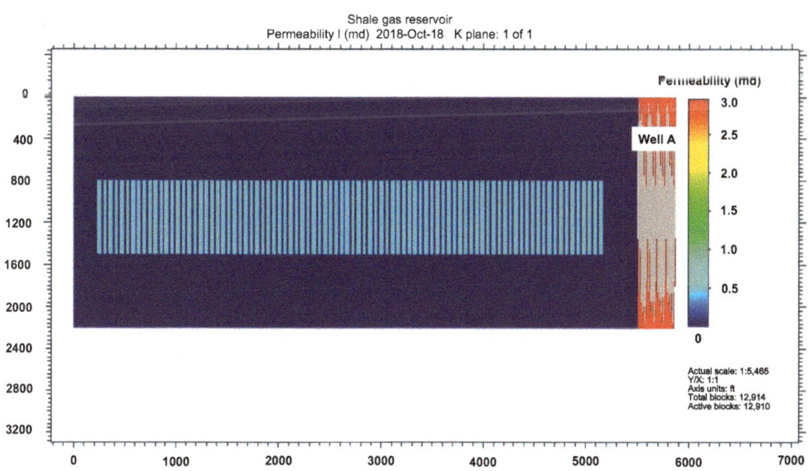

FIGURE 7.2 Reservoir model showing permeability in matrix cells in dark blue, EPA as light blue, and main hydraulic fractures by EDFM cells in red. *EDFM*, Embedded discrete fracture model; *EPA*, enhanced permeability area.

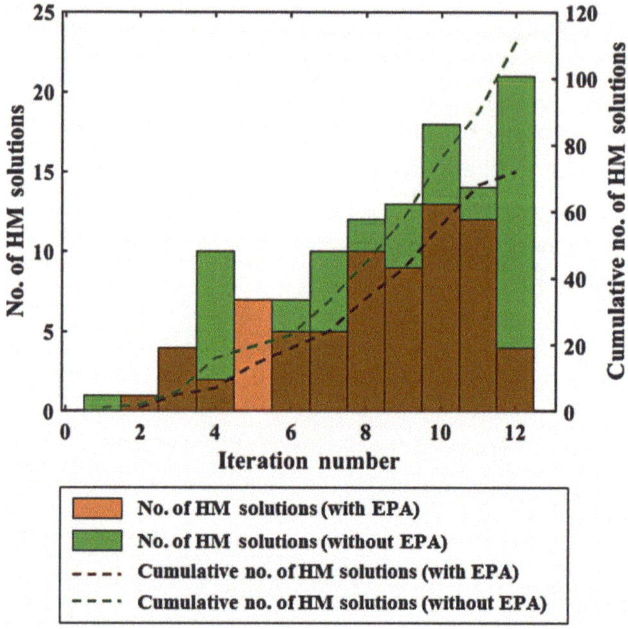

FIGURE 7.3 Comparison of a number of history-matching solutions versus iteration numbers between cases with and without EPA (Tripoppoom, 2019). *EPA*, Enhanced permeability area.

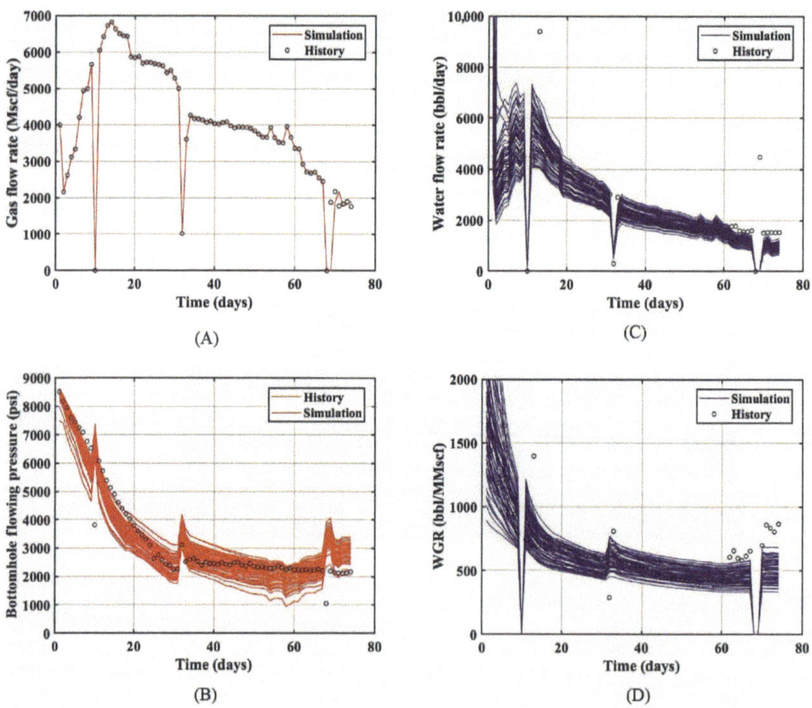

FIGURE 7.4 Comparison between simulation results and historical data of 72 history-matching solutions for the case with EPA: (A) gas flow rate; (B) flowing BHP; (C) water flow rate; and (D) WGR (Tripoppoom, 2019). *BHP*, Bottomhole pressure; *EPA*, enhanced permeability area; *WGR*, water–gas ratio.

light blue area uses the value from SRV permeability, and the red area derives fracture permeability value from fracture conductivity divided by fracture width input. For example, if fracture conductivity is 6 md ft and fracture width is 2 ft, the fracture permeability value is 3 md.

7.3 History matching

The assisted history-matching (AHM) workflow continues as previously explained in Chapter 6, Proxy Comparison Between Neural Network and *k*-Nearest Neighbors. In this case, the workflow also stopped at the maximum iteration of 12. Seventy-two history-matching solutions were found from a total of 325 simulation runs or around 22% compared to 34% for the case with only hydraulic fractures (without EPA). The lower number of solutions found is obviously from a higher number of uncertain parameters including an additional two EPA parameters. The number of history-matching solutions versus iteration numbers is shown in Fig. 7.3. The simulation results of

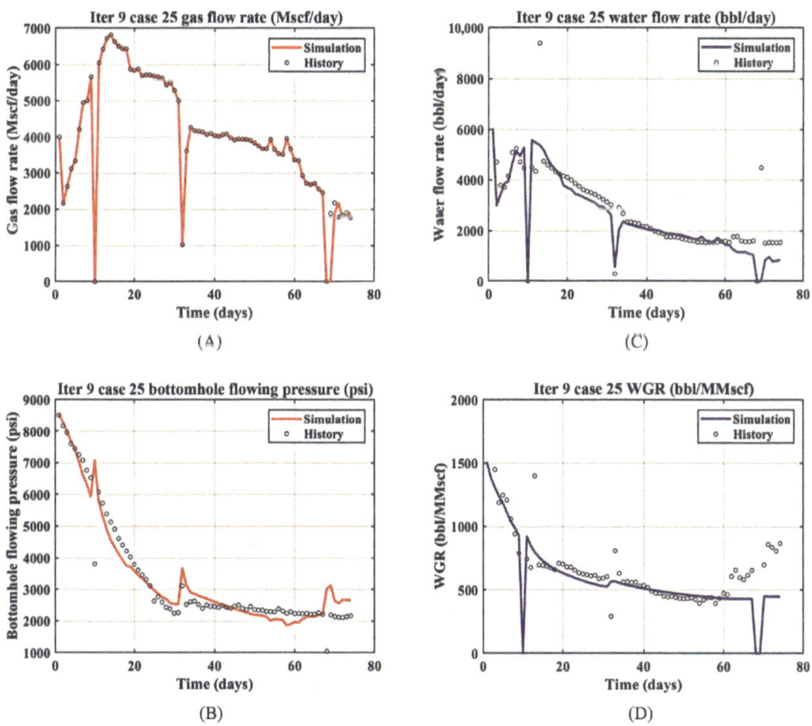

FIGURE 7.5 Comparison between simulation results and historical data for the best match case with EPA: (A) gas flow rate; (B) flowing BHP; (C) water flow rate; and (D) WGR (Tripoppoom, 2019). *BHP*, Bottomhole pressure; *EPA*, enhanced permeability area; *WGR*, water–gas ratio.

all 72 solutions are shown in Fig. 7.4. In addition, the best match cases both with EPA and without EPA (Chapter 6: Proxy Comparison Between Neural Network and *k*-Nearest Neighbors) are shown in Figs. 7.5 and 7.6, respectively, and Table 7.2 summarizes the uncertain properties of the best cases for both scenarios.

Similarly, we obtained posterior distribution of uncertain parameters including the additional uncertain parameters for EPA including SRV permeability and SRV length. The comparison of each uncertain parameter between cases with EPA and without EPA is illustrated in Fig. 7.7. We can see that when there is EPA around the main hydraulic fractures, matrix permeability, fracture half-length, fracture conductivity, fracture water saturation, and fracture width all shift to reduced values. This is because more areas contribute to the fluid flow via the EPA around main hydraulic fractures. For relative permeability, we observed slightly higher values of

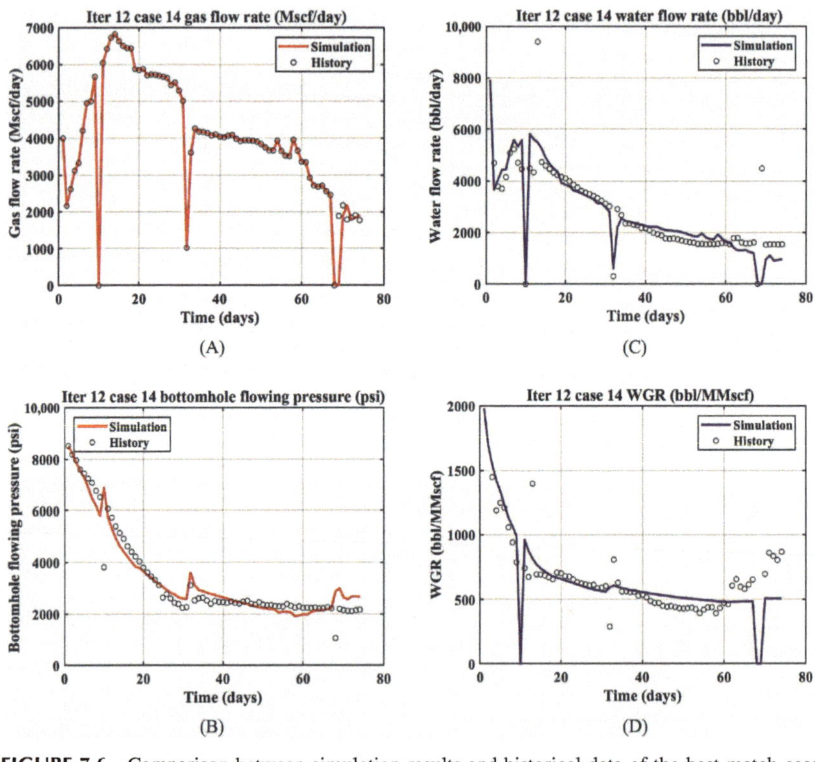

FIGURE 7.6 Comparison between simulation results and historical data of the best match case (iteration 12 case 14) for hydraulic fracture only or without EPA (Chapter 6: Proxy Comparison Between Neural Network and *k*-Nearest Neighbors): (A) gas flow rate; (B) flowing BHP; (C) water flow rate; and (D) WGR (Tripoppoom, 2019). *BHP*, Bottomhole pressure; *EPA*, enhanced permeability area; *WGR*, water–gas ratio.

TABLE 7.2 Summary of 10 uncertain parameters for the best matches for cases both with enhanced permeability area (EPA) (the best case is from case 25 of iteration 9) and without EPA (the best case is from case 14 of iteration 12) (Tripoppoom, 2019).

Uncertain parameters	Unit	With EPA (iteration 9 case 25)	Without EPA (iteration 12 case 14)
Matrix permeability	md	0.0000024	0.000031
Fracture half-length	ft	379	442
Fracture conductivity	md ft	53	59
Fracture water saturation	–	0.81	0.81
Exponent of k_{ro}	–	1.71	1.87
Endpoint of k_{rw}	–	0.92	0.89
Exponent of k_{rw}	–	1.68	1.01
Fracture width	ft	3.79	4.98
SRV permeability	md	0.55	n/a
SRV length	ft	18.4	n/a

SRV, Simulated rock volume.

the k_{rg} exponent for the case with EPA. This means the gas relative permeability is likely to have more curvature because EPA exists around the main hydraulic fractures, as illustrated in Fig. 7.8, while there is no significant difference for water relative permeability parameters, including k_{rw} endpoint and k_{rw} exponent.

7.4 Probabilistic forecasting

After we obtained history-matching solutions, we can perform production forecasts in the probabilistic manner. We constrained the prediction by using a constant bottomhole pressure of 500 psi, as shown in Fig. 7.9. The predicted gas and water productions for both cases, with and without EPA, are shown in Figs. 7.10–7.13, respectively. Also, the empirical cumulative distribution function of gas estimated ultimate recovery (EUR) for cases with and without EPA is illustrated in Fig. 7.14. We can observe that gas EUR is predicted around 25% lower than the case without EPA in Chapter 6, Proxy Comparison Between Neural Network and k-Nearest Neighbors, comparing P50 between the cases with and without EPA shows, respectively, 2302 and 3062 MMscf. Similarly, the water

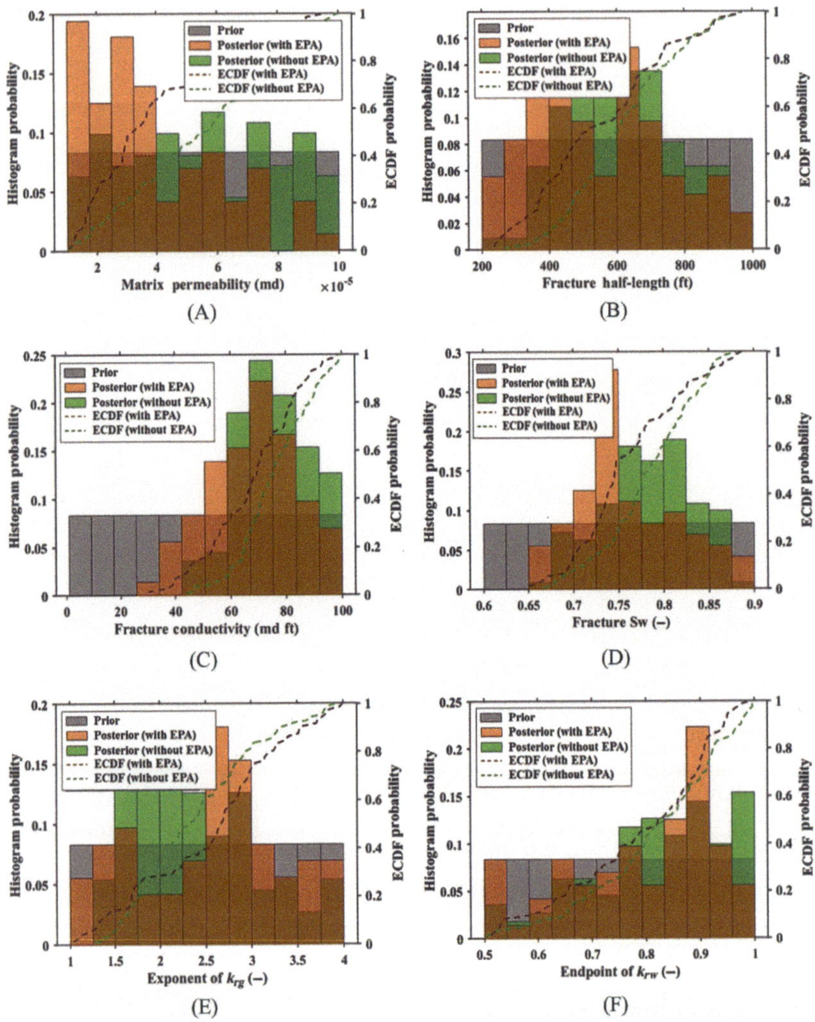

FIGURE 7.7 Comparison between prior and posterior distribution of uncertain parameters for the case with EPA (72 solutions) and the case without EPA (111 solutions): (A) matrix permeability; (B) fracture half-length; (C) fracture conductivity; (D) fracture water saturation; (E) exponent of k_{rg}; (F) endpoint of k_{rw}; (G) exponent of k_{rw}; (H) fracture width; (I) SRV permeability; and (J) SRV length (Tripoppoom, 2019). *EPA*, Enhanced permeability area; *SRV*, simulated rock volume.

production for the case with EPA is lower than the case without EPA, as depicted by Figs. 7.12 and 7.13. The reason is that the cases with lower fracture geometry were obtained from history matching when EPA existed around the hydraulic fractures.

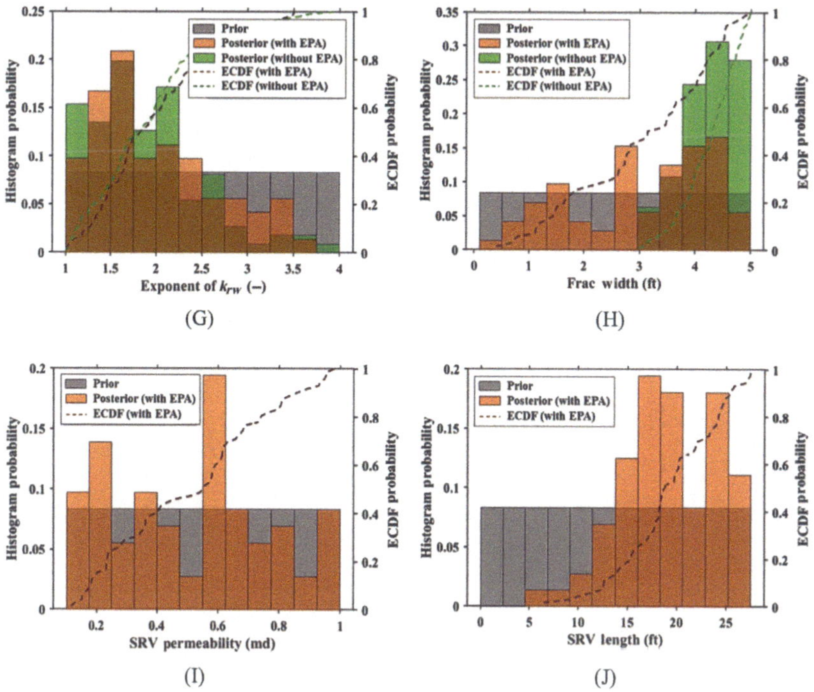

FIGURE 7.7 (Continued).

To sum up, when the EPA existed around hydraulic fractures, the history-matching solutions shifted to lower values for hydraulic fracture geometries and matrix permeability. These changes triggered a discrepancy between the production forecasts for cases with and without EPA. The lower gas EUR production forecast was obtained for the case with EPA. This means that if we assume a simple model with only hydraulic fractures but in a realistic situation where there is EPA due to the small fracture networks created around the main hydraulic fractures, we would overpredict the fracture geometry and gas EUR prediction, in this case by around 25%–30%.

The pressure distributions at different times from best match cases of two scenarios, with and without EPA, are illustrated in Figs. 7.15 and 7.16, respectively. We can see that the pressure between each hydraulic fracture cluster for the case without EPA is higher than the case with EPA at each time step. However, the fracture half-length for the case without EPA is longer than the case with EPA. Therefore this causes the case without EPA to drain more gas, as illustrated by a wider area of depleted pressure than the case with EPA, resulting in a higher gas EUR, as discussed in Fig. 7.14.

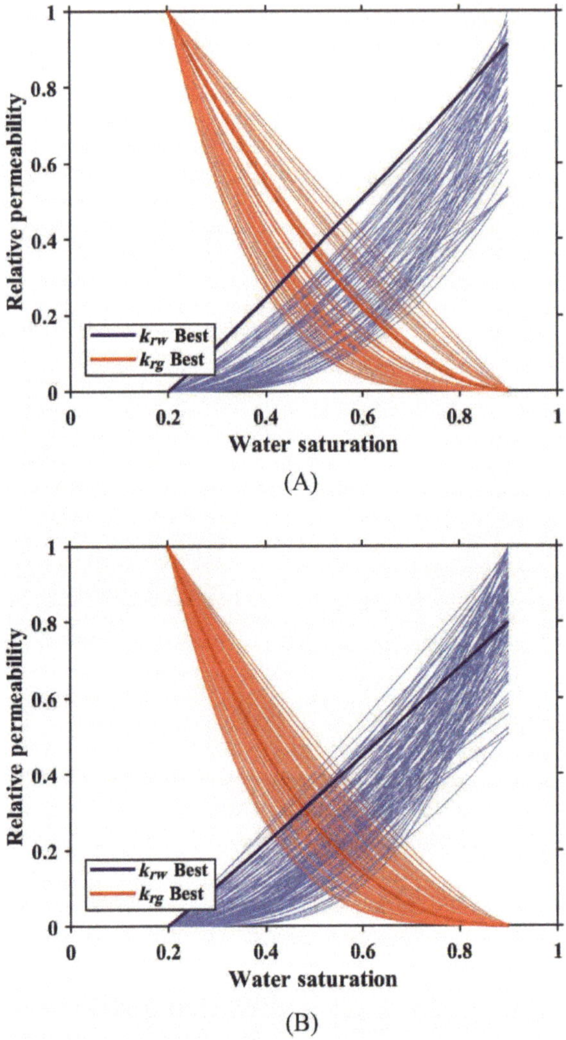

FIGURE 7.8 Relative permeability curves of history-matching solutions and the best case: (A) with and (B) without EPA. *EPA*, Enhanced permeability area.

7.5 Remarks

We applied the AHM workflow to a shale-gas well using the neural network–Markov chain Monte Carlo algorithm. We investigated the impact of the existence of EPA on history-matching solutions and production forecasts. With EPA, lower fracture geometry solutions were obtained in the history-matching solutions compared with the case of

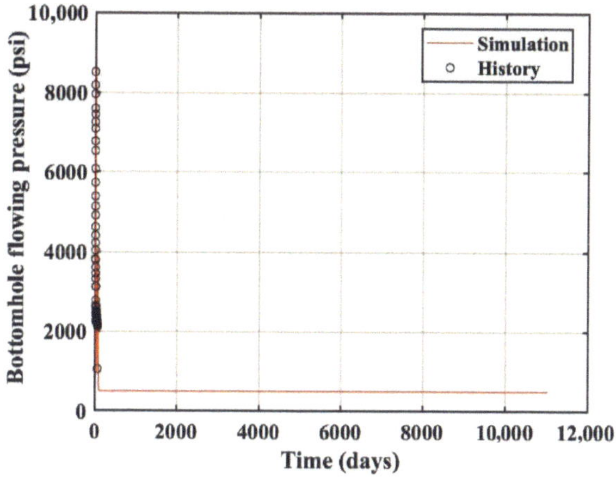

FIGURE 7.9 Production forecast constrained by a constant bottomhole pressure of 500 psi.

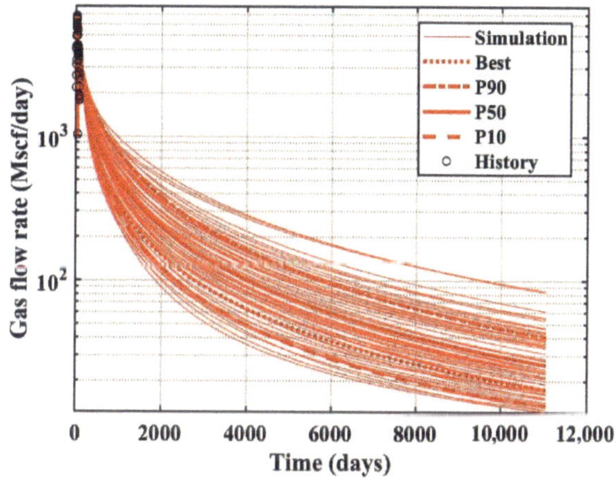

(A)

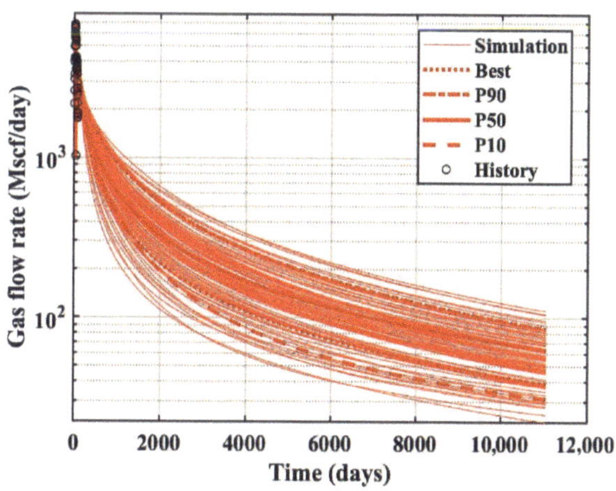

(B)

FIGURE 7.10 Gas flow rate production forecast of all 72 history-matching solutions from AHM workflow with best match case and P10–P50–P90 cases: (A) with and (B) without EPA. *AHM*, Assisted history matching; *EPA*, enhanced permeability area.

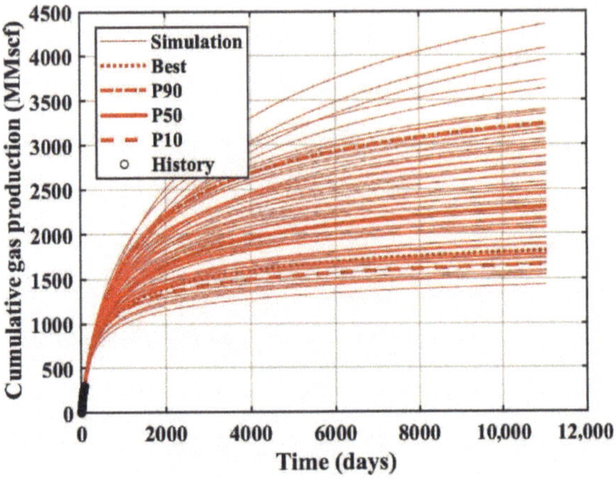

(A)

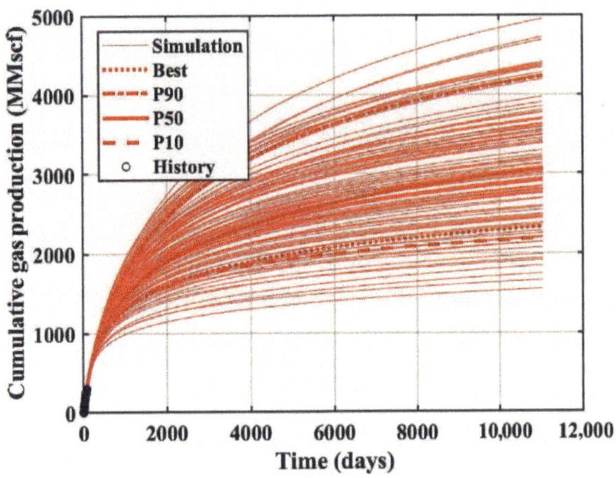

(B)

FIGURE 7.11 Cumulative gas production forecast of all 72 history-matching solutions from AHM workflow with best match case and P10–P50–P90 cases: (A) with and (B) without EPA (Tripoppoom, 2019). *AHM*, Assisted history matching; *EPA*, enhanced permeability area.

hydraulic fractures only (without EPA). This causes the production forecast of the case with EPA to be significantly lower than the one with only hydraulic fractures (without EPA). This means that if a simple model with only hydraulic fractures was assumed while in actual operations where there is EPA due to the small fracture networks created around main hydraulic fractures, we would overpredict the fracture geometry and gas EUR prediction.

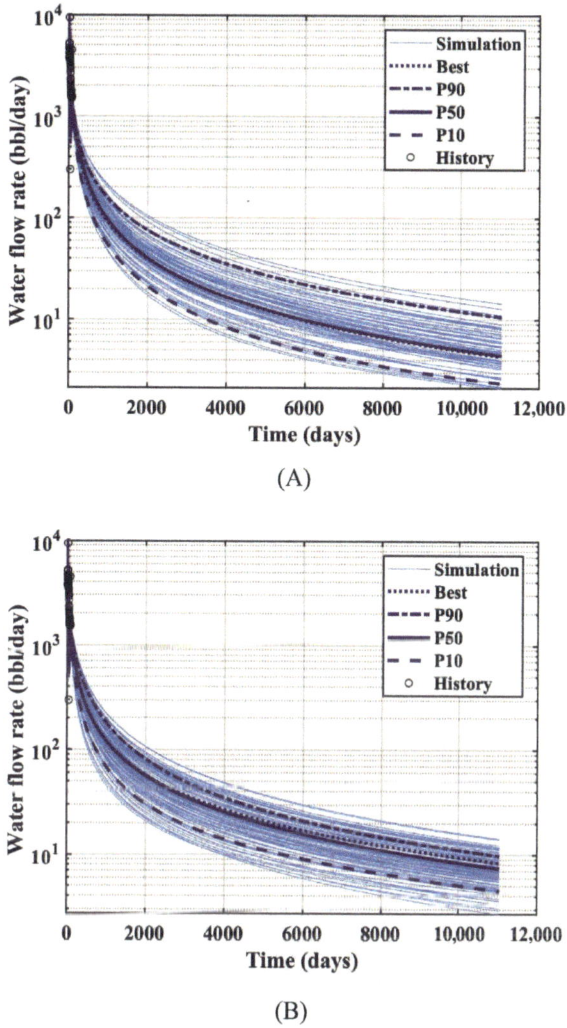

FIGURE 7.12 Water flow rate production forecast of all 72 history-matching solutions from AHM workflow with best match case and P10–P50–P90 cases: (A) with and (B) without EPA. *AHM*, Assisted history matching; *EPA*, enhanced permeability area.

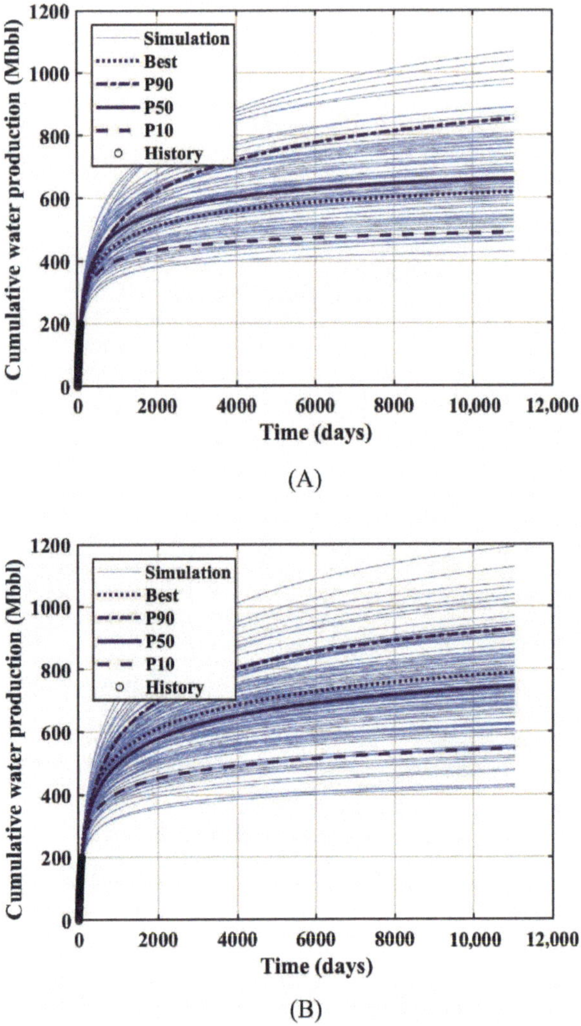

FIGURE 7.13 Cumulative water production forecast of all 72 history-matching solutions from AHM workflow with best match case and P10–P50–P90 cases: (A) with and (B) without EPA. *AHM*, Assisted history matching; *EPA*, enhanced permeability area.

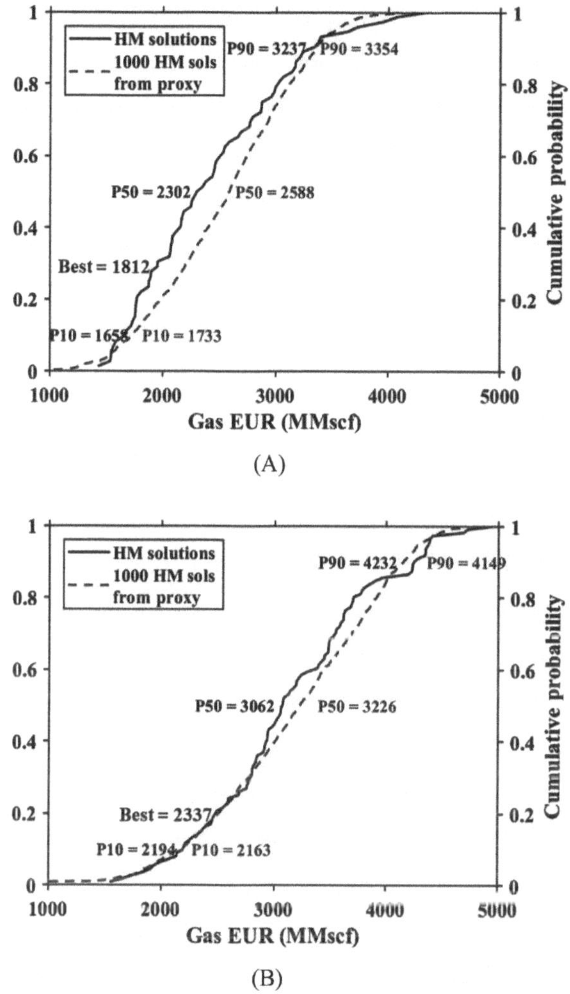

FIGURE 7.14 ECDF of gas EUR for 72 and 1000 history-matching solutions generated by proxy model (neural network): (A) with and (B) without EPA. *ECDF*, Empirical cumulative distribution function; *EPA*, enhanced permeability area; *EUR*, estimated ultimate recovery.

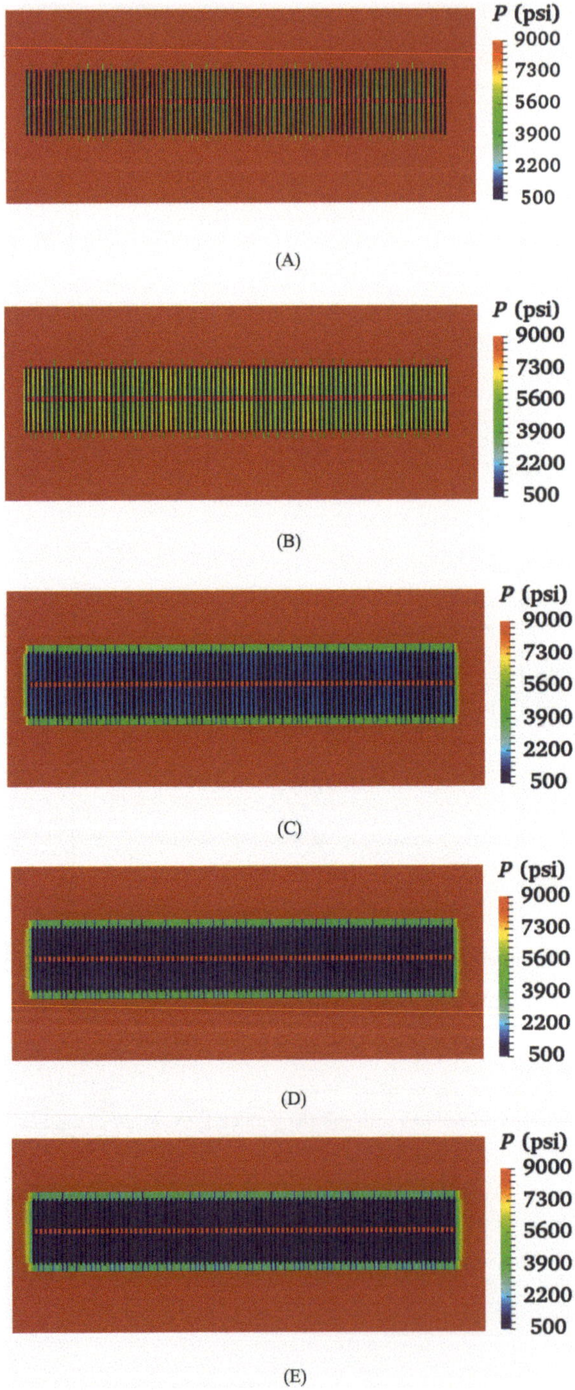

FIGURE 7.15 Pressure distributions of the case of hydraulic fractures and EPA: (A) 30 days; (B) 74 days (end of history); (C) 10 years after history; (D) 20 years after history; and (E) 30 years after history. *EPA*, Enhanced permeability area. *Modified from Tripoppoom, S., 2019. Assisted History Matching Workflow for Unconventional Reservoirs (M.S. thesis). The University of Texas at Austin.*

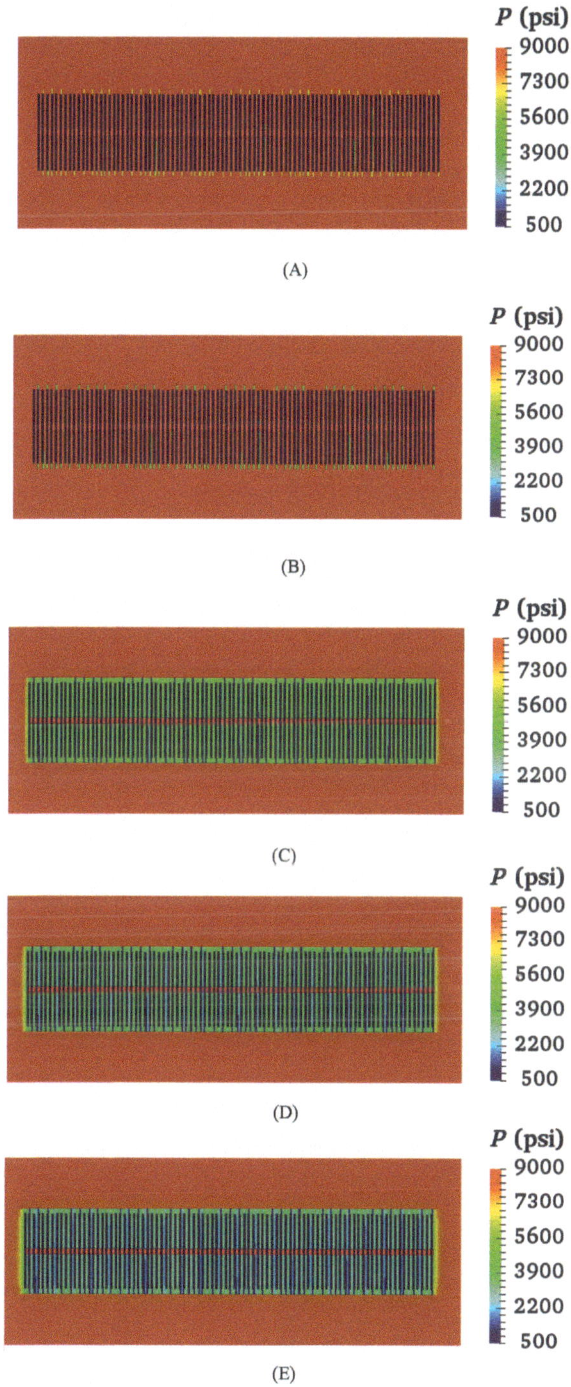

FIGURE 7.16 Pressure distributions of the case of with hydraulic fractures only (without EPA): (A) 30 days; (B) 74 days (end of history); (C) 10 years after history; (D) 20 years after history; and (E) 30 years after history. *EPA*, Enhanced permeability area. *Modified from Tripoppoom, S., 2019. Assisted History Matching Workflow for Unconventional Reservoirs (M.S. thesis). The University of Texas at Austin.*

References

Chang, H., Zhang, D., 2018. History matching of stimulated reservoir volume of shale-gas reservoirs using an iterative ensemble smoother. SPE Journal 23 (2), 346–366.

Cipolla, C.L., Maxwell, S.C., Mack, M.G., 2012. Engineering guide to the application of microseismic interpretations. In: Paper SPE 152165, Presented at SPE Hydraulic Fracturing Technology Conference, The Woodlands, TX, February 6–8, 2012.

Du, C.M., Zhang, X., Zhan, L., Gu, H., Hay, B., Tushingham, K., et al., 2010. Modeling hydraulic fracturing induced fracture networks in shale gas reservoirs as a dual porosity system. In: Paper SPE 132180, Presented at International Oil & Gas Conference and Exhibition, Beijing, China, June 8–10, 2010.

Mayerhofer, M.J., Lolon, E., Warpinski, N.R., Cipolla, C.L., Walser, D.W., Rightmire, C.M., 2010. What is stimulated reservoir volume? SPE Production & Operations 25 (1), 89–98.

Rubin, B., 2010. Accurate simulation of non Darcy flow in stimulated fractured shale reservoirs. In: Paper SPE 132093, Presented at SPE Western Regional Meeting, Anaheim, CA, May 27–29, 2010.

Shakiba, M., Sepehrnoori, K., 2015. Using embedded discrete fracture model (EDFM) and microseismic monitoring data to characterize the complex hydraulic fracture networks. In: Paper SPE 175142, Presented at SPE Annual Technical Conference and Exhibition, Houston, TX, September 28–30, 2015.

Shrivastava, K., Hwang, J., Sharma, M., 2018. Formation of complex fracture networks in the Wolfcamp Shale: calibrating model predictions with core measurements from the hydraulic fracturing test site. In: Paper SPE 191630, Presented at SPE Annual Technical Conference and Exhibition, Dallas, TX, September 24–26, 2018.

Taleghani, A.D., Olson, J.E., 2013. How natural fractures could affect hydraulic-fracture geometry. SPE Journal 19 (1), 161–171.

Tripoppoom, S., 2019. Assisted History Matching Workflow for Unconventional Reservoirs (M. S. thesis). The University of Texas at Austin.

Wu, K., Olson, J.E., 2015. Simultaneous multifracture treatments: fully coupled fluid flow and fracture mechanics for horizontal wells. SPE Journal 20 (2), 337–346.

Yin, J., Xie, J., Datta-Gupta, A., Hill, A.D., 2011. Improved characterization and performance assessment of shale gas wells by integrating stimulated reservoir volume and production data. In: Paper SPE 148969, Presented at SPE Eastern Regional Meeting, Columbus, Ohio, August 17–19, 2011.

Chapter 8

Shale-gas well with and without natural fractures

8.1 Introduction

One of the challenges in unconventional reservoirs is the existence of natural fractures. Natural fractures do not only affect fracture propagation during hydraulic fracturing (Taleghani and Olson, 2013; Wu and Olson, 2016; Agrawal et al., 2019) but also have an effect on the production performance in the long term (Yu et al., 2018), including future development decisions (Yu et al., 2019). Moreover, the hydraulic fracture and reservoir properties obtained from history matching may not be representative due to the neglect of natural fracture existence.

Natural fractures can be characterized from regional natural fracture information such as outcroppings, seismic, core data (Shrivastava et al., 2018), and image logs; however, this information still has uncertainty. With the proposed workflow of the neural network-Markov chain Monte Carlo (NN-MCMC) algorithm and an efficient embedded discrete fracture model (EDFM) to model discrete fractures, natural fractures can be included during history matching as uncertain parameters. The benefits of including natural fractures are first that the posterior distributions of hydraulic fractures and reservoir properties are more representative of what happens when natural fracture existence is considered, and second, the uncertainty of natural fractures is calibrated with production data not deterministically, but probabilistically.

In this chapter, we used a shale-gas well with 724 days of production data. Gas flow rate, water flow rate, and wellhead pressure (WHP) were available for history matching. We performed two scenarios consisting of with and without natural fractures to demonstrate the effect of natural fracture existence on hydraulic fractures and reservoir parameters' posterior distribution. Also, the production forecast of two scenarios was obtained and discussed.

8.2 Reservoir model

For this study, we constructed a homogeneous reservoir model due to the fact that the static model integrating geological and petrophysical inputs was not available. The model dimensions are 6600 ft × 2950 ft in the x and y

Assisted History Matching for Unconventional Reservoirs. DOI: https://doi.org/10.1016/B978-0-12-822242-3.00010-9
© 2021 Elsevier Inc. All rights reserved.

directions and the model dimension in the z direction is equal to fracture height. In other words, we assumed a 2D model with the reservoir response within fracture height which is equal to model thickness. The grid block dimension is 50 ft × 50 ft × fracture height. In cases where we have more information regarding vertical heterogeneity, the whole thickness with different properties in the vertical direction can be modeled with more than one layer. However, in this study, that information was not available. Therefore, we continued the study with as simple a model as the information allowed.

The initial reservoir pressure is 7397 psi. The reservoir model description is summarized in Table 8.1. Also, the well and fracture information are summarized in Table 8.2. The horizontal well has a lateral length of 5800 ft. As shown in Fig. 8.1, the well was completed with 20 stages of hydraulic fractures with a total cluster number of 58. For the fracture modeling, we used

TABLE 8.1 Summary of basic reservoir properties used in this study (Tripoppoom et al., 2019b).

Reservoir description	Value	Unit
Model dimension ($x \times y$)	6000 × 2950	ft
Number of grid blocks ($x \times y \times z$)	120 × 59 × 1	–
Initial reservoir pressure	7397	psi
Reservoir temperature	200	°F
Residual water saturation	20%	–
Total compressibility	3×10^{-6}	psi^{-1}
Matrix porosity	6%	–

TABLE 8.2 Summary of well and hydraulic fractures information (Tripoppoom et al., 2019b).

Well and fractures description	Value	Unit
True vertical depth of a lateral section	8228	ft
Well length	5800	ft
Number of stages	20	–
Number of clusters	58	–
Cluster spacing	100	ft

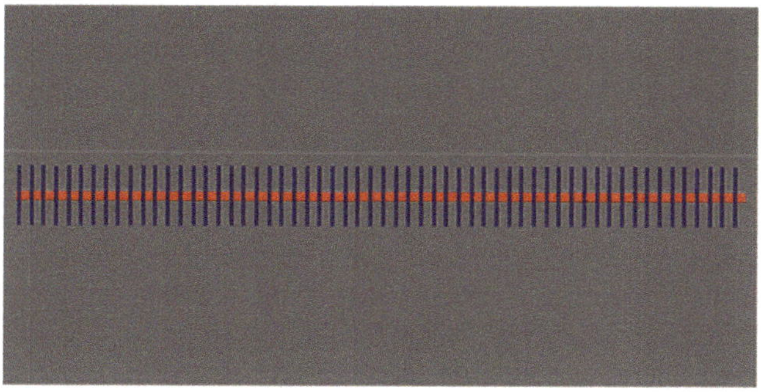

FIGURE 8.1 Reservoir model including 58 hydraulic fractures (in blue) modeled by EDFM method in a horizontal well (in red) (Tripoppoom et al., 2019b). *EDFM*, Embedded discrete fracture model.

TABLE 8.3 Langmuir isotherm parameters for modeling the gas desorption effect (Tripoppoom et al., 2019b).

Langmuir isotherm parameters	Value	Unit
Inverse-pressure parameter for the Langmuir isotherm model, CH_4	0.000534	1/psi
Maximum moles of adsorbed component per unit mass of rock, CH_4	0.060148	gmole of component/ lb of rock
Rock density	162.318	lb/ft^3

the EDFM method to handle fractures nonintrusively relative to the main reservoir grid.

For shale-gas reservoirs, we included the gas desorption effect using a Langmuir isotherm, as shown in Table 8.3. Also, we considered the pressure-dependent permeability for both matrix and fractures due to a geomechanical effect during reservoir depletion. The permeability reduction happens when reservoir pressure is depleted. In this study, the permeability modulus of 0.055 and 0.056 for matrix and fractures, from laboratory work, were used.

8.3 History matching

We had a historical production of 724 days including gas flow rate, water flow rate, and WHP as published in Tripoppoom et al. (2019a). For history matching, we used gas flow rate, water flow rate, and bottomhole pressure (BHP) calculated from WHP due to wellbore pressure loss. We assumed the constant pressure loss from wellhead to bottomhole to account for reservoir depth and a mostly constant water–gas ratio (WGR) value after water flow back period. However, in cases where more complicated WGRs and fluctuating gas flow rate are observed, more complicated conversion of the function of gas flow rate, WGR, and WHP should be used for WHP-BHP conversion.

8.3.1 Case 1: hydraulic fractures only (no natural fractures)

For the case with only hydraulic fractures, we identified the uncertain parameters and their ranges, as summarized in Table 8.4. We can categorize the uncertain parameters into two groups. In the first group are reservoir properties, including matrix permeability, matrix water saturation, and relative permeability curve parameters. In the second group are hydraulic fracture properties, including fracture height, fracture half-length, fracture conductivity, fracture water saturation, and fracture width. We determined the initial ranges of each parameter based on our initial knowledge or prior knowledge about this well and reservoir. For example, for matrix permeability, we used information from core data and flow back analysis. However, the information from core data normally indicates a low value of matrix permeability, such as 0.001 md. Therefore, we increased the maximum matrix permeability to 0.04 md because we expected the apparent matrix permeability to show an effect from natural fractures. Based on our experience, the success of history matching significantly depends on choosing reasonable parameter values to mimic the production data we are matching. We cannot achieve reasonable matching quality with the wrong parameter range inputs, even though an accurate and efficient history-matching algorithm is used. In cases where

TABLE 8.4 Summary of the uncertain parameters of reservoir and fracture parameters (Tripoppoom et al., 2019b).

Uncertain parameters	Unit	Min value	Max value
Matrix permeability (km)	md	0.0001	0.04
Fracture height	ft	16	115
Fracture half-length (X_f)	ft	200	1000
Fracture conductivity	md ft	1	100
Fracture water saturation	–	0.50	0.90
Matrix water saturation	–	0.30	0.50
Exponent of k_{rg}	–	1	4
Endpoint of k_{rw}	–	0.5	1
Exponent of k_{rw}	–	1	4
Fracture width	ft	0.1	1

good history-matching quality cannot be achieved, the first thing to review is parameter inputs and adjust them.

We then used Latin hypercube design to populate 50 initial cases scattering the entire parameter domain. We will achieve a better quality in the construction of NN using this method rather than using random sampling, such as with the Monte Carlo method. The simulation results of 50 initial cases from Latin hypercube design are shown in Fig. 8.2. We observed that the simulation results cover the historical data and their shapes are similar to production history. This means we can possibly find history-matching solutions in our initial uncertain parameter ranges. On the other hand, if the simulation results do not cover the historical data or the shapes of the production profile are totally different, we need to review our initial uncertain parameter range inputs and revise them.

Instead of constructing one proxy model for each response parameter, we used multiple proxy models to mitigate the nonlinearity of the problem. Therefore, we had to define the proxy points to construct proxy models (NN in this case study for proxy-based MCMC algorithm). The global objective

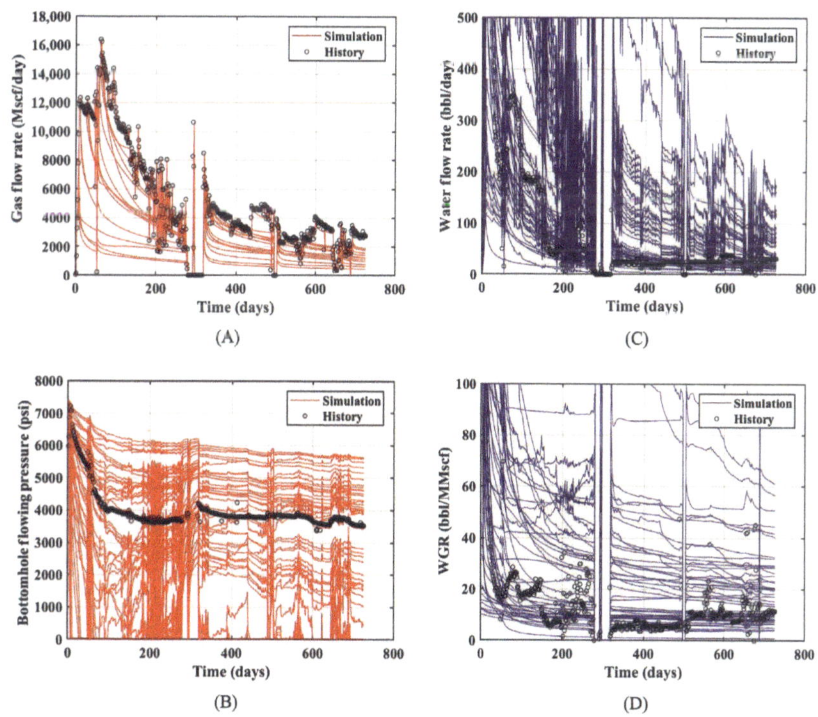

FIGURE 8.2 Comparison between simulation results and historical data of the cases from Latin hypercube design (iteration 1): (A) gas flow rate; (B) flowing BHP; (C) water flow rate; and (D) WGR (Tripoppoom et al., 2019b). *BHP*, Bottomhole pressure; *WGR*, water–gas ratio.

function will be calculated from the many points we selected based on assigned weight values. In an ideal case, we would choose proxy points for as long as the computational effort allows. However, adding unnecessary proxy points would hinder the performance of the history-matching algorithm. This means we need to achieve an acceptable accuracy with optimized computational effort. The criterion is to choose proxy points that capture the major trend of production data. In this case study, we used a total of 12 proxies from different dates in production data. This included 6 proxies for BHP at day 12, 100, 250, 350, 530, and 680 (Fig. 8.3A) and 6 proxies for WGR at 12, 50, 120, 180, 400, and 600 (Fig. 8.3B). All 12 proxies captured the major trend of BHP and WGR. We can use more than 12 proxy points; however, one thing to be aware of is that this will increase the computing time for

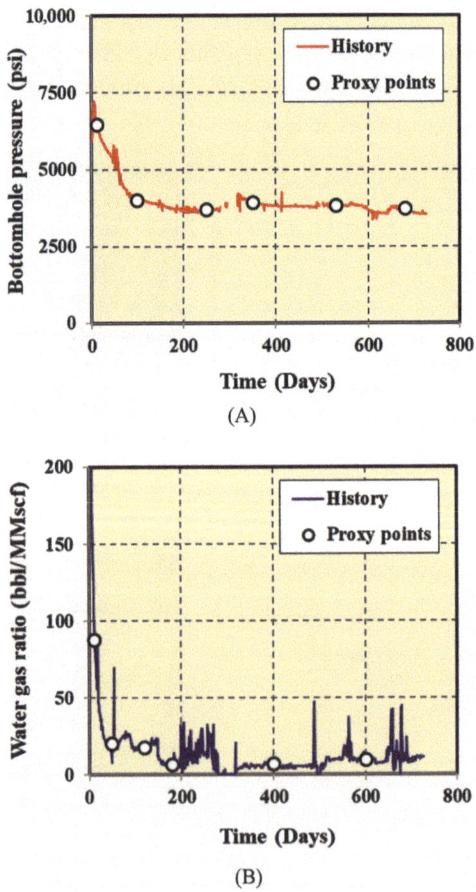

FIGURE 8.3 Multiple proxies of 12 including (A) 6 proxies of BHP and (B) 6 proxies of WGR (Tripoppoom et al., 2019b). *BHP*, Bottomhole pressure; *WGR*, water–gas ratio.

NNs in every step of MCMC and in every history-matching iteration. Therefore, including redundant proxy points that do not give more valuable information about the production trend may not be a good option. On the other hand, if we include fewer proxy points, we may obtain history-matching solutions with less actual matching quality; for example, if we have only one proxy point for bottomhole pressure at the early time and another point at the end of history. The NN-MCMC algorithm will not evaluate any points in between these two points. This certainly affects the history-matching quality between these two proxy points.

We used NN as a proxy model to predict the values without running actual simulations while executing the NN-MCMC algorithm. The NN-MCMC algorithm progressed iteratively. In each iteration the algorithm proposed 25 new cases as history-matching solution candidates to be validated with the actual reservoir simulator. Then, new simulation results were included in the training data set, and then the NNs were re-trained. The stopping criteria of the workflow is either the proxy convergence or the maximum number of iterations. In this study, we used a proxy convergence value of 5% difference between the current and previous proxy. The workflow stopped at a maximum of 12 iterations, in this case.

After the assisted history-matching (AHM) workflow stopped and all simulation runs were completed, we screened the history-matching solutions from all runs of all iterations. We defined history-matching criteria of 10% and 40% for BHP and WGR responses based on our decision for the least acceptable matching quality. The history-matching solutions must pass these two criteria, or the intersection of two response parameters, as shown in Fig. 8.4. The criteria for WGR is higher due to a more acceptable range of water phase. In addition, their naturally low values created a higher value of relative error percentage. For example, if the historical WGR is 5 bbl/MMscf and the simulation WGR is 10 bbl/MMscf, because we used relative error to calculate the WGR error objective function, the discrepancy between historical data and simulation result would be 100%, even though the absolute value difference is only 5 bbl/MMscf.

Fig. 8.5 illustrates the number of history-matching solutions versus iteration number. We obtained 89 solutions from a total of 325 simulation runs, which is around 27%. One thing to highlight is that more solutions were

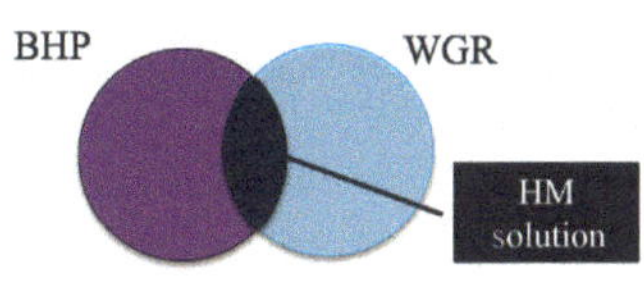

FIGURE 8.4 History-matching solutions are screen by the criteria of 10% and 40% for each objective function of BHP and WGR, respectively (Tripoppoom et al., 2019b). *BHP*, Bottomhole pressure; *WGR*, water–gas ratio.

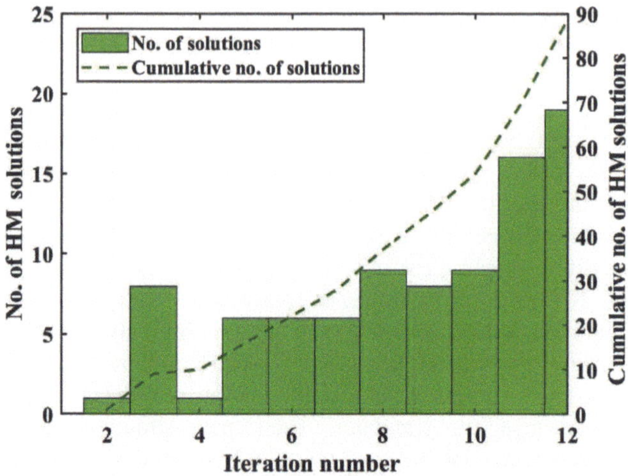

(A)

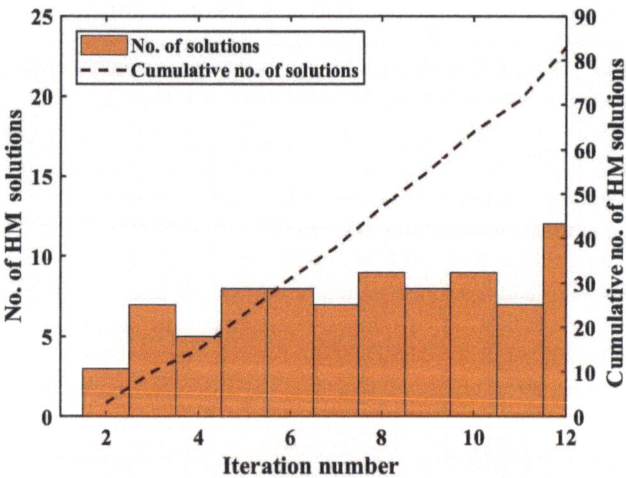

(B)

FIGURE 8.5 Number of history-matching solutions versus iteration number: (A) case 1 hydraulic fractures only (no NF) and (B) case 2 hydraulic fractures and natural fractures (with NF). *NF*, Natural fracture.

found in later iterations because the quality of NN is better—more actual data points from more simulation runs are included in the NN training. All simulation results of 89 history-matching solutions are shown in Fig. 8.6, and the simulation results of the best match case for the hydraulic fractures only case [no natural fracture (NF)] are illustrated in Fig. 8.7, including gas flow rate, BHP, water flow rate, and WGR.

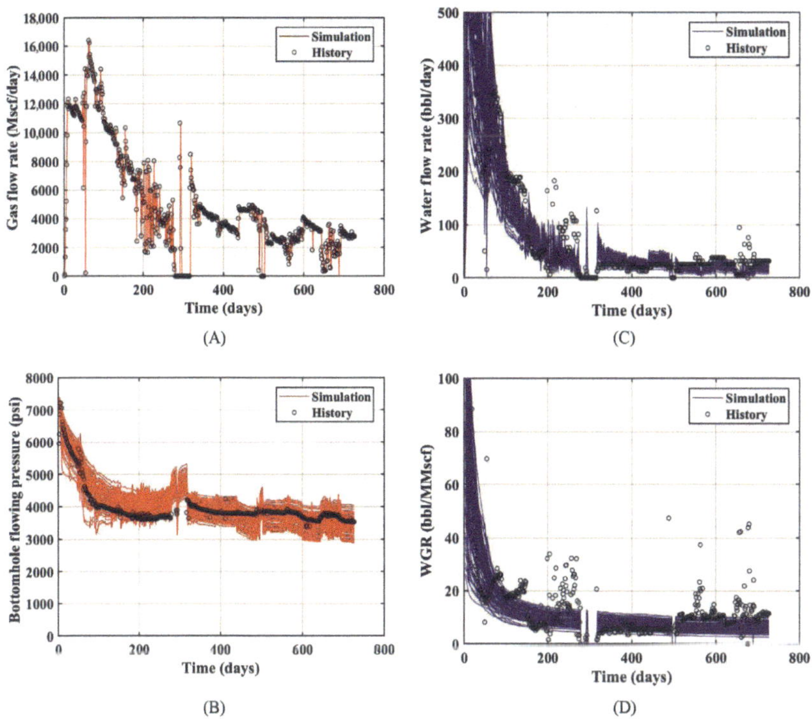

FIGURE 8.6 Comparison between simulation results of 89 history-matching solutions and historical data of case 1 with hydraulic fractures only (no NF): (A) gas flow rate; (B) flowing BHP; (C) water flow rate; and (D) WGR (Tripoppoom et al., 2019b). *BHP*, Bottomhole pressure; *NF*, natural fracture; *WGR*, water–gas ratio.

One thing we can obtain from the history-matching workflow is the posterior distribution of uncertain parameters from 89 history-matching solutions. As shown in Fig. 8.8, the prior uniform distribution we specified at the beginning has changed to the posterior distribution. This means we can characterize reservoir and fracture properties from historical production data in a probabilistic manner. While the initial range of matrix permeability is 0.0001–0.04 md, the exponent of k_{rg} 1–4 and the endpoint of k_{rw} 0.5–1 can make combinations with other uncertain parameters to produce history-matching solutions. We can characterize other uncertain parameters; for example, fracture height is mostly in the range of 40–100 ft, fracture half-length in the lower side, and fracture conductivity in the higher side near 50–80 md ft tend to produce history-matching solutions. In addition, fracture water saturation in the range of 0.55–0.8, matrix water saturation of 0.35–0.45, the exponent of k_{rw} in the higher end of 4, and fracture width below 0.8 ft are likely to match the values of historical WGR trend.

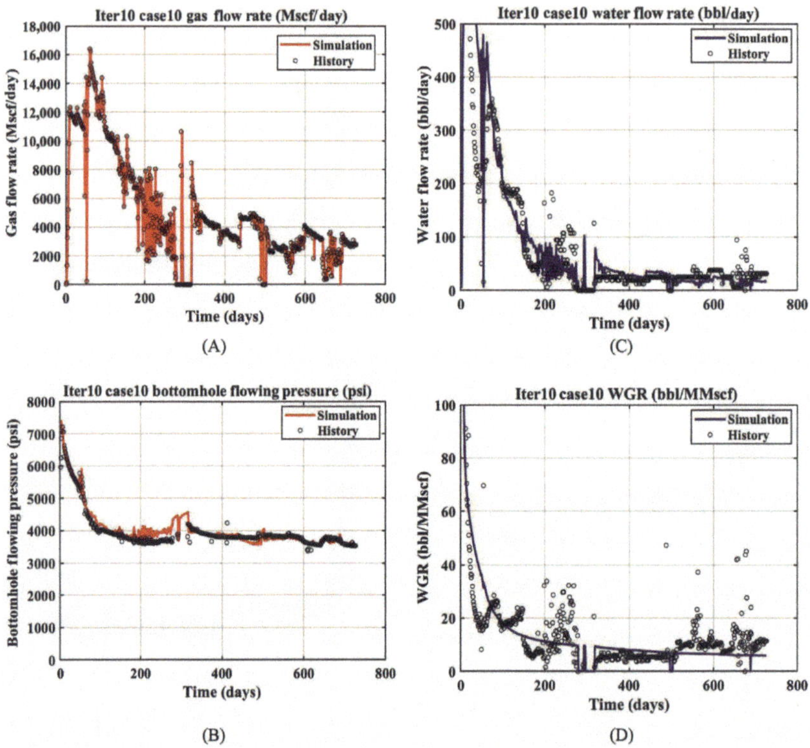

FIGURE 8.7 Comparison between simulation results of the best match case (iteration 10 case 10) and historical data of the case with hydraulic fractures only (no NF): (A) gas flow rate; (B) flowing BHP; (C) water flow rate; and (D) WGR (Tripoppoom et al., 2019b). *BHP*, Bottomhole pressure; *NF*, natural fracture; *WGR*, water–gas ratio.

8.3.2 Case 2: hydraulic fractures and natural fractures (with NF)

The second case we investigated is the case of hydraulic fractures and natural fractures. The objective is to evaluate the effect of natural fracture existence on the history-matching solutions and production forecast. All reservoir model information and uncertain parameter ranges are the same as in case 1 (Tables 8.1–8.4). In this case, we added uncertain parameters for natural fracture properties including the number of natural fractures, natural fracture length, and natural fracture conductivity, as shown in Table 8.5. For the number and length of natural fractures, these two parameters were linked to natural fracture density or, in this study, natural fracture length per unit area, and natural fracture surface area per unit area. The number and the length of natural fracture ranges were from our prior knowledge of natural fracture density in this reservoir. Also, for natural fracture height, we assumed natural fracture height is equal to hydraulic fracture height; in other words, natural fractures also fully penetrate the whole model thickness as the

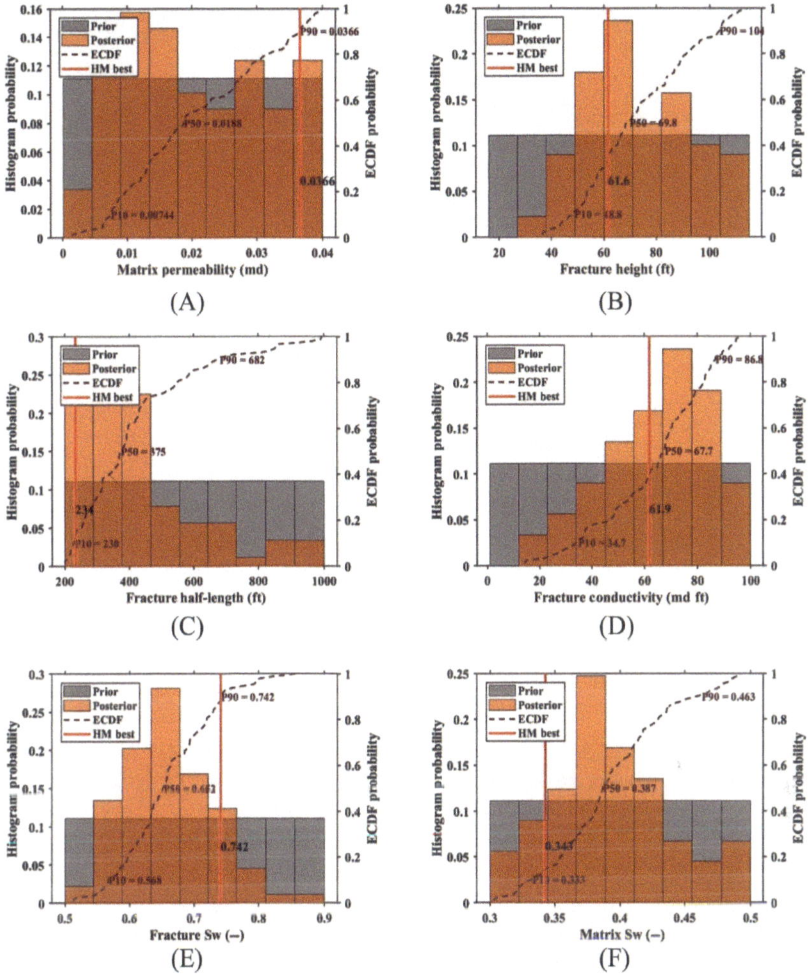

FIGURE 8.8 Prior and posterior distribution of case 1 with hydraulic fractures only (no NF) with P10–P50–P90 values and best match case: (A) matrix permeability; (B) fracture height; (C) fracture half-length; (D) fracture conductivity; (E) fracture water saturation; (F) matrix water saturation; (G) exponent of k_{rg}; (H) endpoint of k_{rw}; (I) exponent of k_{rw}; and (J) fracture width (Tripoppoom et al., 2019b).

2D model assumption. Nevertheless, in cases when more information regarding natural fractures is available, more complicated natural fractures can also be modeled such as natural fracture geometry in 3D space; however, a lot more computational effort is required. For natural fracture conductivity a maximum is assumed that is 10 times lower than the maximum of hydraulic fracture conductivity. These three uncertain parameters normally have more impact on flow behavior and overshadow other natural fracture parameters.

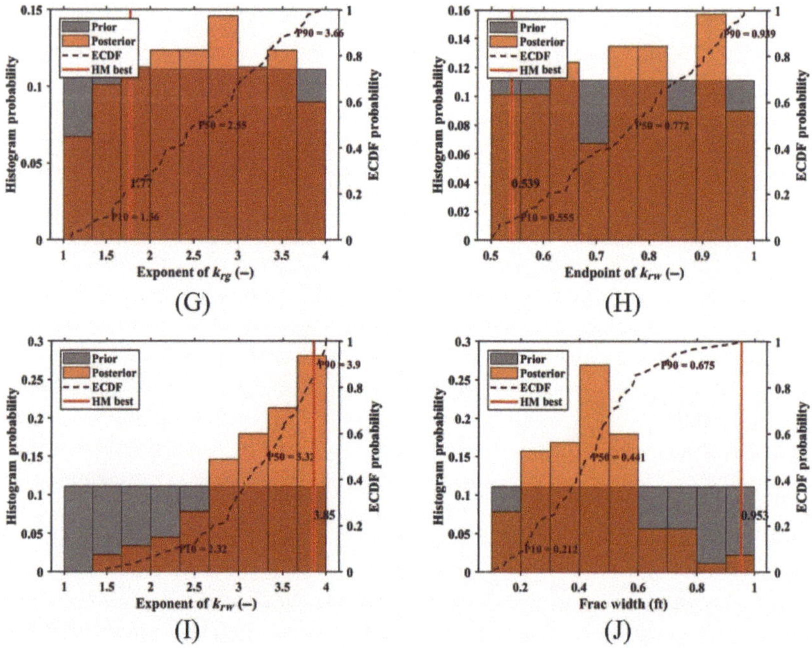

FIGURE 8.8 (Continued).

TABLE 8.5 Summary of the uncertain parameters of natural fractures (Tripoppoom et al., 2019b).

Uncertain NF parameters	Unit	Min value	Max value
Total number of NF	–	200	1200
NF length (same for both NF sets)	ft	200	800
NF conductivity	md ft	1	10

NF, Natural fracture.

Therefore, we used deterministic values for all other natural fracture parameters such as theta, dip angle, and natural fracture width, as summarized in Table 8.6 We specified two sets of natural fractures with different theta. To illustrate, Fig. 8.9 shows hydraulic fractures with natural fractures modeled by EDFM. With the effect from natural fractures, our hypothesis is that the natural fractures intersect with themselves and hydraulic fractures could enhance more connection to the production well. This means the existence of natural fractures could affect the history-matching solutions of other fracture

TABLE 8.6 Summary of the fixed parameters of natural fractures (Tripoppoom et al., 2019b).

Fixed NF parameters	Unit	Value
Number of NF set	–	2
NF height	ft	As hydraulic fracture height
NF theta	degree	NF set 1: 45 NF set 2: 135
NF dip angle	degree	90

NF, Natural fracture.

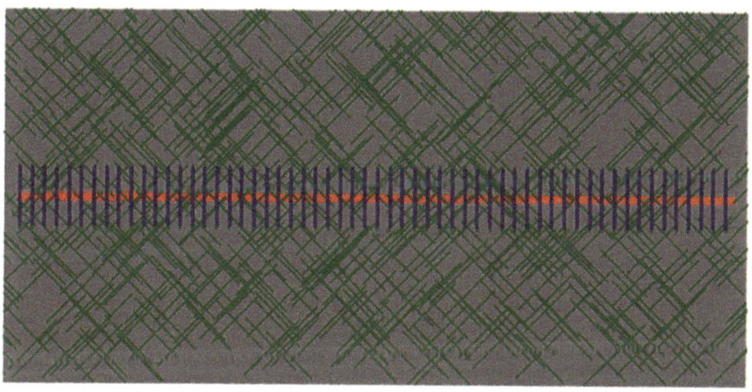

FIGURE 8.9 The horizontal well (in red) with 58 hydraulic fractures (in blue) and two sets of natural fractures (in green) modeled by EDFM (Tripoppoom et al., 2019b). *EDFM*, Embedded discrete fracture model.

and reservoir uncertain parameters and the production forecast, but we still do not know how much this impact would be until the history matching is performed.

The history-matching process progressed the same as case 1. In this case, the history-matching workflow also stopped at the maximum iteration of 12. With the same history-matching solution screening criteria the workflow found 83 history-matching solutions from a total of 325 simulation runs (around 26%) compared to 89 history-matching solutions of case 1 with only hydraulic fractures (Fig. 8.5B). The lower number of history-matching solutions is obviously from the higher degree of freedom including natural fracture uncertain parameters.

All simulation results of the 83 history-matching solutions are shown in Fig. 8.10, including gas flow rate, BHP, water flow rate, and WGR.

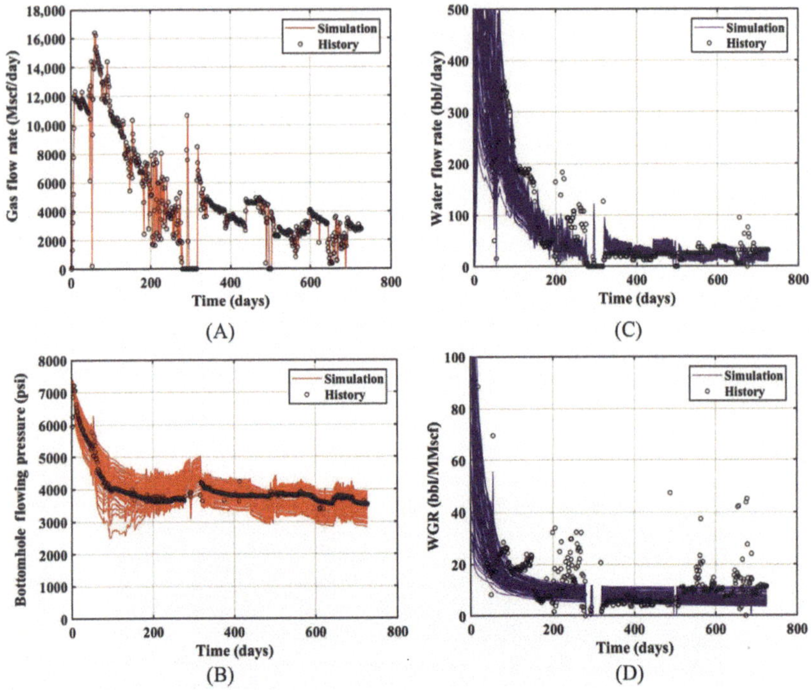

FIGURE 8.10 Comparison between simulation results of 83 history-matching solutions and historical data of case 2 with hydraulic fractures and natural fractures (with NF): (A) gas flow rate; (B) flowing BHP; (C) water flow rate; and (D) WGR (Tripoppoom et al., 2019b). *BHP*, Bottomhole pressure; *NF*, natural fracture; *WGR*, water–gas ratio.

Similarly, we obtained posterior distribution of uncertain parameters, especially natural fracture parameters, together with best match case values and P10–P50–P90 values, as shown in Fig. 8.11. In terms of natural fracture density, we illustrated in Fig. 8.12 the natural fracture length per unit area and total natural fracture surface area per unit area. Natural fracture length per unit area is calculated from the multiplication of the number of natural fractures and natural fracture length, and divided by total model area. For natural fracture surface area per unit area, the natural fracture height (equal to hydraulic fracture height or model thickness) is multiplied by natural fracture length per unit area to obtain this parameter. We can see that the lower values of natural fracture density from posterior distribution yield more history-matching solutions. One thing to be aware of is that even with the same natural fracture density and the same natural fracture parameters, the production performance responses could be different due to different seeds of natural fractures resulting in different reservoir connectivity. Therefore, the stochastic prediction by NNs of the same natural fracture

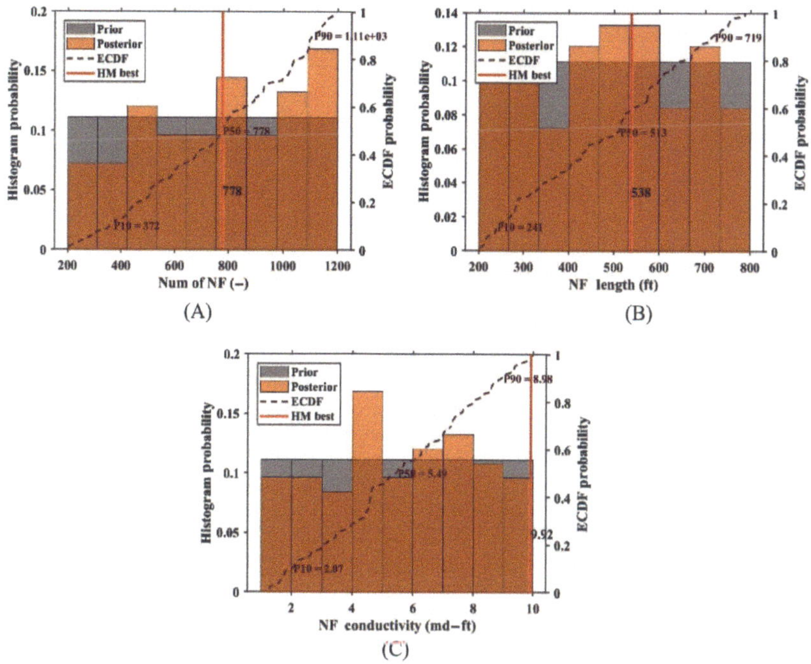

FIGURE 8.11 Prior and posterior distribution of natural fractures parameters of case 2 with hydraulic fractures and natural fractures (with NF) with P10–P50–P90 values and best match case: (A) number of NF; (B) NF length; and (C) NF conductivity (Tripoppoom et al., 2019b). *NF*, Natural fracture.

input can be adopted to improve the research regarding natural fracture history matching.

To illustrate the effect of including natural fractures into history matching, a comparison between prior and posterior distribution of uncertain parameters for both case 1 (No NF) and case 2 (With NF) is presented in Fig. 8.13. When natural fractures exist together with hydraulic fractures, the posterior distribution of some uncertain parameters significantly differs in comparison with the case with only hydraulic fractures. While there is no significant change for matrix permeability and fracture half-length, we observed a significantly lower fracture height, lower fracture conductivity, and higher fracture water saturation because more fluid flow is enhanced by natural fractures. For relative permeability, we see the lower exponent of k_{rg} and lower exponent of k_{rw}, in other words, less curvature on both k_{rg} and k_{rw}. These changes possibly compensate for the lower flow from other uncertain parameters. Lastly, the uncertainty of natural fractures can be reduced from the prior distribution by using the AHM workflow. As shown in Fig. 8.13, we observe that natural fracture conductivity and natural fracture

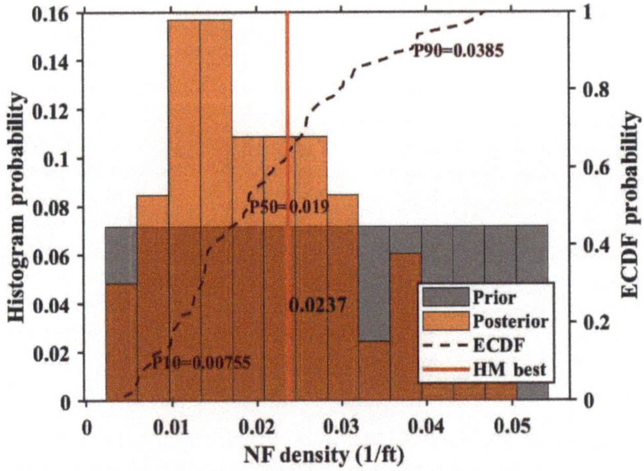

(A)

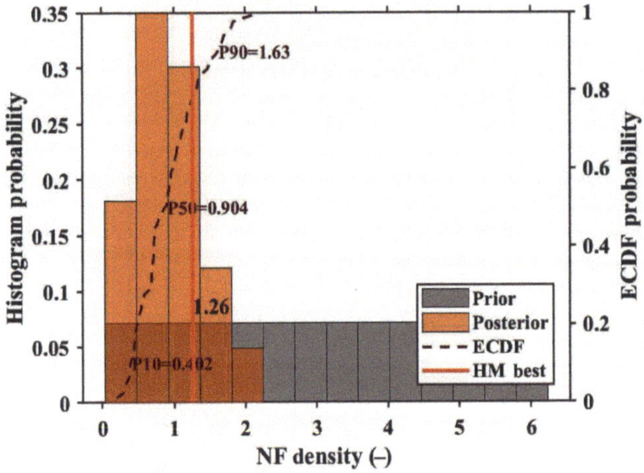

(B)

FIGURE 8.12 Natural fractures density: (A) natural fractures length per unit area; and (B) natural fractures surface area per unit area.

length did not change significantly from prior distribution while we see a significant change in the posterior distribution of the number of natural fractures from the initial prior distribution. The higher end values tend to provide more history-matching solutions. This implies that, for this case, the number of natural fractures has more impact on the production than natural fracture length and conductivity because any value of natural fracture length and conductivity could yield history-matching solutions.

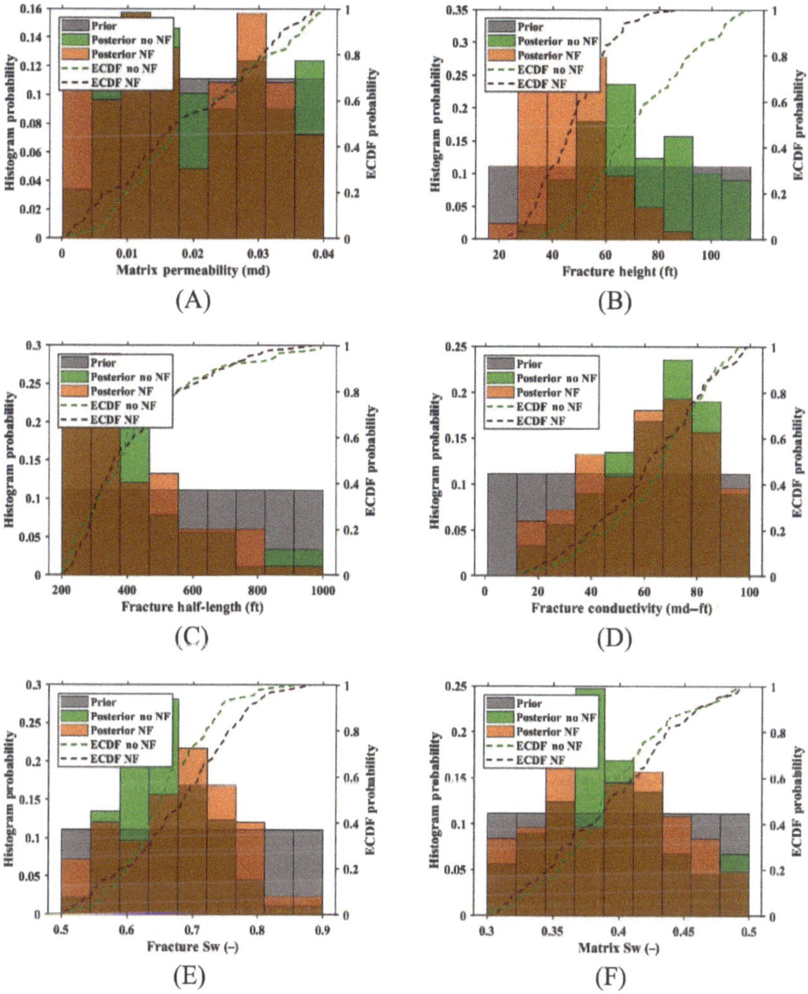

FIGURE 8.13 Prior and comparison of posterior distribution between case 1 with hydraulic fractures only (no NF) and case 2 with hydraulic fractures and natural fractures (with NF): (A) matrix permeability; (B) fracture height; (C) fracture half-length; (D) fracture conductivity; (E) fracture water saturation; (F) matrix water saturation; (G) exponent of k_{rg}; (H) endpoint of k_{rw}; (I) exponent of k_{rw}; and (J) fracture width (Tripoppoom et al., 2019b). *NF*, Natural fracture.

In addition, the pressure distributions of the best match case for both scenarios are depicted in Figs. 8.14 and 8.15. We can see the difference in the pressure profile of uniform and nonuniform depleted areas in the x−y plane for case 1 and case 2, respectively. One thing to note is that these depleted areas in x−y between two cases are not obviously different because the fracture half-length posterior distribution of the two cases (Fig. 8.13C) is not

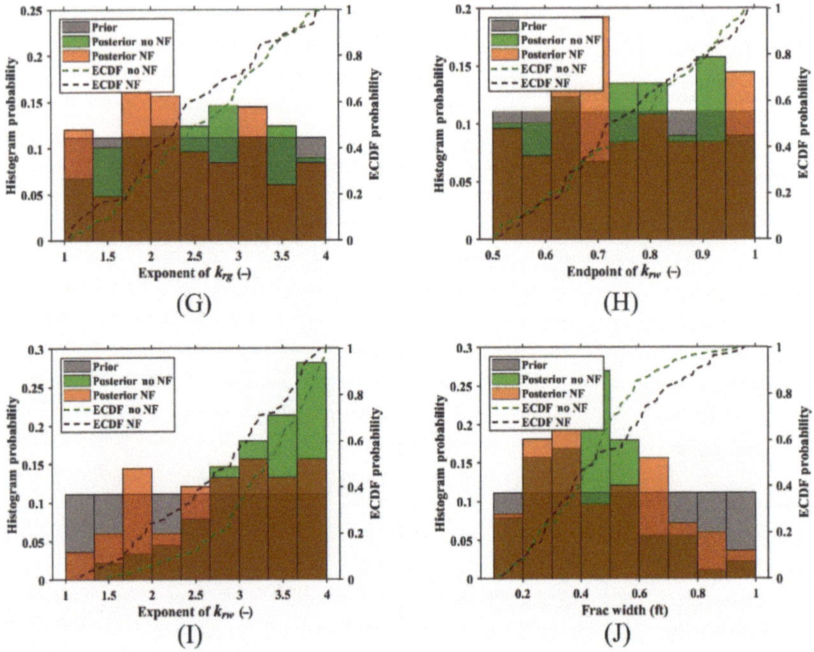

FIGURE 8.13 (Continued).

significantly different based on history matching. In fact, the main difference should be from fracture height (Fig. 8.13B) or in the z direction, which is previously discussed.

8.4 Discussion

In this section, we emphasize the performance of the NN-MCMC algorithm in the proposed AHM workflow by presenting the history-matching results and analysis of case 2 with hydraulic fractures and natural fractures (with NF).

Fig. 8.16 shows the parallel coordinate plot of history-matching solutions and nonhistory-matching solutions. Each line represents one realization of uncertain parameters. We can see that only some combinations of uncertain parameters can be history-matching solutions; for example, ones with lower fracture half-lengths and higher fracture conductivity in the ranges tend to be history-matching solutions. Another thing to be highlighted is that the NN-MCMC algorithm already explored and validated other combinations, but they could not reproduce the history match solutions, shown as the gray lines in Fig. 8.16. This confirms the NN-MCMC algorithm efficiency in exploring throughout the possible uncertain parameter domain.

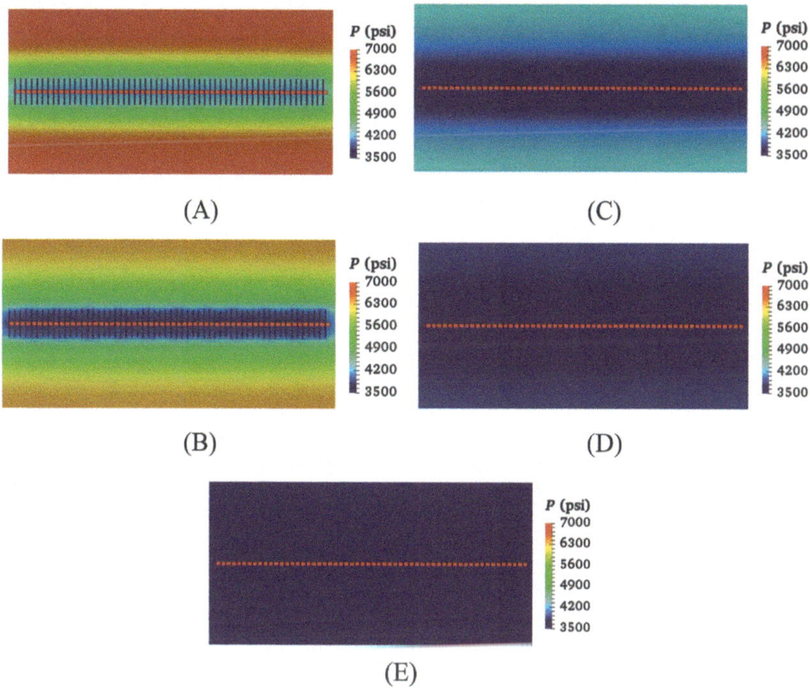

FIGURE 8.14 Pressure distribution of the best case for case 1 hydraulic fractures only (no NF): (A) after 365 days; (B) after 724 days (end of history); (C) after 10 years; (D) after 20 years; and (E) after 30 years (Tripoppoom, 2019). *NF*, Natural fracture.

As the NN-MCMC algorithm performed automatically and proposed new cases in each iteration to be validated with reservoir simulator, Fig. 8.17 shows the objective function values from simulation results versus simulation number index. We observed that objective function values are improving as the algorithm progresses (higher simulation number index) and more solutions are found in the later iterations due to the improvement of NN. Also, the proposed values for each uncertain parameter versus simulation index are shown in Fig. 8.18. The values at later iterations can be seen to converge to some specific value range, especially for fracture height. In addition, Fig. 8.19 displays the cross plot between proxy estimation (using NN) versus simulation results from reservoir simulator for various errors including global error, BHP error, and WGR error. We can see the improvement of NN quality by comparing it to the unit slope line. As iteration numbers increase, the proposed points are closer to the unit slope line. For each uncertain parameter the cross plots are demonstrated in Fig. 8.20. Similar improvement is observed for each uncertain parameter. To sum up, the NN-MCMC algorithm in the proposed AHM

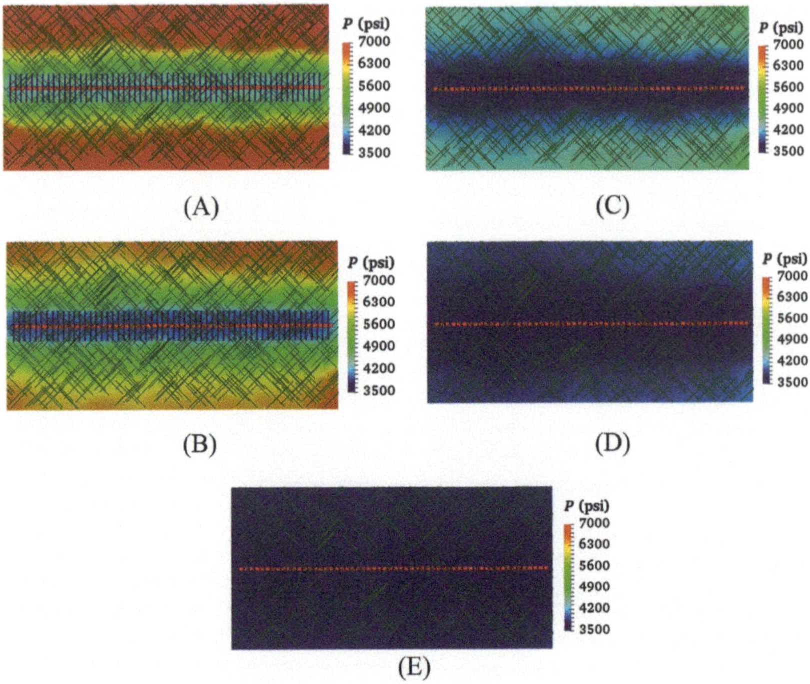

FIGURE 8.15 Pressure distribution of the best case for case 2 hydraulic fractures and natural fractures (with NF): (A) after 365 days; (B) after 724 days (end of history); (C) after 10 years; (D) after 20 years; and (E) after 30 years (Tripoppoom, 2019). *NF*, Natural fracture.

workflow can perform history matching automatically and find multiple history-matching solutions efficiently.

For total CPU time, Fig. 8.21 shows the total elapsed time for both cases by portions including proxy-based MCMC algorithm, EDFM preprocessor, simulation runs, and other tasks such as code initialization, automatic simulation files generation, and automatic simulation results reading. The computer we used is a 3.7 GHz with 4-core processor. We also ran the simulation with 4-parallel computation. For case 1 with hydraulic fractures only, the total elapsed time from iteration 1 to 12 is 27 hours with most running time contributing to the proxy-based MCMC algorithm exploring a hundred thousand steps in each iteration. For case 2 with hydraulic fractures and natural fractures the total elapsed time is higher at 54 hours with most of the portion allocated to the EDFM preprocessor. This is because EDFM handled hundreds to thousands of discrete natural fractures for each simulation run which is significantly more than the first case where only 58 hydraulic fractures were included in each simulation run. In sum, we finished automatic history matching for both cases within 4 days.

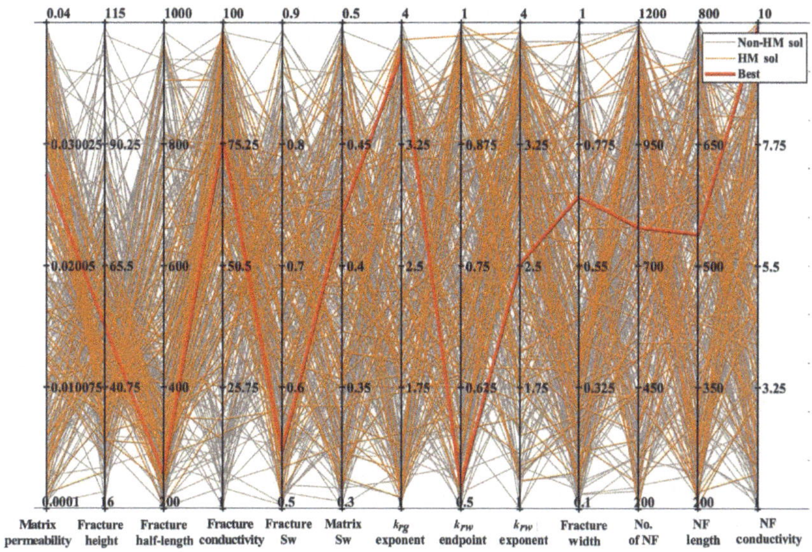

FIGURE 8.16 Parallel coordinate plot of all cases proposed by the NN-MCMC algorithm of case 2 with hydraulic fractures and natural fractures (with NF) (Tripoppoom et al., 2019b). *NF*, Natural fracture; *NN*, neural network; *MCMC*, Markov chain Monte Carlo.

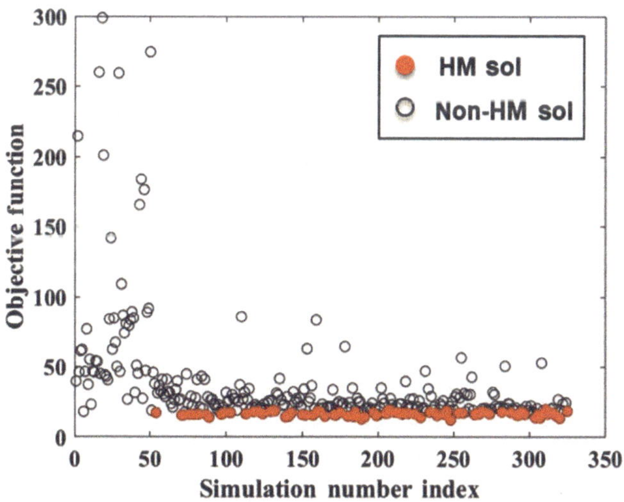

FIGURE 8.17 Objective function value or global error versus simulation number index of case 2 with hydraulic fractures and natural fractures (Tripoppoom et al., 2019b).

8.5 Probabilistic production forecast

After we obtained multiple realizations of history-matching solutions from the AHM workflow, we performed a probabilistic production

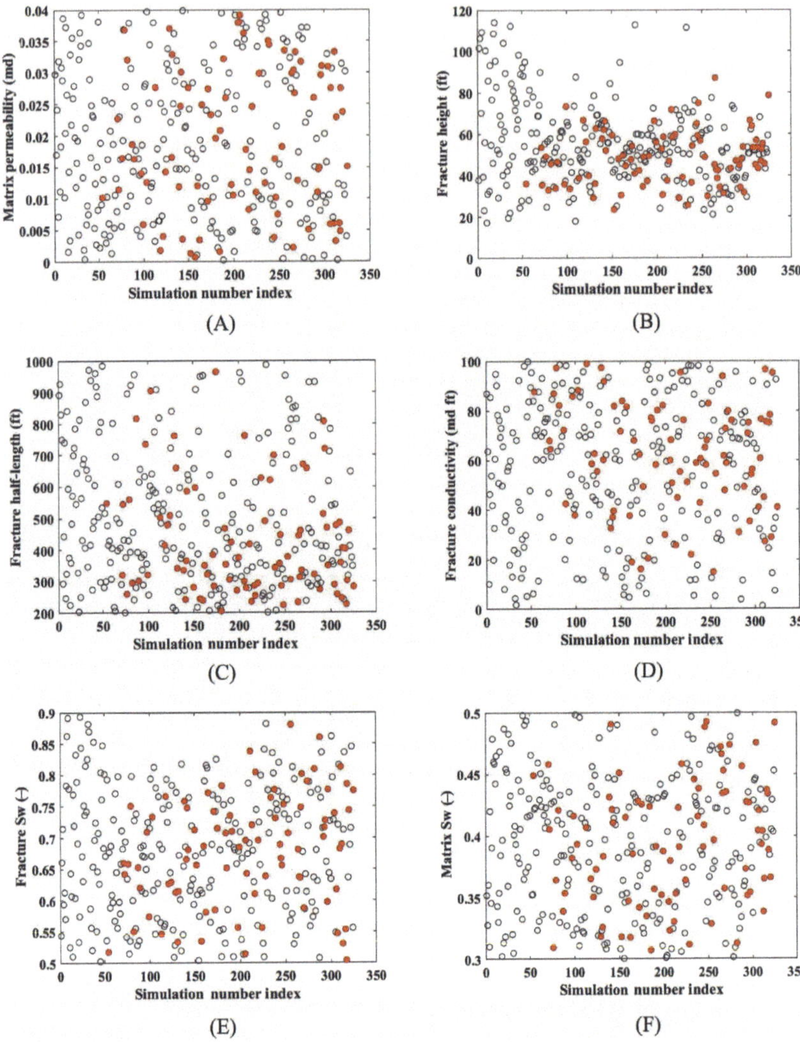

FIGURE 8.18 Uncertain parameter values versus simulation number index: (A) matrix permeability; (B) fracture height; (C) fracture half-length; (D) fracture conductivity; (E) fracture water saturation; (F) matrix water saturation; (G) exponent of k_{rg}; (H) endpoint of k_{rw}; (I) exponent of k_{rw};; (J) fracture width; (K) number of NF; (L) NF length; and (M) NF conductivity (Tripoppoom, 2019). *NF*, Natural fracture.

forecast. In this study, we forecasted the production for 30 years. To perform a reasonable production forecast, we assumed a constant BHP of 3500 psi for prediction (Fig. 8.22A). The reason is that the BHP had been constantly around 3500 psi for several months near the end of the

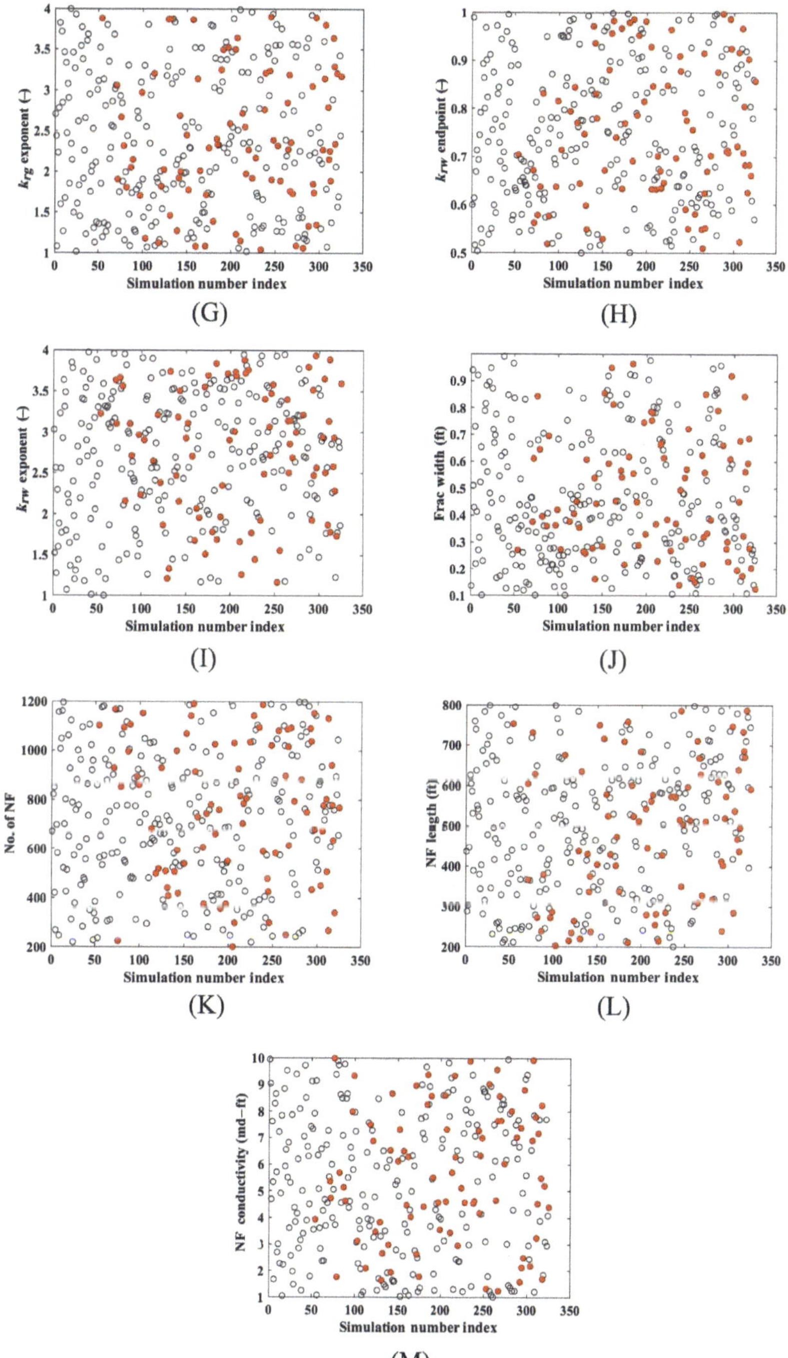

FIGURE 8.18 (Continued).

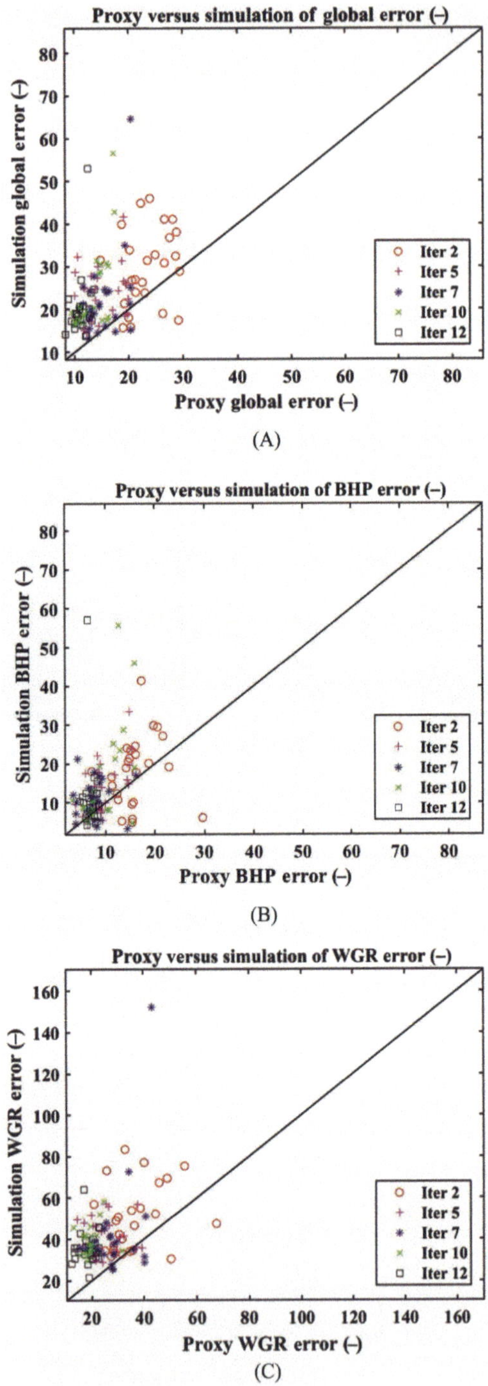

FIGURE 8.19 Cross plot between proxy and simulation values of case 2 hydraulic fractures and natural fractures: (A) global error (calculated from 12 proxies); (B) BHP error (calculated from 6 BHP proxies); and (C) WGR error (calculated from 6 WGR proxies) (Tripoppoom et al., 2019b). *BHP*, Bottomhole pressure; *WGR*, water–gas ratio.

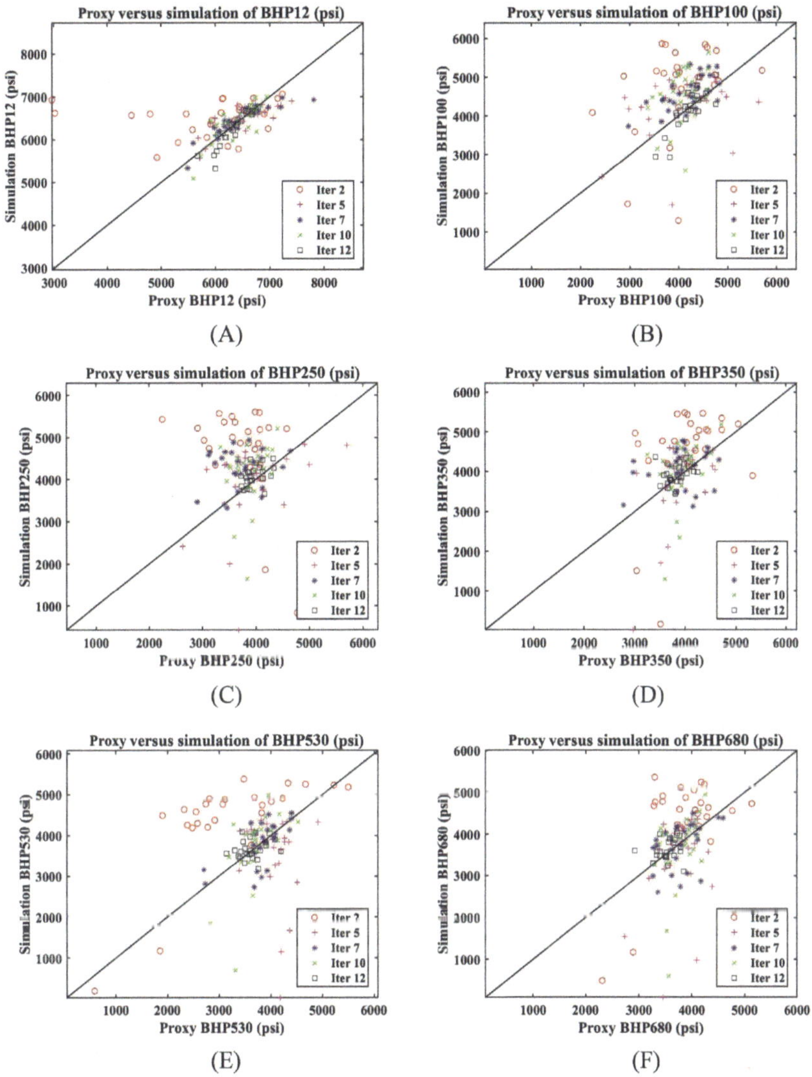

FIGURE 8.20 Cross plot between proxy and simulation values of 12 proxies of case 2 with hydraulic fractures and natural fractures: (A) BHP at day 12; (B) BHP at day 100; (C) BHP at day 250; (D) BHP at day 350; (E) BHP at day 530; (F) BHP at day 680; (G) WGR at day 12; (H) WGR at day 50; (I) WGR at day 120; (J) WGR at day 180; (K) WGR at day 400; and (L) WGR at day 600 (Tripoppoom, 2019). *BHP*, Bottomhole pressure; *WGR*, water–gas ratio.

historical period. We performed a production forecast for cases both with and without natural fractures. The simulation results of the production forecast for both cases are plotted in Figs. 8.22 and 8.23 together with the best match and P10–P50–P90 cases.

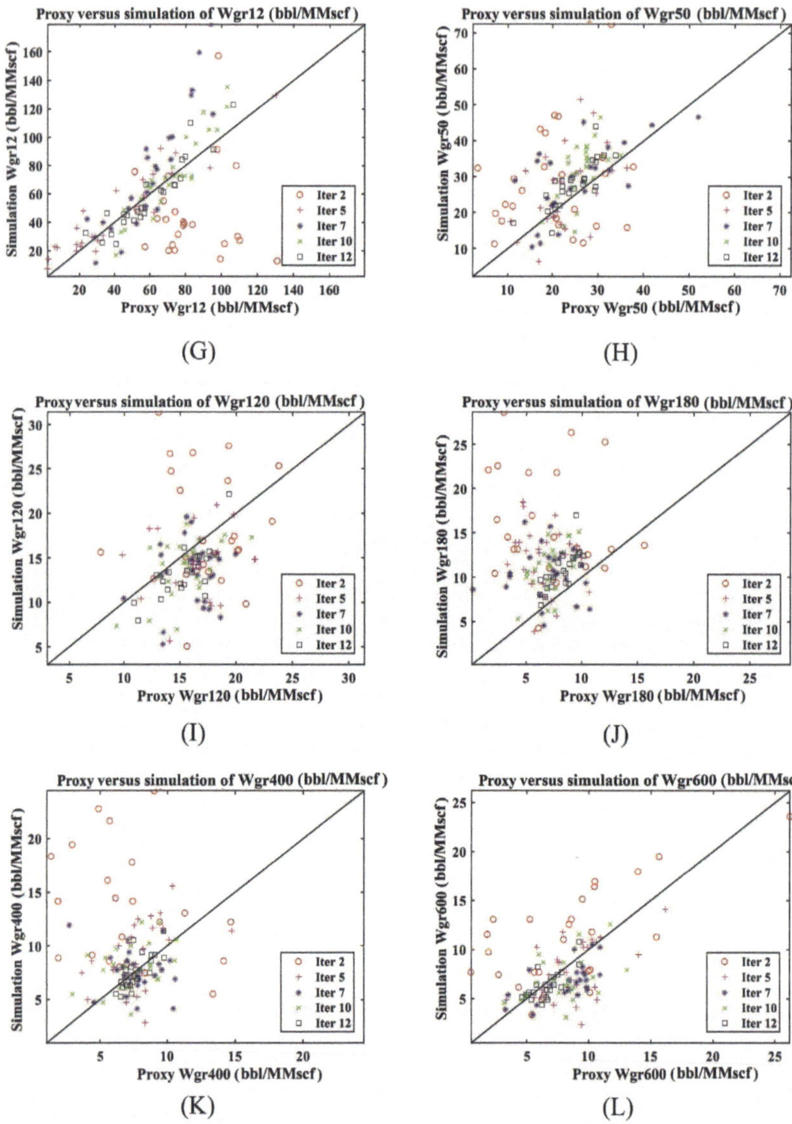

FIGURE 8.20 (Continued).

To compare the gas estimated ultimate recovery (EUR) between both cases, the empirical cumulative distribution function (ECDF) curves of gas EUR are shown in Fig. 8.24. With the existence of natural fractures, the ECDF of gas EUR is lower than the case with only hydraulic fractures. This is because a lower fracture height posterior distribution is obtained during history matching when natural fractures exist (other uncertain

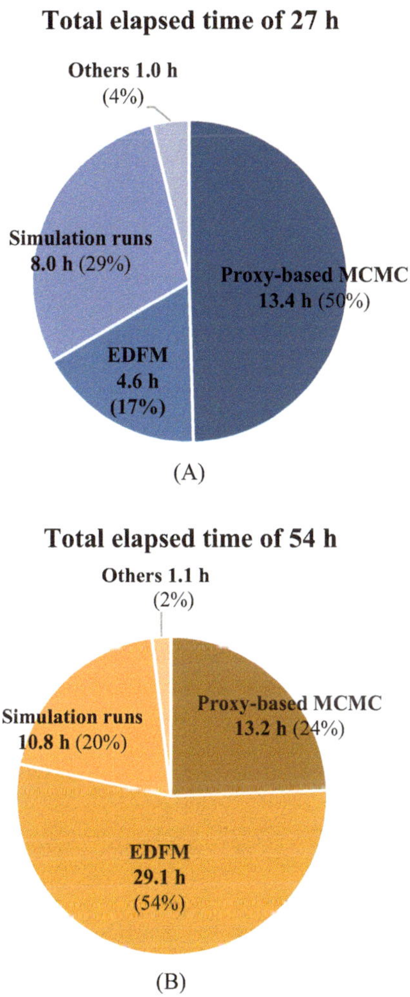

FIGURE 8.21 Total elapsed time including proxy-based MCMC, EDFM simulation runs, and others: (A) case 1 with hydraulic fractures only (no NF) and (B) case 2 with hydraulic fractures and natural fractures (with NF). *EDFM*, Embedded discrete fracture model; *MCMC*, Markov chain Monte Carlo; *NF*, natural fracture.

parameters such as fracture half-length and fracture conductivity are similar for both cases with and without natural fractures). If the same fracture height posterior distribution of the case with only hydraulic fractures was used, adding natural fractures into the system would certainly increase production. However, to match historical production data when natural fractures exist, a lower fracture height posterior was obtained, as illustrated in Fig. 8.13B.

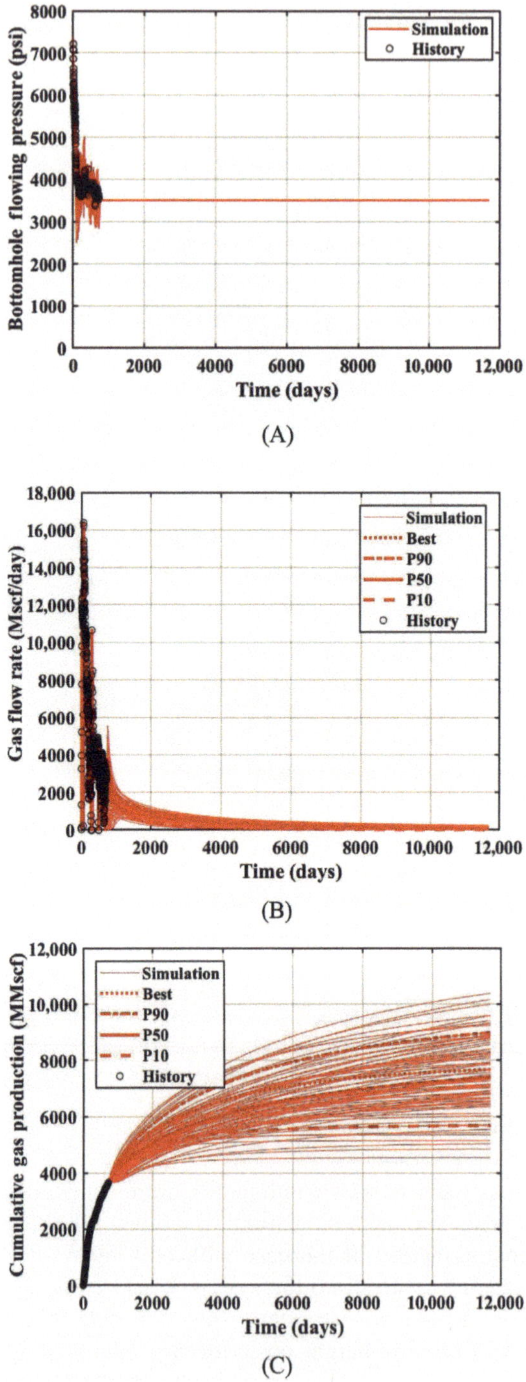

FIGURE 8.22 Production forecast of history-matching solutions of case 1 with hydraulic fractures only (no NF): (A) BHP; (B) gas flow rate; and (C) cumulative gas production (Tripoppoom, 2019). *BHP*, Bottomhole pressure; *NF*, natural fracture.

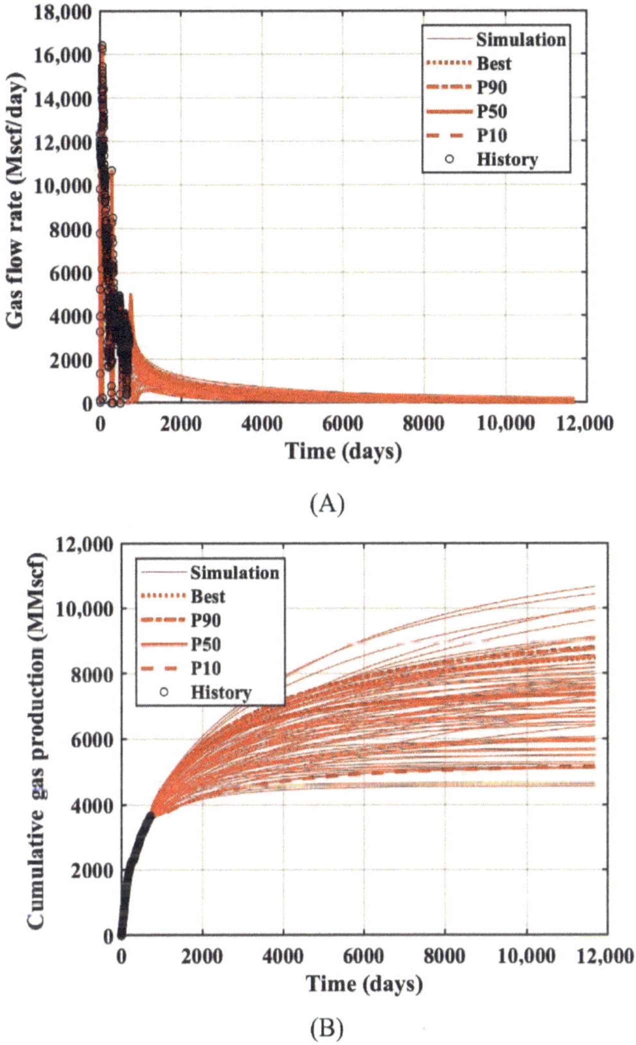

FIGURE 8.23 Production forecast of history-matching solutions of case 2 with hydraulic fractures and natural fractures: (A) gas flow rate and (B) cumulative gas production (Tripoppoom, 2019).

In other words, we would overpredict the gas production if we assumed this well with only hydraulic fractures in spite of the fact that there are natural fractures in the system. The degree of gas EUR overprediction would depend on natural fracture properties and reservoir properties. These would certainly be different for each reservoir. In this case, we observe around a 10% discrepancy in predicted gas EUR.

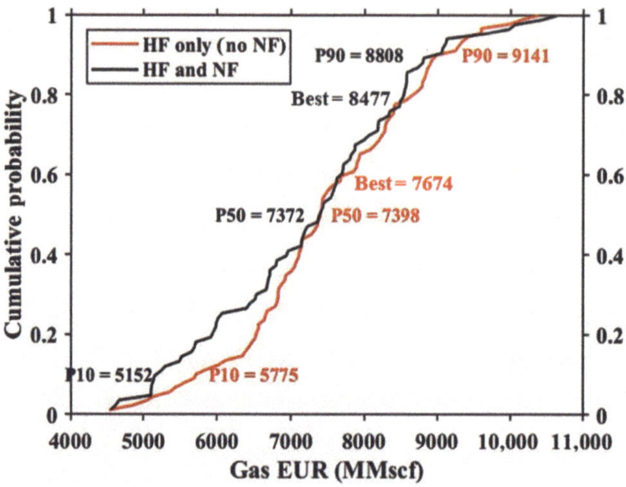

FIGURE 8.24 The comparison of gas EUR empirical cumulative distribution function between case 1 with hydraulic fractures only and case 2 with hydraulic fractures and natural fractures (Tripoppoom, 2019). *EUR*, Estimated ultimate recovery.

To conclude, if there are natural fractures in an actual reservoir, they should be included in the reservoir model for history matching, since we cannot represent the whole system of natural fractures and hydraulic fractures by a simplified model with hydraulic fractures only.

8.6 1000 History-matching solutions from neural networks

In cases where we would like a smoother posterior distribution of uncertain parameters, we can use the latest proxy model or NN to generate more history-matching solutions. To illustrate, Fig. 8.25 shows the posterior distribution and ECDF of 1000 history-matching solutions for both case 1 with hydraulic fractures only (no NF) and case 2 with hydraulic fractures and natural fractures (with NF). We can observe that the distributions are similar to the history-matching cases in cases 1 and 2 where there were, respectively, 89 and 83 history-matching solutions (Fig. 8.13) but with a smoother distribution. Furthermore, NNs can be created for gas EUR prediction based on the production forecast results of 89 and 83 history-matching solutions in each case. Then, we can estimate gas EUR for 1000 history-matching solutions from the proxy (NN), as plotted in Fig. 8.26. The gas EUR ECDF of the 1000 solutions from proxy yields approximately the same range compared to the actual history solutions. However, there is some slight difference between the actual history-matching solutions and the 1000 solutions from

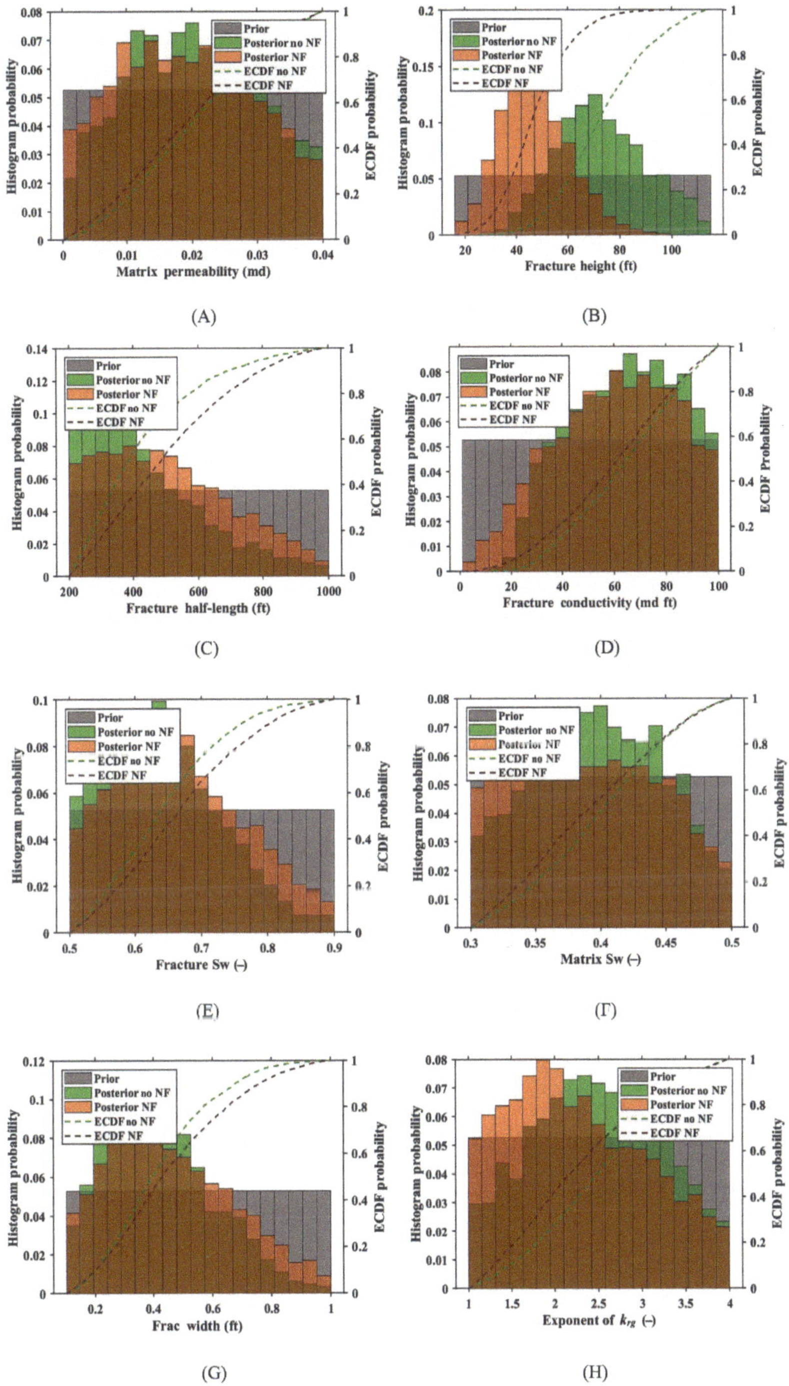

FIGURE 8.25 Comparison of prior and posterior distribution of 1000 history-matching solutions from proxy between case 1 with hydraulic fractures only (no NF) and case 2 with hydraulic fractures and natural fractures (with NF): (A) matrix permeability; (B) fracture height; (C) fracture half-length; (D) fracture conductivity; (E) fracture water saturation; (F) matrix water saturation; (G) exponent of k_{rg}; (H) endpoint of k_{rw}; (I) exponent of k_{rw};; (J) fracture width; (K) number of NF; (L) NF length; and (M) NF conductivity (Tripoppoom, 2019). *NF*, Natural fracture.

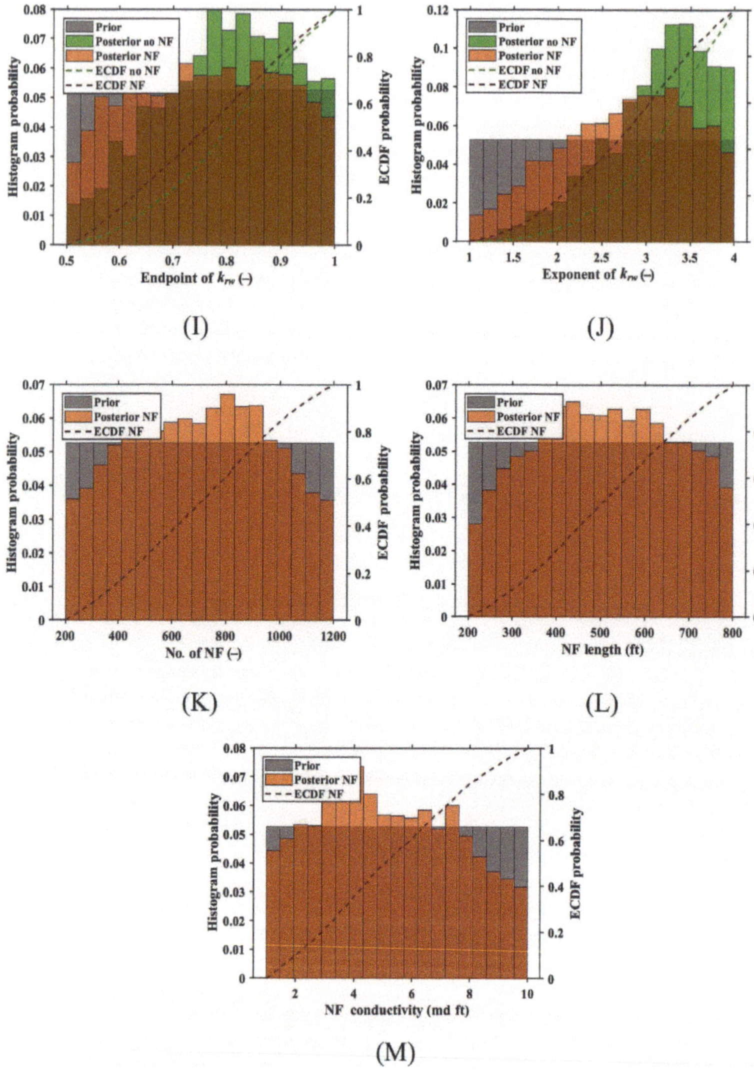

FIGURE 8.25 (Continued).

proxy. The reasons are first, these 1000 solutions from proxy have not been validated with the actual simulator, whether they are all history-matching solutions or not, as it is infeasible to perform, and second, the gas EUR proxy or NN in this study could contribute to the EUR prediction inaccuracy, as we did not run all production forecasts of all 1000 cases, directly.

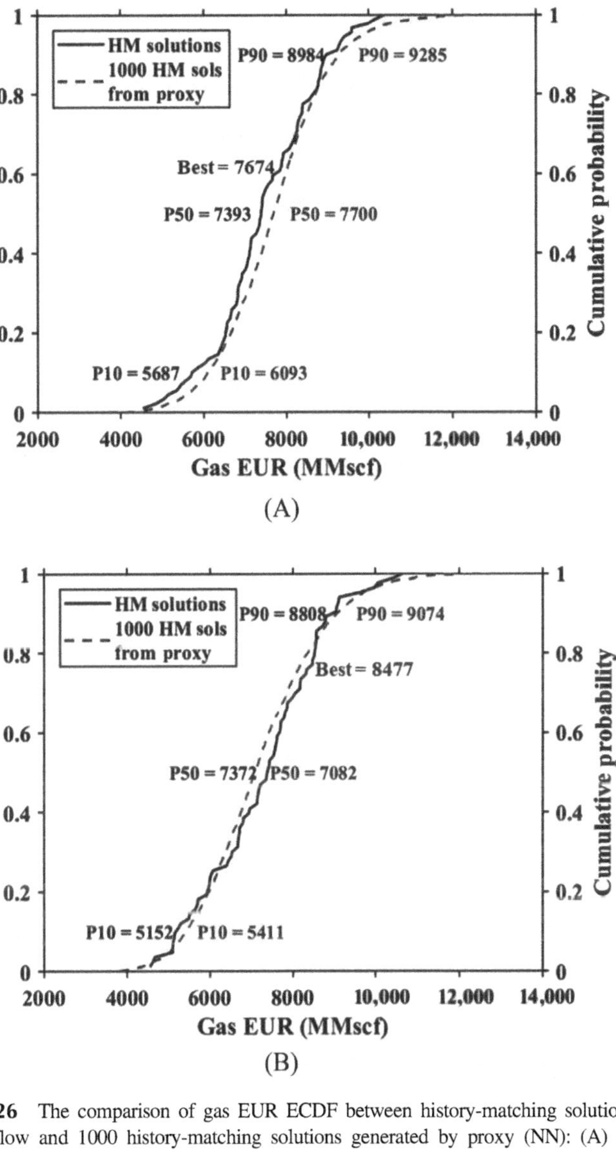

FIGURE 8.26 The comparison of gas EUR ECDF between history-matching solutions from the AHM workflow and 1000 history-matching solutions generated by proxy (NN): (A) case 1 with hydraulic fractures only (no NF); and (B) case 2 with hydraulic fractures and natural fractures (with NF) (Tripoppoom, 2019). *AHM*, Assisted history matching; *ECDF*, empirical cumulative distribution function; *EUR*, estimated ultimate recovery; *NF*, natural fracture; *NN*, neural network.

Therefore we have to use the outputs and analysis of proxy solutions with caution and should use them only to compare to the actual history-matching solutions from the AHM workflow.

8.7 Benefits from the study

The benefits from this work are that we can characterize fracture properties such as fracture height, fracture half-length, and fracture conductivity. More importantly, we can quantify their uncertainties from production data, as shown in Figs. 8.8, 8.11, and 8.13. With this information, they can be integrated to calibrate fracture propagation models to improve any future hydraulic fracturing design and infill well placement optimization. For example, we can obtain dimensionless fracture conductivity (F_{CD}), as shown in Fig. 8.27 from this AHM workflow by calculating from fracture conductivity, fracture half-length, and matrix permeability values of each realization. The results of F_{CD} could be used to reflect the effectiveness of hydraulic fracturing for this well and proppant transport performance. In this case, we can interpret that the created hydraulic fractures are still finite in fracture conductivity based on F_{CD} values and can be improved.

We can also characterize reservoir properties and their uncertainties such as matrix permeability, matrix water saturation, and relative permeability, which are sometimes challenging to acquire by other methods for shale reservoirs. To illustrate, the relative permeability curves for shale reservoirs are sometimes infeasible to implement in the laboratory. However, with the AHM workflow, the relative permeability curves can be characterized, and their uncertainty can be captured. As shown in Fig. 8.28, relative permeability curves of all 83 history-matching solutions together with the best match case are plotted. Furthermore, in case we can characterize those reservoir properties by other methods, the reservoir properties information from the workflow can be integrated with other techniques, such as petrophysical

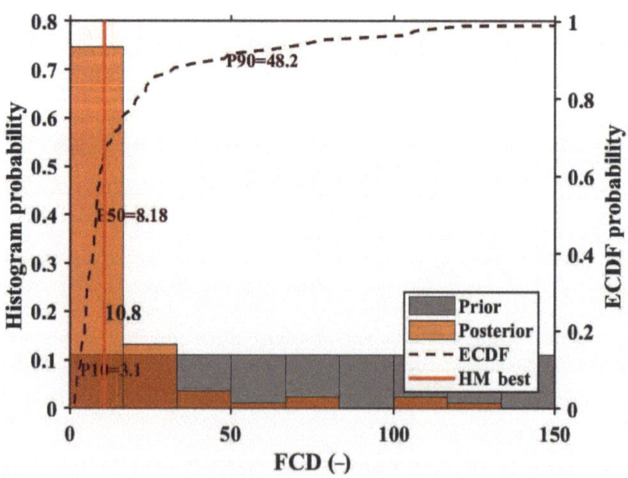

FIGURE 8.27 F_{CD} of 83 history-matching solutions of case 2 with hydraulic fractures and natural fracture shown with the best match case. F_{CD}, Dimensionless fracture conductivity.

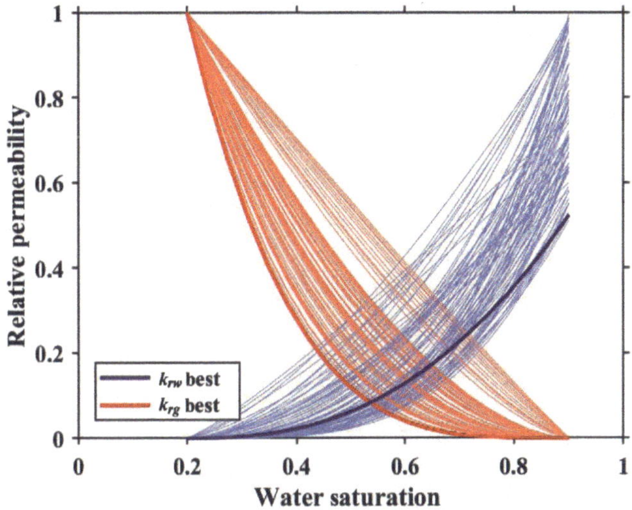

FIGURE 8.28 Relative permeability curves of all 83 history-matching solutions of case 2 with hydraulic fractures and natural fracture shown with the best match case.

interpretation and core data, to produce a more comprehensive analysis. Lastly, we can reduce the uncertainty of natural fractures from the AHM workflow and include them as the part of realization uncertainty as natural fractures play an important role in the long-term production forecast. Furthermore, the probabilistic production forecast, capturing uncertainty during history matching, can be obtained from the physical-based numerical model rather than using an empirical method such as decline curve analysis alone.

8.8 Remarks

We utilized the AHM workflow to perform history matching with a real shale-gas well automatically. Two scenarios, including hydraulic fractures only and hydraulic fractures with natural fractures, were investigated. From the AHM workflow, we can obtain multiple history-matching solutions and characterize hydraulic fractures, natural fractures, and reservoir properties in a probabilistic manner. The effect of natural fracture existence on the posterior distribution of uncertain parameters can be evaluated. In this case, we found that the case with natural fractures has posterior distributions significantly lower in fracture height, lower in fracture conductivities, and higher in fracture water saturation than the case without natural fractures. Also, we observed less curvature on the relative permeability curves of k_{rg} and k_{rw}. This information can be used further for hydraulic fracturing design improvement and well placement optimization. After that, we performed a production forecast for both scenarios and found that there would be some difference in the gas production forecast. If we simplified the system of hydraulic fractures with natural fractures by only modeling hydraulic fractures, we would have over-predicted gas EUR.

References

Agrawal, S., Shrivastava, K., Sharma, M.M., 2019. Effect of shear slippage on the interaction of hydraulic fractures with natural fractures. In: Paper SPE 194361, Presented at SPE Hydraulic Fracturing Technology Conference and Exhibition, The Woodlands, TX, February 5–7, 2019.

Shrivastava, K., Hwang, J., Sharma., M., 2018. Formation of complex fracture networks in the Wolfcamp Shale: calibrating model predictions with core measurements from the hydraulic fracturing test site. In: Paper SPE 191630, Presented at SPE Annual Technical Conference and Exhibition, Dallas, TX, September 24–26, 2018.

Taleghani, A.D., Olson, J.E., 2013. How natural fractures could affect hydraulic-fracture geometry. SPE Journal 19 (1), 161–171.

Tripoppoom, S., 2019. Assisted History Matching Workflow for Unconventional Reservoirs (M.S. thesis). The University of Texas at Austin.

Tripoppoom, S., Yu, W., Huang, H., Sepehrnoori, K., Song, W., Dachanuwattana, S., 2019a. A practical and efficient iterative history matching workflow for shale gas well coupling multiple objective functions, multiple proxy-based MCMC and EDFM. Journal of Petroleum Science and Engineering 176, 594–611.

Tripoppoom, S., Yu, W., Sepehrnoori, K., Miao, J. 2019b. Application of assisted history matching workflow to shale gas well using EDFM and neural network-Markov chain Monte Carlo algorithm. In: Paper URTeC 2019-659, Presented at Unconventional Resources Technology Conference, Denver, CO, July 22–24, 2019.

Wu, K., Olson, J.E., 2016. Numerical investigation of complex hydraulic-fracture development in naturally fractured reservoirs. SPE Production & Operations 31 (4), 300–309.

Yu, W., Xu, Y., Liu, M., Wu, K., Sepehrnoori, K., 2018. Simulation of shale gas transport and production with complex fractures using embedded discrete fracture model. AIChE Journal 64 (6), 2251–2264.

Yu, W., Hu, X., Liu, M., Wang, W., 2019. Investigation of the effect of natural fractures on multiple shale-gas well performance using non-intrusive EDFM technology. Energies 12 (5), 932.

Chapter 9

Shale-oil well with and without natural fractures

9.1 Introduction

We applied the assisted history-matching (AHM) workflow in a case study of a real shale-oil well in the Permian Basin. The well had been producing for 834 days. The available information for HM was oil flow rate, gas flow rate, water flow rate, and wellhead pressure or tubing head pressure (THP). First, we investigated the scenario with only hydraulic fractures (HFs). Then, we studied a scenario featuring HFs with a single case of natural fractures (NFs). This is to see the impact of NF existence on HM solutions. Lastly, we included both HFs and NFs for HM.

9.2 Reservoir model

For this study, a static model from geologists and petrophysicists was not available. Instead, we built a homogeneous model with model dimensions of 10,000 × 3300 × 327.5 ft. Because the petrophysical interpretation was not available for this well, we assumed a gross thickness of 327.5 ft from the well logging of a nearby well in the same formation. The grid cell dimensions are 20 × 100 × 41 ft. For fluid properties, we used the pressure-volume temperature (PVT) laboratory results available in this well. We used a black oil model to simulate three-phase flow. The initial reservoir pressure was 9911 psi which is above the bubble-point pressure (2628 psi). The reservoir model description is summarized in Table 9.1. The well and fracture information are summarized in Table 9.2. For well and fracture description, it is a horizontal well with a lateral length of 9401 ft. As shown in Fig. 9.1, the well was completed with 50 stages of HFs with a total cluster number of 193. The well has been producing through 7″ casing and 4.5″ liner. We assumed a 75% cluster efficiency which is usually observed in actual well performance. This is analogous to the distributed temperature sensing information available from the nearby well. Also, we assumed 70% of the HFs go above the well and 30% below the well. For fracture modeling, we used the embedded discrete fracture model (EDFM) technique to model HFs nonintrusively relative to the main reservoir grid.

Assisted History Matching for Unconventional Reservoirs. DOI: https://doi.org/10.1016/B978-0-12-822242-3.00012-2
© 2021 Elsevier Inc. All rights reserved.

TABLE 9.1 Summary of basic reservoir properties used in this study (Tripoppoom et al., 2020).

Reservoir description	Value	Unit
Model dimension ($x \times y \times z$)	10,000 × 3300 × 327.5	ft
Number of grid blocks ($x \times y \times z$)	500 × 33 × 8	–
Grid cell dimension ($x \times y \times z$)	20 × 100 × 41	ft
Reservoir top depth	11,649	ft
Initial reservoir pressure	9911	psi
Reservoir temperature	180	°F
Matrix porosity	9.5	%
Residual water saturation	20	%
Bubble-point pressure (Pb)	2628	psi
Oil gravity	45	°API
Solution gas–oil ratio at Pb	1021	scf/stb

TABLE 9.2 Summary of well and hydraulic fractures information (Tripoppoom et al., 2020).

Well and fractures description	Value	Unit
True vertical depth of a lateral section	11,877	ft
Reservoir top true vertical depth	11,649	ft
Well length	9401	ft
Number of stages	50	–
Number of clusters	193	–
Cluster efficiency	75%	–
Fracture height above: below well	0.7:0.3	–
Fracture width	0.01	ft
Production string	7″ Casing and 4.5″ liner	–

9.3 History matching

We had historical production for 834 days including oil flow rate, gas flow rate, water flow rate, operating condition, and THP. The THP was

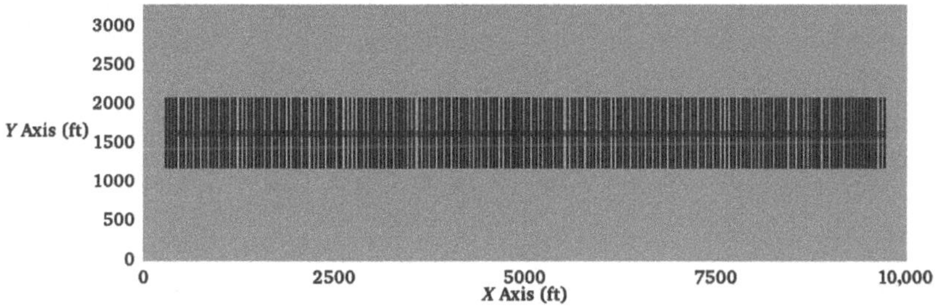

FIGURE 9.1 Reservoir model (top view) including 193 hydraulic fractures (in blue) modeled by EDFM in a horizontal well (in red) (Tripoppoom et al., 2020). *EDFM*, Embedded discrete fracture model.

converted to bottomhole pressure (BHP) using a simple wellbore model accounting for hydrostatic pressure in wellbore (well depth), different water-cut values, and different oil flow rates. For this study, we used BHP and the flow rates for oil, gas, and water. This study focuses on HM of multiphase flow including oil, gas, and water. We investigated three cases including HFs only, HFs with a single realization of NFs, and HFs with a full realization of NFs.

9.3.1 Case 1: hydraulic fractures only

For the case with only HFs, we identified the uncertain parameters and their ranges, as summarized in Table 9.3. We can group the uncertain parameters into two groups. First is reservoir properties, consisting of matrix permeability, matrix water saturation, and relative permeability curve parameters. Second is HF properties, consisting of fracture height, fracture half-length, fracture conductivity, and fracture water saturation. We integrated information from diagnostic fracture injection test (DFIT) analysis, well log from nearby wells and core data, microseismic data, and HM experience to determine the initial ranges of each parameter, as shown in Table 9.3. This step of identifying parameters and their ranges is crucial for the HM process because these initial values determine the success of the HM solutions found during the workflow. If the HM later was not successful, one main reason could be that the initial ranges of information we identified based on prior knowledge were not reasonable. This means we should review all information we had and adjust the ranges.

We generated 50 initial cases by using Latin hypercube design. The simulation results of 50 initial cases are shown in Fig. 9.2. The simulation results of all cases cover the historical production. This means the history match

TABLE 9.3 Summary of the uncertain parameters of reservoir and fracture properties (Tripoppoom et al., 2020).

Uncertain parameters	Unit	Min value	Max value	Sources of information
Matrix permeability (k_m)	md	0.0001	0.02	• Nearby well core data • DFIT analysis
Fracture height	ft	100	310	• Historical microseismic data • Reservoir thickness
Fracture half-length (X_f)	ft	200	1000	• Historical microseismic data
Fracture conductivity	md ft	10	200	• DFIT analysis
Fracture water saturation	–	0.6	0.8	• History-matching experience
Matrix water saturation	–	0.35	0.60	• Well logging interpretation
Exponent of k_{ro}	–	1	4	• History-matching experience
Endpoint of k_{rw}	–	0.5	1	• History-matching experience
Exponent of k_{rw}	–	1	4	• History-matching experience

DFIT, Diagnostic fracture injection test.

solutions are in our initial uncertain parameter ranges and there is no need to revise the uncertain parameter ranges at this step.

The next step is to define the number of proxies used to calculate the objective function which will be used later in the neural network-Markov chain Monte Carlo (NN-MCMC) algorithm. We used multiple proxies to calculate the global objective function and each objective function will be calculated from many dates of historical production data, as shown in Eq. (2.21). The criteria for selecting the number of proxy points to be used is limited by the computational effort, but a general rule of thumb for the optimal number of proxies is that the selected proxies should capture the major production data trend. In this case study, we used a total of 14 proxies from different dates in production data. This included seven proxies for oil flow rate at days 39, 165, 311, 405, 470, 630, and 824 (Fig. 9.3A); four proxies for water cut at days 8, 200, 478, and 800 (Fig. 9.3B); and three proxies for derivative of oil flow rate at days 221, 420, and 763 (Fig. 9.3C). All 14

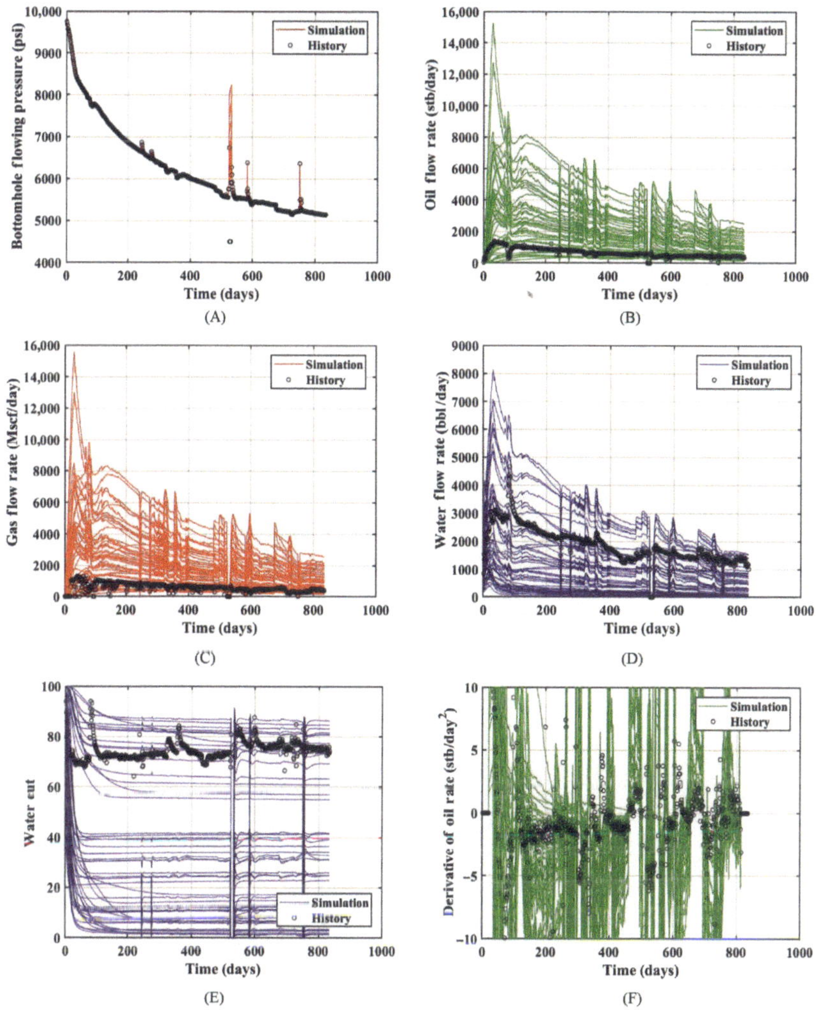

FIGURE 9.2 Comparison between simulation results and historical data of the cases from Latin hypercube design (iteration 1): (A) flowing BHP as constraint; (B) oil flow rate; (C) gas flow rate; (D) water flow rate; (E) water cut; and (F) derivative of oil rate in 40 days (Tripoppoom et al., 2020). *BHP*, Bottomhole pressure.

proxies captured the major trends of oil flow rate, water cut, and derivative of oil flow rate. The reason we included the derivative of oil rate as one of the objective functions is that some cases during the HM process can give the right slope of oil rate but the incorrect values. This indicates the cases are approaching an HM solution and the HM algorithm can adjust some uncertain parameter values to shift the oil rate values to match the production

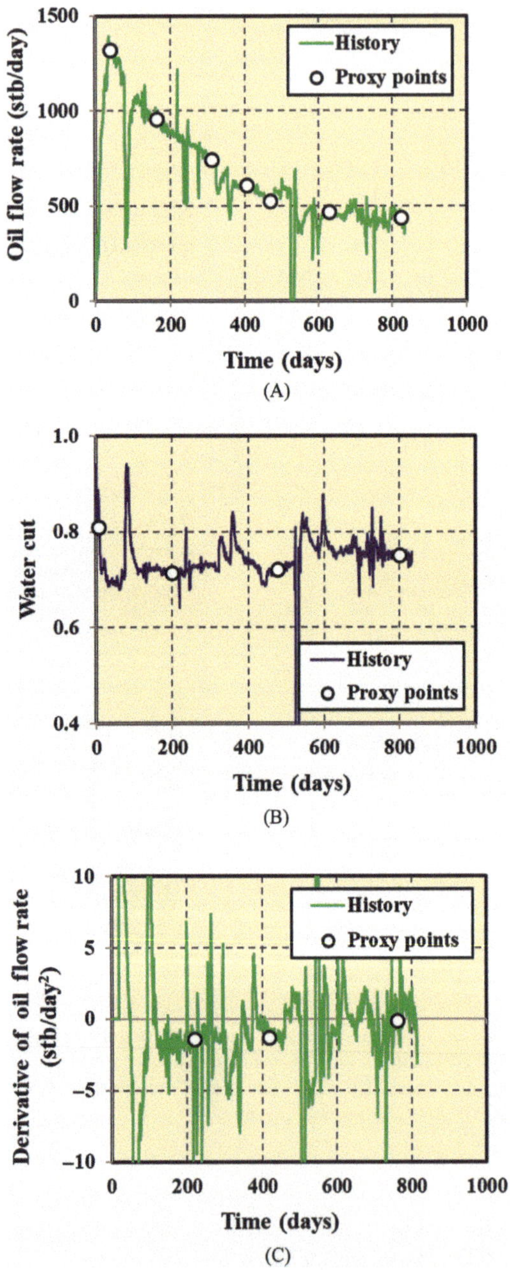

FIGURE 9.3 Multiple proxies of 14 including (A) 7 proxies of oil flow rate; (B) 4 proxies of water cut; and (C) 3 proxies of derivative of oil flow rate (Tripoppoom et al., 2020).

data. Therefore, by using the derivative of oil rate, we can prioritize cases with the right slope in the candidate pool during each iteration of the HM workflow. As a side note, we previously attempted HM by not including the derivative of oil flow rate as a subobjective function, and we observed poorer HM results than the one which included the derivative of oil flow rate.

After that, we calculated the objective function separately for oil flow rate, water cut, and derivative of oil flow rate from those multiple proxies. Finally, we obtained one global objective function from all 14-proxy values with assigned weighted values on each proxy. NNs were used as proxy model to predict each proxy value without running the actual simulation during the NN-MCMC algorithm because it is infeasible to run all samplings of 10^5-10^6 by NN-MCMC with reservoir simulator in each iteration of the HM workflow.

The NN-MCMC algorithm progressed in an iterative manner. In each iteration, the algorithm proposed 25 new cases as HM solution candidates. These candidates were validated with the actual reservoir simulator. Next, new simulation results from those candidates were included in the data set and then the re-training of the NNs was done. The stopping criteria of the workflow typically is either proxy convergence or the maximum number of iterations. In this study, we used a 5% discrepancy between current and previous proxies to define proxy convergence. The workflow terminated at the maximum of 12 iterations in this case.

After we retrieved all simulation runs from each iteration, the next step is to screen for HM solutions from all iterations. However, before that, we have to define the HM solution criteria, which is based on an engineering decision of the least acceptable matching quality (Li et al., 2019). In this study, we used the criteria of 15%, 15%, and 75% for objective functions of oil flow rate, water cut, and derivative of oil flow rate. The HM solutions should pass all three criteria and can be viewed as the intersection of these three criteria, as shown in Fig. 9.4. The criteria for the derivative of oil flow rate was relatively higher because the derivative values in this case are near zero. Small differences from the actual values can cause high percentage values. Fig. 9.5 shows the number of HM solutions versus iteration number.

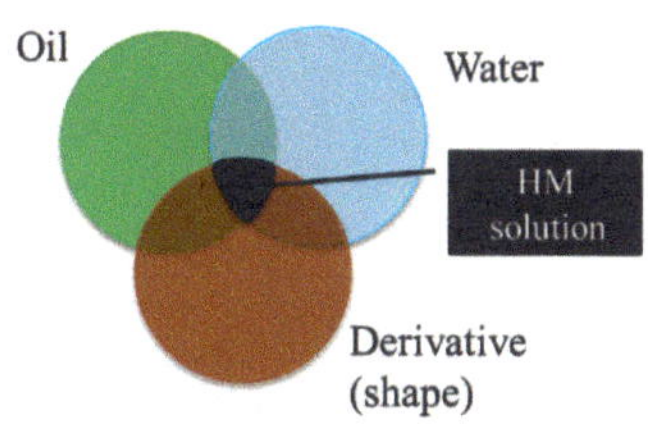

FIGURE 9.4 History-matching solutions criterion of oil flow rate error, water-cut error, and derivative of oil flow rate error (Tripoppoom et al., 2020).

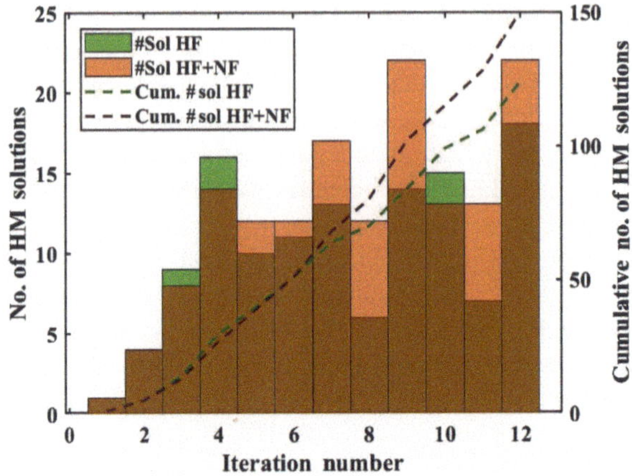

FIGURE 9.5 Number of history-matching solutions versus iteration number for case 1 HF and case 2 HF + NF (Tripoppoom et al., 2020). *HF*, Hydraulic fracture; *NF*, natural fracture.

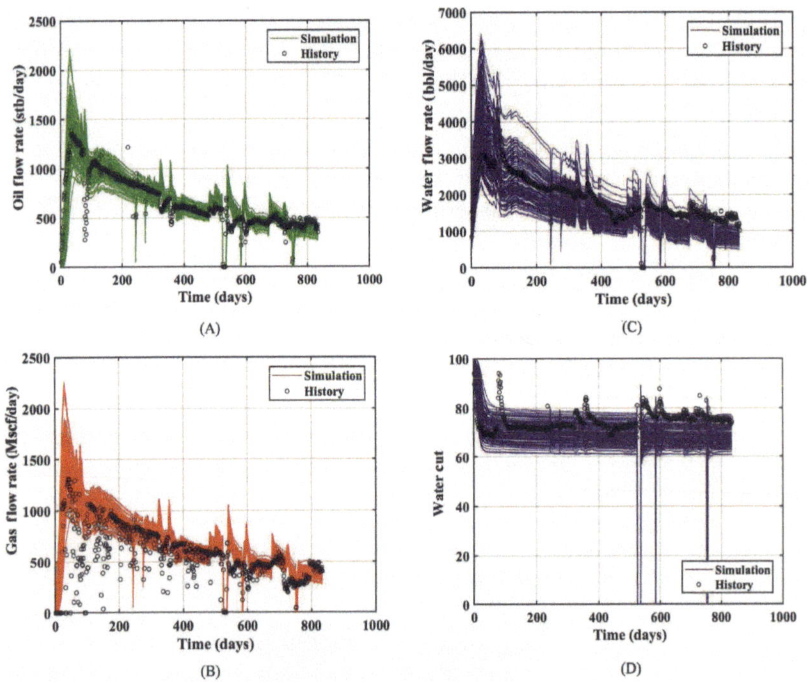

FIGURE 9.6 Comparison between simulation results of 124 history-matching solutions and the production data of case 1 HF only: (A) oil flow rate; (B) gas flow rate; (C) water flow rate; and (D) water cut (Tripoppoom et al., 2020). *HF*, Hydraulic fracture.

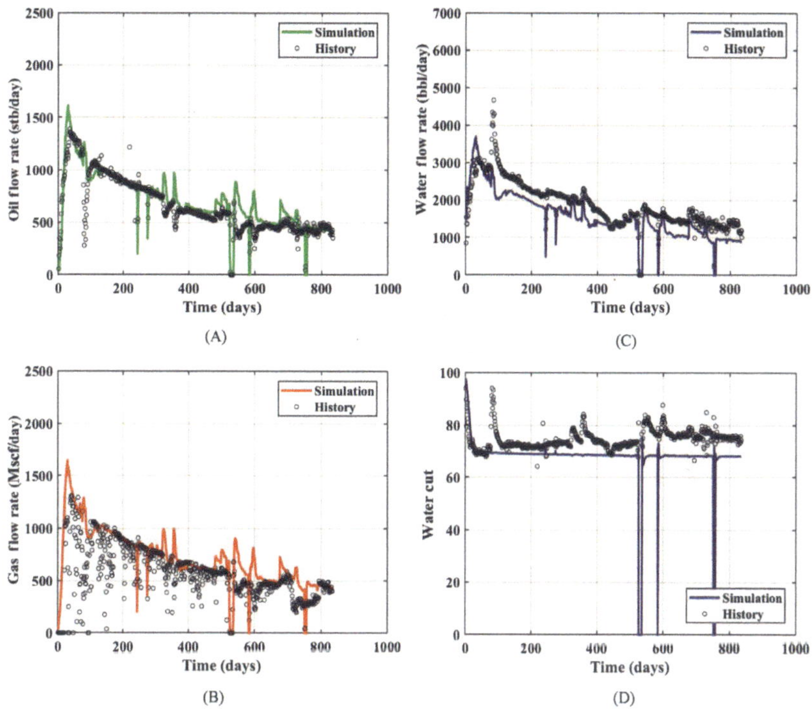

FIGURE 9.7 Comparison between simulation results of the best match case and production data of case 1 HF only: (A) oil flow rate; (B) gas flow rate; (C) water flow rate; and (D) water cut (Tripoppoom et al., 2020). *HF*, Hydraulic fracture.

We obtained 124 solutions from a total of 325 simulation runs, which is around 38%. One thing to highlight is that more solutions were found at the later iteration because the quality of NNs was better—more actual data points from more simulation runs are included in the NNs training. All simulation results of 124 HM solutions are shown in Fig. 9.6, including oil, gas, water flow rates, and water cut.

For the best match case in this chapter, we performed an extra iteration to generate realizations around P50 values of each uncertain parameter (plus and minus 10% of P50 values). Then, the realization with the lowest global error in this extra iteration will be used as the best match case. As a result, we obtained the best match case that was more aligned with the P50 values or the most likely realization. Unlike the best match case that was screened from the lowest global error from all iterations as used in previous chapters, the realization may locate near P10 or P90 of posterior distribution of some uncertain parameters. The best match case HM results are illustrated in Fig. 9.7.

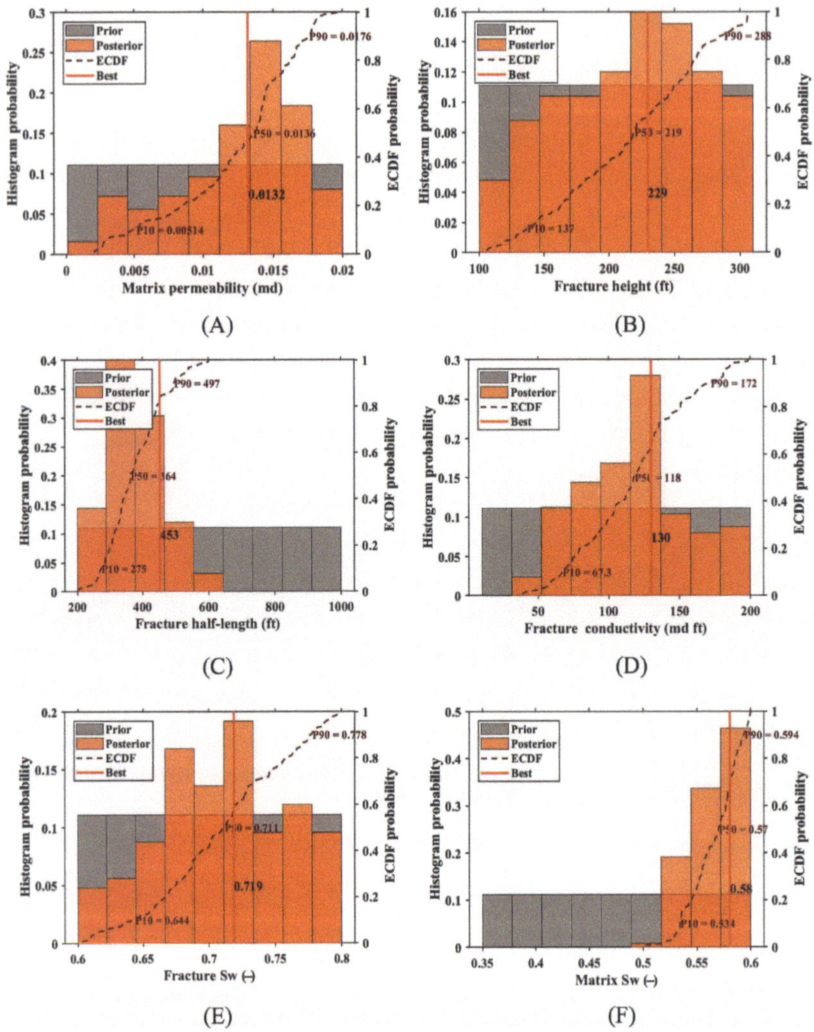

FIGURE 9.8 Comparison between prior and posterior distribution of uncertain parameters (137 solutions) of case 1 HF only (ECDF represents empirical cumulative distribution function): (A) matrix permeability; (B) fracture height; (C) fracture half-length; (D) fracture conductivity; (E) fracture water saturation; (F) matrix water saturation; (G) exponent of k_{ro}; (H) endpoint of k_{rw}; and (I) exponent of k_{rw} (Tripoppoom et al., 2020). *HF*, Hydraulic fracture.

From the HM workflow, we obtained the posterior distribution of uncertain parameters from 124 HM solutions. As shown in Fig. 9.8, the prior uniform distribution we specified at the beginning has been transformed to the posterior distribution. In other words, those reservoir and fracture properties

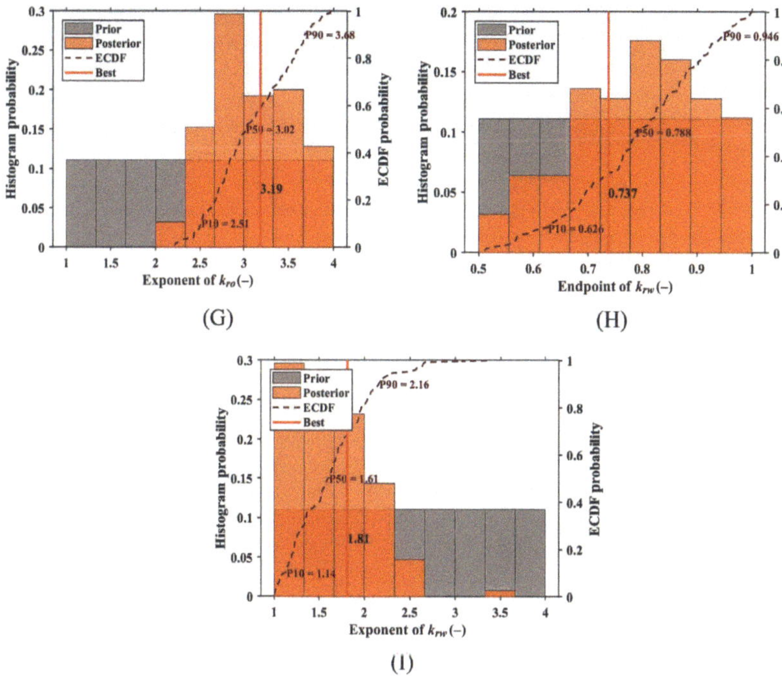

FIGURE 9.8 Continued

TABLE 9.4 Summary of single realization of natural fractures (NFs) used in case 2 (Tripoppoom et al., 2020).

NFs properties		NF1	NF2
Number of fractures	–	500	500
Theta	degree	45	135
Length	ft	100	100
Dip angle	degree	90	90
Permeability	md	10	10
Aperture	ft	0.1	0.1
Fracture conductivity	md ft	1	1

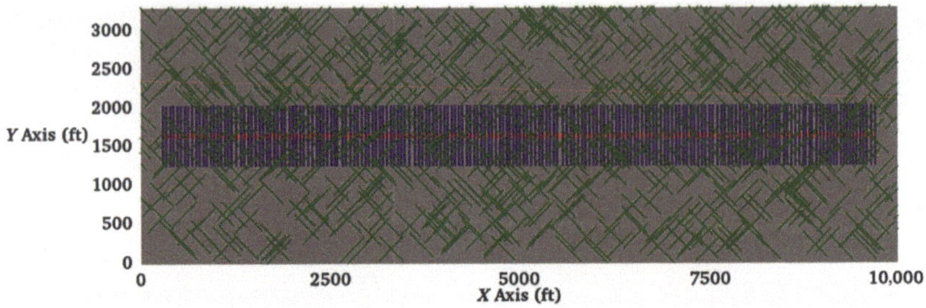

FIGURE 9.9 Reservoir model (top view) including 193 hydraulic fractures (in blue) and natural fractures (in green) modeled by EDFM in a horizontal well (in red) (Tripoppoom et al., 2020). *EDFM*, Embedded discrete fracture model.

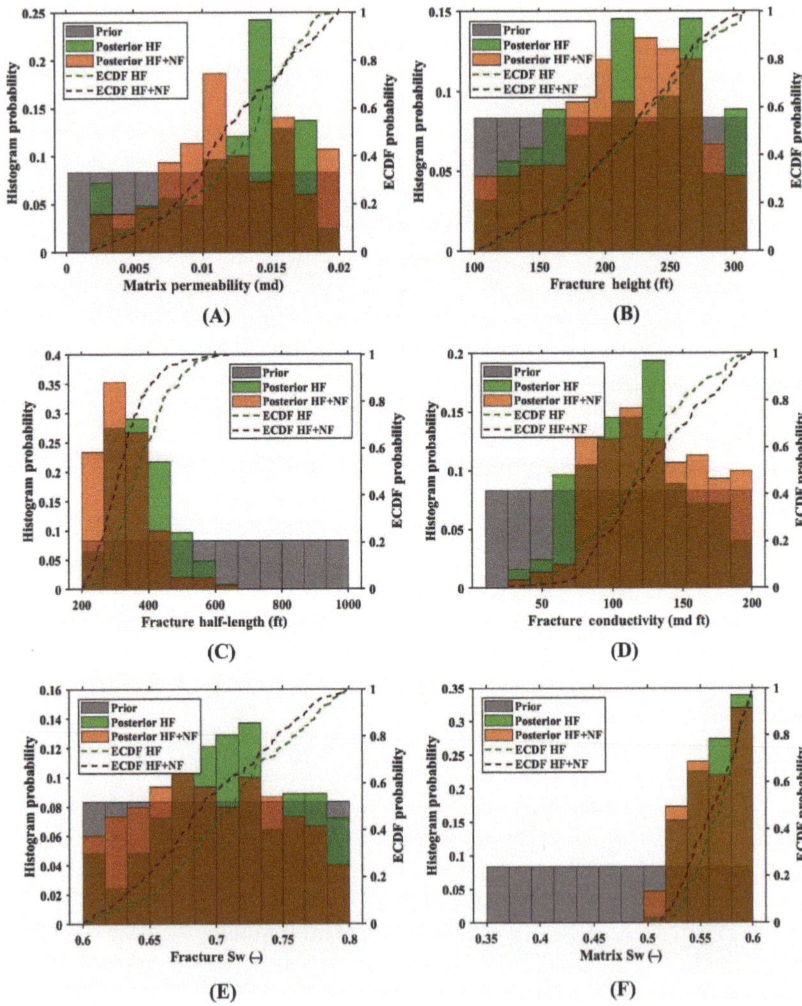

FIGURE 9.10 Comparison between prior and posterior distribution of uncertain parameters of case 1 HFs only and case 2 HFs and a single realization of NFs (HF + NF): (A) matrix permeability; (B) fracture height; (C) fracture half-length; (D) fracture conductivity; (E) fracture water saturation; (F) matrix water saturation; (G) exponent of k_{ro}; (H) endpoint of k_{rw}; and (I) exponent of k_{rw} (Tripoppoom et al., 2020). *HF*, Hydraulic fracture; *NF*, natural fracture.

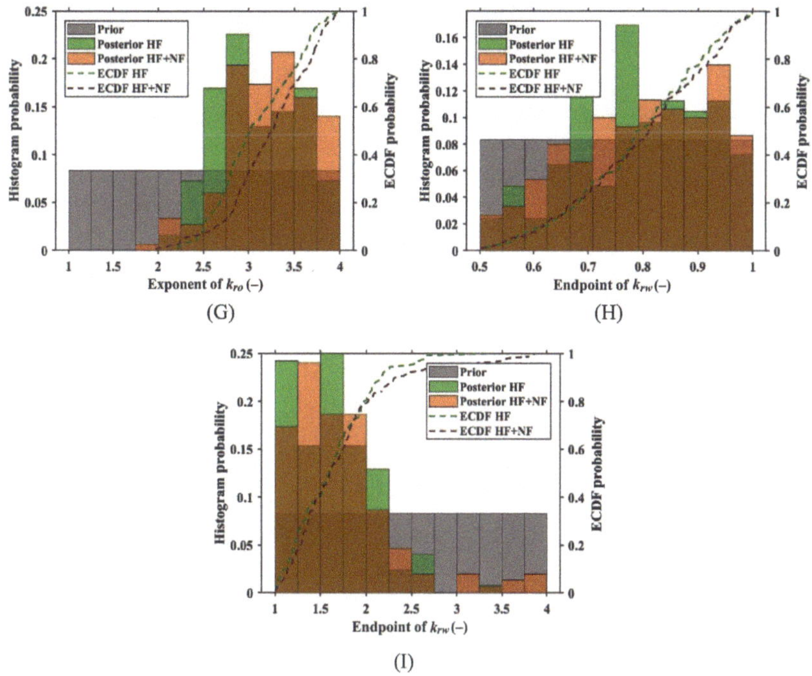

FIGURE 9.10 Continued

were characterized from historical production data in the probabilistic manner. For a fracture height of 100–300 ft and a fracture water saturation of 0.6–0.8, any values in these ranges with a combination of other parameters still provided an HM solution. However, the uncertainty range of the other uncertain parameters can be narrowed down; for example, matrix permeability near the value of 0.02 md is likely to be the HM solution for the case without NFs—fracture half-length in the range of 200–400 ft tends to provide HM solutions. Matrix water saturation of a value above 0.5, exponent of k_{ro} above 2, and exponent of k_{rw} near 1 can match the values of water cut of 70% for the late-time production data. The best match case that aligned with P50 value is also shown in Fig. 9.8 as the red line.

9.3.2 Case 2: hydraulic fractures and single realization of natural fractures

The second case is the scenario with HFs and a single realization of NFs. The objective is to evaluate how NF's existence impacts the HM solutions.

All reservoir model information and uncertain parameter ranges are the same as in case 1 (Tables 9.1–9.3). The NF properties used in this case are summarized in Table 9.4. We specified two sets of NFs with different theta angles. The number and length of the NFs used are from our prior knowledge of NF density in this reservoir, such as NF length per area. Fig. 9.9 shows HFs with NFs modeled by EDFM. Our hypothesis is that the NFs intersect with themselves and that HFs could provide more connection to the production well. Therefore, we first performed a quick comparison between the cases with and without NFs using middle case values. This is to have an idea whether NFs in this well significantly affect the production forecast or not. Moreover, the existence of NFs could affect the HM solutions of other

TABLE 9.5 Summary of the uncertain parameters of reservoir and fracture properties used in case 3 (Tripoppoom et al., 2020).

Uncertain parameters	Unit	Min value	Max value	Sources of information
Matrix permeability (km)	md	0.0001	0.001	• Core data • DFIT analysis
Fracture height	ft	100	310	• Historical microseismic data • Reservoir thickness
Fracture half-length (X_f)	ft	200	800	• Historical microseismic data • History-matching experience from cases 1 and 2
Fracture conductivity	md ft	10	200	• DFIT analysis
Fracture water saturation	–	0.65	0.8	• History-matching experience • History-matching experience from cases 1 and 2
Matrix water saturation	–	0.35	0.65	• Well logging interpretation
k_{ro} exponent	–	1	4	• History-matching experience
k_{rw} endpoint	–	0.5	1	• History-matching experience
k_{rw} exponent	–	1	4	• History-matching experience
Total number of NF	–	400	1500	• Regional NFs density
NF length	ft	200	800	• Regional NFs density
NF conductivity	md ft	1	10	• Regional NFs density

DFIT, Diagnostic fracture injection test; *NF*, natural fracture.

fracture and reservoir uncertain parameters. However, the magnitude of the effect from NFs on posterior distribution is still unknown until we perform HM.

The HM workflow continued similarly to case 1. In this case, the HM workflow also terminated at the maximum iteration of 12. With the same HM solution screening criteria, the workflow found 150 HM solutions from a total of 325 simulation runs (around 46%) compared to 124 HM solutions of case 1 with only HFs (Fig. 9.5).

The comparison between prior and posterior distribution of uncertain parameters for both case 1 (HF) and case 2 (HF + NF) are shown in Fig. 9.10. The effect from the existence of NFs interpreted from posterior distribution is that matrix permeability and fracture half-length have shifted to lower values due to more flow contributing from NFs. The fracture conductivity is higher and fracture water saturation is lower to compensate for the lower area of HFs geometry to match the same production data. For relative permeability parameters the exponent of k_{ro} is higher than case 1, causing more curvature for k_{ro}. To sum

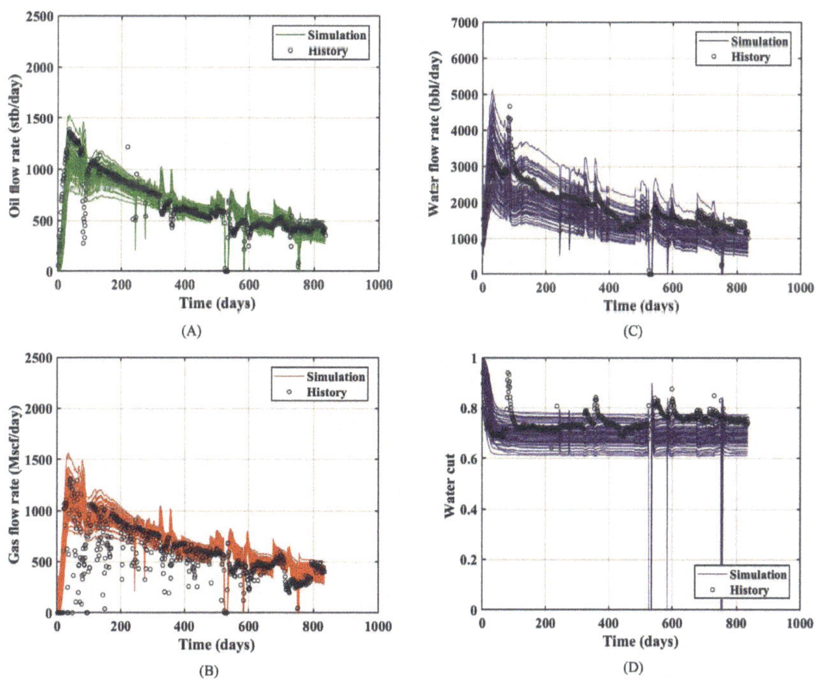

FIGURE 9.11 Comparison between simulation results of 71 history-matching solutions and production data of case 3 HFs and full realization of NFs (HF + full NF): (A) oil flow rate; (B) gas flow rate; (C) water flow rate; and (D) water cut (Tripoppoom et al., 2020). *HF*, Hydraulic fracture; *NF*, natural fracture.

up, by modeling NFs together with HFs, the characterization of some fracture and reservoir properties was changed. However, the limitation in this case is that we assumed we knew the information about NFs as a single realization. In fact, NFs are also one of the shale-reservoir uncertainties and should be included as uncertain parameters, as we will illustrate in the next case.

9.3.3 Case 3: hydraulic fractures and full realization of natural fractures

From case 2, we investigated the effect from a single realization of NFs to the posterior distribution; however, NFs also had high uncertainty and needed to be included as uncertain parameters. Thus, case 3 included both HFs and a full realization of NFs in the HM process. From the new information and the knowledge we gained from cases 1 and 2, the initial ranges of uncertain parameters were revised as summarized in Table 9.5. The sources of information to determine the range for each uncertain parameter are also shown in Table 9.5. The

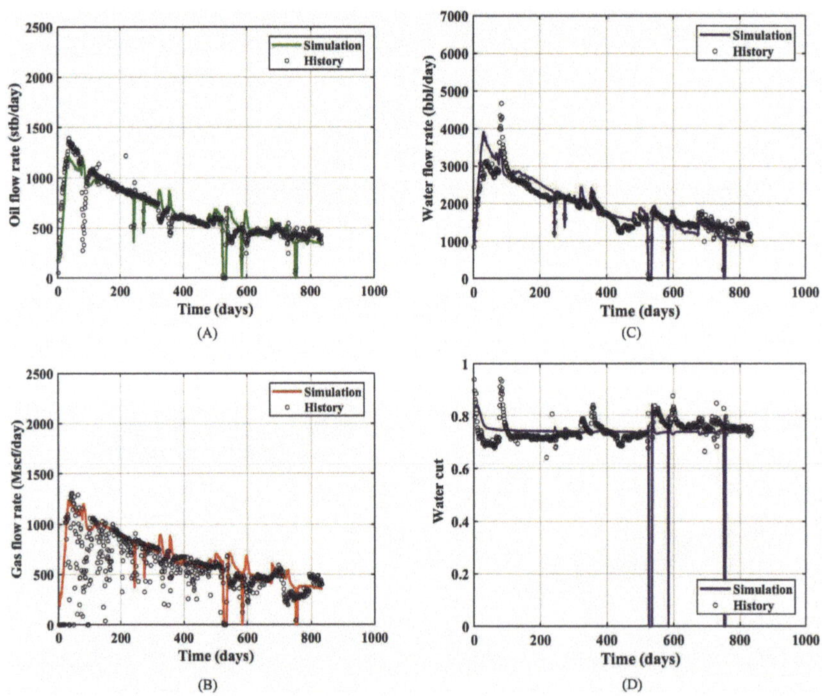

FIGURE 9.12 Comparison between simulation results of the best match case and production data of case 3 HFs and full realization of NFs (HF + full NF): (A) oil flow rate; (B) gas flow rate; (C) water flow rate; and (D) water cut (Tripoppoom et al., 2020). *HF*, Hydraulic fracture; *NF*, natural fracture.

matrix permeability maximum value is constrained at a max of 0.001 md due to information from core data. The fracture half-length maximum value is reduced from 1000 to 800 ft due to no HM solutions being found above 800 ft for both cases 1 and 2. Matrix water saturation is extended from 0.6 to 0.65, and fracture water saturation is now 0.65–0.8 because the posterior distribution of cases 1 and 2 suggested that higher values can give more HM solutions. In addition, the uncertain parameters for NFs are added based on prior knowledge in this reservoir including the number of NFs, NF length, and NF conductivity. These three parameters normally dominate the flow behavior compared with other NF parameters. Therefore, other NF parameters such as theta, dip angle, and NF width are deterministic values and are the same as Table 9.4 from case 2.

The HM workflow progressed the same as cases 1 and 2. For case 3, the HM workflow also stopped at the maximum iteration of 12, while NNs convergence did not reach 5%. With the same HM criteria as the previous two cases, 71 HM solutions were found from 325 simulation runs (22%). The lower number of HM solutions is because we have a higher number of uncertain parameters, in other words, a higher degree of freedom including NF uncertain parameters. All simulation results of 71 HM solutions are shown in Fig. 9.11, including oil, gas, water flow rates, and water cut.

For the best match case, we used the same best match case definition as we used in case 1 HFs only. We performed another iteration to find the best match case around P50 values, and Fig. 9.12 illustrates the simulation results of this best match case. The comparison between the prior and posterior distribution of each uncertain parameter is shown in Fig. 9.13. The best match case parameter values are also plotted in Fig. 9.13. The effect of changing the range of matrix permeability had some effect on other parameters' posterior distribution such as fracture height, fracture conductivity, and exponent of k_{ro}. However, with the revised range, other HM solutions are found for a matrix water saturation above 0.60. Lastly, the uncertainty of NFs can be decreased from the prior distribution by adopting the AHM workflow. As shown in Figs. 9.13J–L, we observed that NF conductivity did not change significantly from prior distribution while we saw that the number of NFs and NF length posterior distribution changed significantly from the initial prior distribution. This implies that the number of NFs and NF length have more influence on the production than NF conductivity because any value of NF conductivity can give HM solutions.

9.4 History-matching results and discussion

In this section, we illustrate the performance of the NN-MCMC algorithm in the proposed AHM workflow by using the HM results of case 3.

Parallel coordinate plots of HM solutions and non-HM solutions are depicted in Fig. 9.14. Each realization of uncertain parameters is plotted as a line. Only some combinations of uncertain parameters are HM solutions; for example, ones with higher fracture height, lower fracture half-length, and higher matrix water saturation in the ranges are likely to be HM solutions. Moreover, this plot emphasizes that the NN-MCMC algorithm

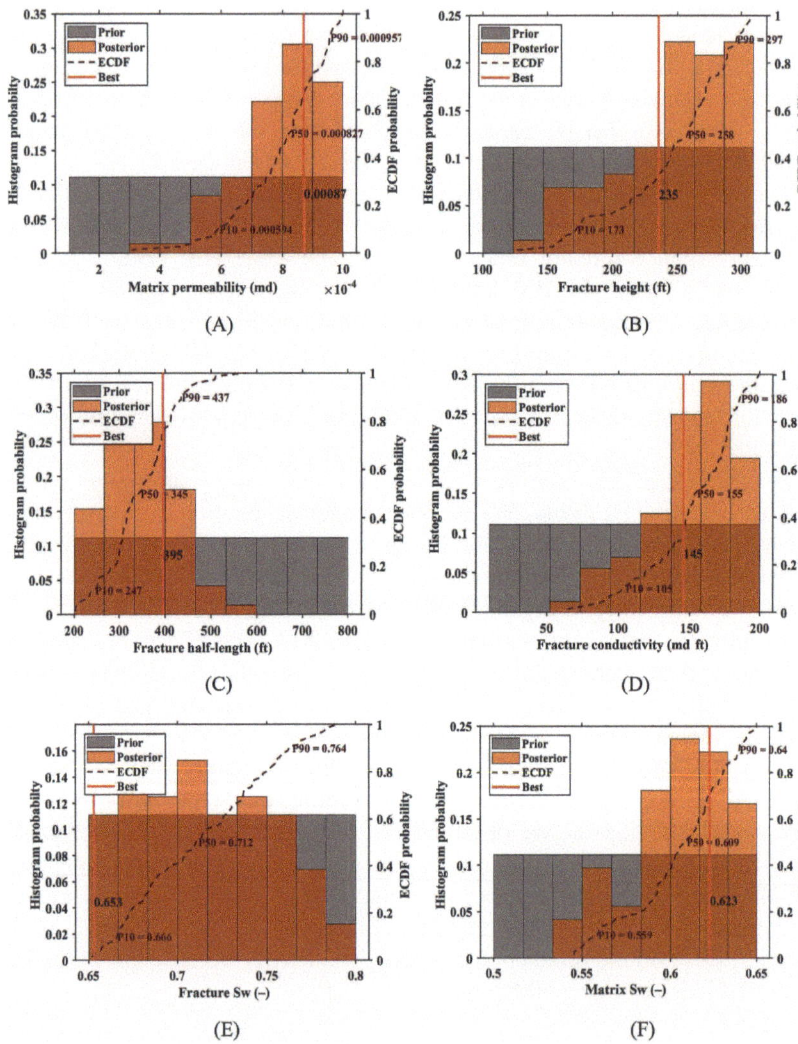

FIGURE 9.13 Comparison between prior and posterior distribution of uncertain parameter (71 solutions) of case 3 HFs and full realization of NFs (HF + full NF): (A) matrix permeability; (B) fracture height; (C) fracture half-length; (D) fracture conductivity; (E) fracture water saturation; (F) matrix water saturation; (G) exponent of k_{ro}; (H) endpoint of k_{rw}; (I) exponent of k_{rw}; (J) number of NF; (K) NF length; and (L) NF conductivity (Tripoppoom et al., 2020). *HF*, Hydraulic fracture; *NF*, natural fracture.

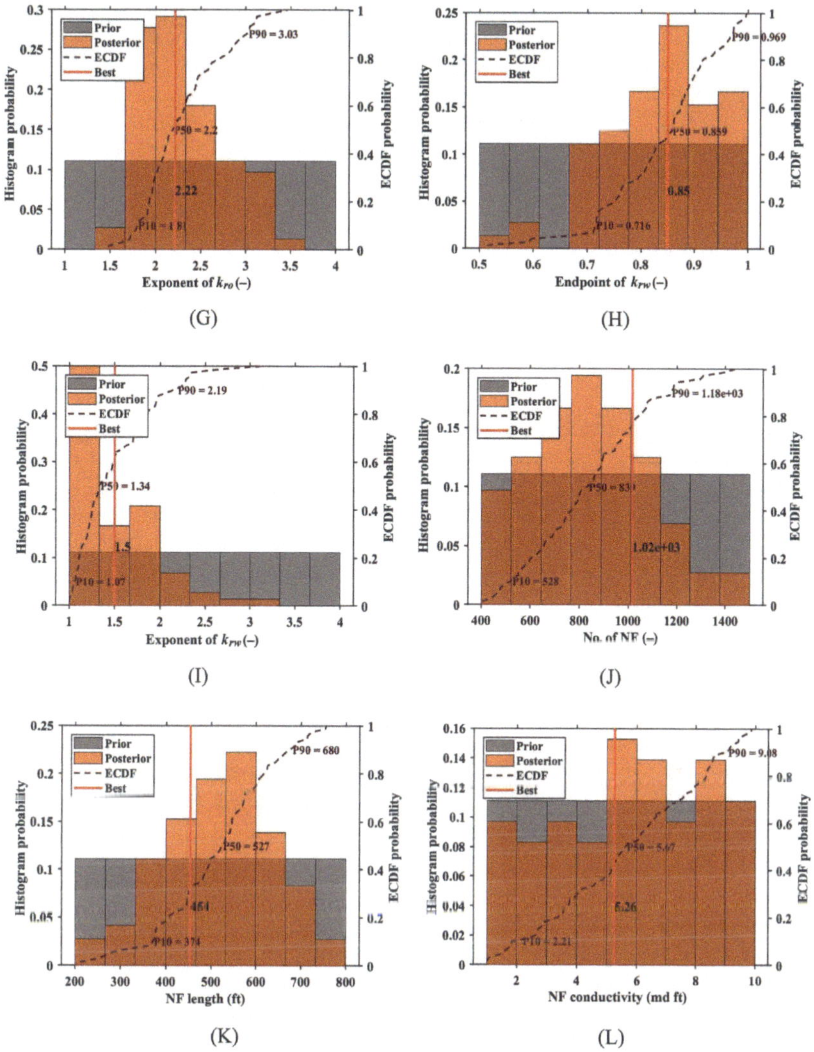

FIGURE 9.13 Continued

already explored and validated other combinations, but they did not pass the criteria to be history match solutions, as shown as the gray lines in Fig. 9.14. Therefore, it confirms the efficiency of the NN-MCMC algorithm; in that, the algorithm explores throughout the possible uncertain parameter domain.

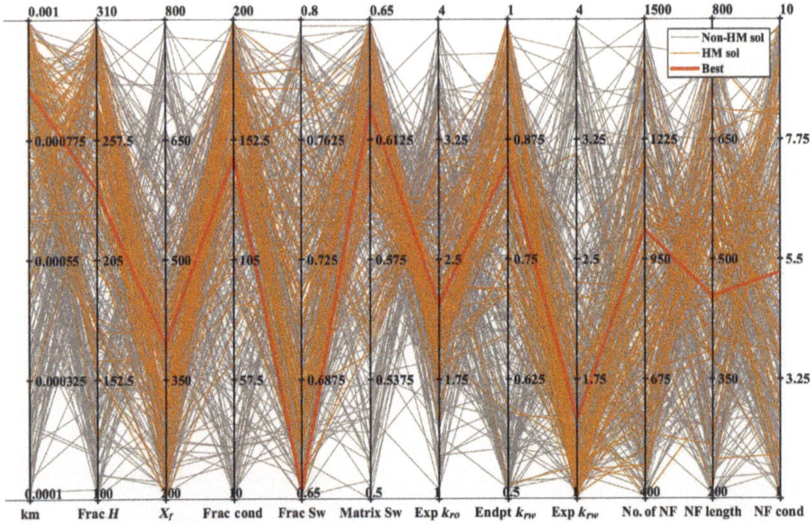

FIGURE 9.14 Parallel coordinate plot of all realizations for case 3 HFs and full realization of NFs (HF + full NF) proposed by the NN-MCMC algorithm including nonhistory-matching solutions, history-matching solutions, and the best case (Tripoppoom et al., 2020). *HF*, Hydraulic fracture; *MCMC*, Markov chain Monte Carlo; *NF*, natural fracture; *NN*, neural network.

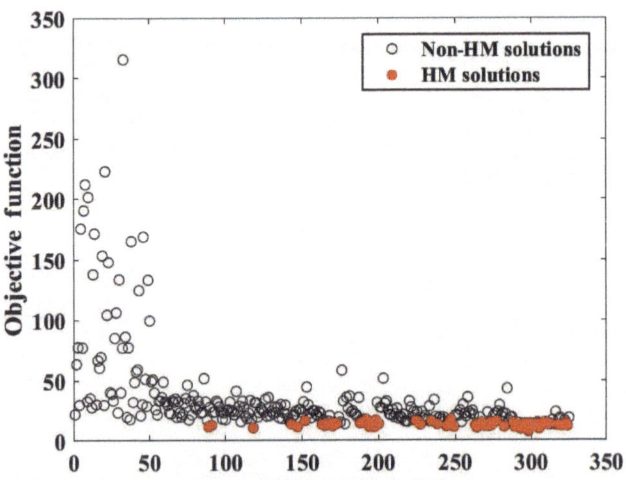

FIGURE 9.15 Objective function value or global error versus simulation number index showing non-HM solutions (*black circles*) and HM solutions (*solid red circles*) (Tripoppoom et al., 2020). *HM*, History matching.

As iteration progresses, we demonstrate the evolution of the HM objective function. Fig. 9.15 shows the objective function values from simulation results versus simulation number index. We observed that the values of the objective function are significantly improved since simulation index 51 which is iteration 2 after the initial iteration of Latin hypercube design. This confirms the performance of the NN-MCMC algorithm to propose the possible candidate as an HM solution. However, in iteration 2 (simulation index of 51−75), there are no HM solutions (*filled red dot*). After that, as the workflow continues, the objective function values gradually improve, and more solutions are found in the later iterations. This is because the accuracy of NNs is getting better as more known points from simulation runs are accumulated.

For each uncertain parameter, Fig. 9.16 illustrates how the proposed cases of each uncertain parameter value versus simulation number index progress by the NN-MCMC algorithm. We found that the proposed points converge to the posterior distribution as the simulation number index increases. This demonstrates the performance and efficiency of the NN-MCMC algorithm. Only certain ranges of values can be HM solutions as shown in the filled red points. In addition, Fig. 9.17 depicts the cross plot between proxy or NNs' estimation versus simulation results for global error and three objective functions for oil flow rate, water cut, and derivative of oil flow rate calculated from 14 NNs. The improvement of NN quality can be observed as the proposed points are closer to the unit slope line when the iteration number is higher. Also, Fig. 9.18 shows the details of each NN or proxy estimate versus simulation results. As the iteration progresses, the NNs prediction quality is better and gets closer to the unit slope line. To conclude, the NN-MCMC algorithm in the proposed AHM workflow performs HM automatically and finds multiple HM solutions efficiently.

For a total running time, based on four parallel licenses for simulation and 3.7 GHz with 4-core processor computer, the workflow completed the automatic HM in around 2 days for the case without NFs and around 5 days for the case with NFs. The elapsed time included all steps in the AHM workflow, including creating the simulation case automatically, training NNs, running the proxy-based MCMC algorithm, running the EDFM preprocessor, running reservoir simulation, and evaluating workflow stopping criteria. The reason that higher running time was required for the case with NFs is because there were more calculations regarding EDFM and more grid blocks to represent NFs.

9.5 Probabilistic production forecast

After we obtain multiple realizations of HM solutions from the AHM workflow, we can perform production forecasting probabilistically. In this study, we forecasted the production for 30 years constrained by declining

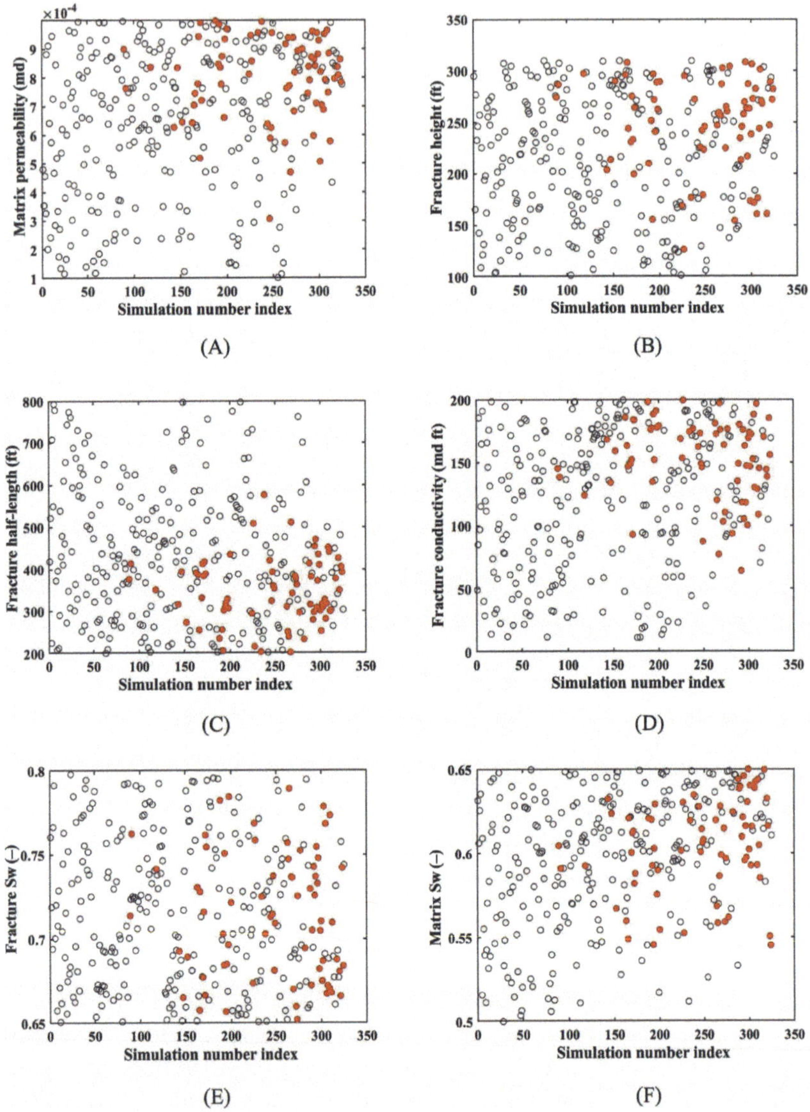

FIGURE 9.16 Uncertain parameter values versus simulation number index showing black circles non-HM solutions (*black circles*) and HM solutions (*solid red circles*) proposed by the NN-MCMC algorithm: (A) matrix permeability; (B) fracture height; (C) fracture half-length; (D) fracture conductivity; (E) fracture water saturation; (F) matrix water saturation; (G) exponent of k_{ro}; (H) endpoint of k_{rw}; (I) exponent of k_{rw}; (J) number of NF; (K) NF length; and (L) NF conductivity (Tripoppoom et al., 2020). *HM*, History matching; *MCMC*, Markov chain Monte Carlo; *NF*, natural fracture; *NN*, neural network.

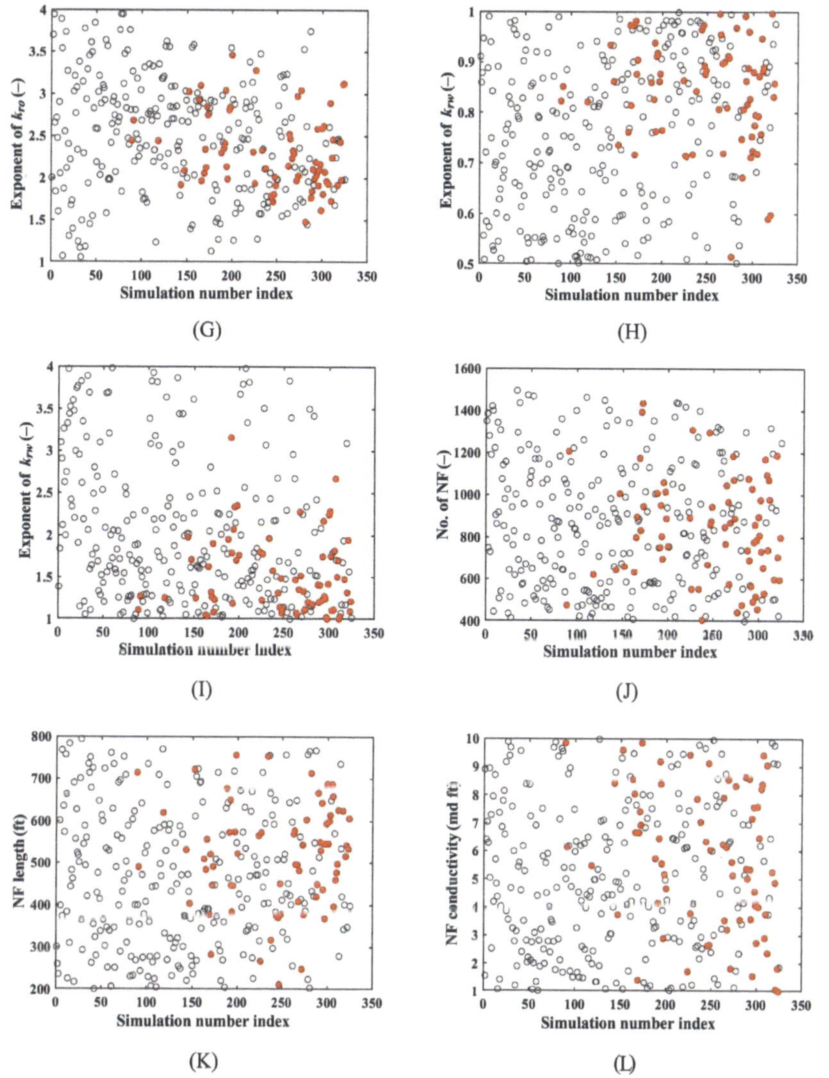

FIGURE 9.16 Continued

BHP until constant at 1000 psi (Fig. 9.19). We performed a production forecast for the HFs only case (case 1) and HFs with full realization of NFs case (case 3). The simulation results of the production forecasts for both cases are shown in Figs. 9.19 and 9.20. P10–P50–P90 of oil and gas estimated ultimate recovery (EUR) can be obtained. The abrupt changes of oil and gas rate seen in Figs. 9.19C and D and 9.20C and D

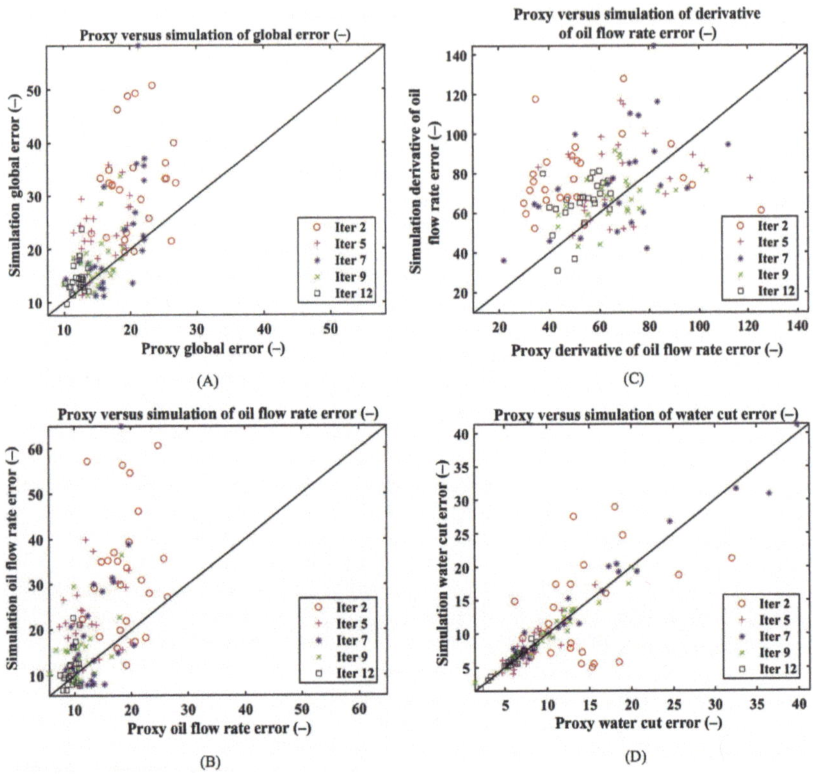

FIGURE 9.17 Cross plots between proxy and simulation values of (A) global error (calculated from 14 proxies); (B) oil flow rate error (calculated from 7 oil flow rate proxies); (C) derivative of oil flow rate error (calculated from 3 derivative of oil flow rate proxies); and (D) water cut error (calculated from 5 water cut proxies) (Tripoppoom et al., 2020).

are because reservoir pressure is below bubble-point pressure and three-phase flow occurred.

To compare the oil and gas EUR of both cases, the empirical cumulative distribution function (ECDF) curves of oil and gas EUR are shown in Fig. 9.21. With the existence of NFs the ECDF of oil EUR is around 30% higher than the case with only HFs at P50 value. This is because NFs provide more connected drainage volume to the well in the long term. On the other hand, the ECDF of gas EUR for the case with NFs is around 15% lower than the case with only HFs at P50 value. The reason is that there is more flux from matrix to the wellbore causing a slower depletion of reservoir pressure around the well. This contributes to less gas being liberated from oil. In other words, we underpredict the oil production and overpredict gas production if we assume this well only contains HFs in spite of the fact that there are NFs in the system. Please note that the magnitude of oil EUR discrepancy between the cases with and without NFs should be different in different areas and reservoirs because the EUR discrepancy is subject to NF properties for each well.

To conclude, if NFs are present in the reservoir, they should be included in the reservoir model for HM; in other words, we cannot simplify the whole system of NFs and HFs by modeling only HFs with an increased matrix permeability as this affects long-term production forecast.

9.6 History-matching solutions from neural networks

To obtain a smoother posterior distribution of uncertain parameters, we can utilize either the latest proxy model or NNs to generate more HM solutions, such as 1000 cases. To illustrate, Fig. 9.22 shows 1000 HM solutions of case 3 (HF + full NF) with the best HM case. We can observe that the distributions are similar to those of 71 HM solutions in case 3 (Fig. 9.13) but with a smoother distribution. Furthermore, another NN can be created for EUR prediction for oil and gas based on the production forecast results of 71 HM solutions. Then, we can estimate oil and gas EUR for 1000 HM solutions from proxy (NNs), as shown in (Fig. 9.23). The oil and gas EUR ECDF of the 1000 solutions from proxy yields approximately the same range of P10–P50–P90 oil and gas EUR. However, there is some difference between 71 HM solutions and 1000 solutions from proxy. The reasons are first, these 1000 solutions from proxy have not been validated with the actual simulator as to whether they are all HM solutions or not as it is infeasible to perform, and second, the oil and gas EUR proxy or NNs in this study could contribute to the EUR prediction inaccuracy as we did not run production forecasts for all 1000 cases directly. Therefore, the solutions from proxy or NNs that have not been validated with the reservoir simulator for both the HM part and production forecast have to be used with caution and should be used only to compare to the actual HM solutions from the AHM workflow.

9.7 Benefits from the study

As there is no direct measurement of effective fracture geometry and fracture properties, the production data can be used to characterize properties such as fracture height, fracture half-length, and fracture conductivity, and, more importantly, quantify their uncertainties, as shown in Figs. 9.8, 9.10, and 9.13. With this information the uncertainty of the fracture propagation model can be calibrated with the information from production HM. This calibrated fracture propagation model based on geomechanical properties can be used to further improve any future hydraulic fracturing design. Also, the information such as effective fracture half-length and fracture height can be the input for infill well placement optimization.

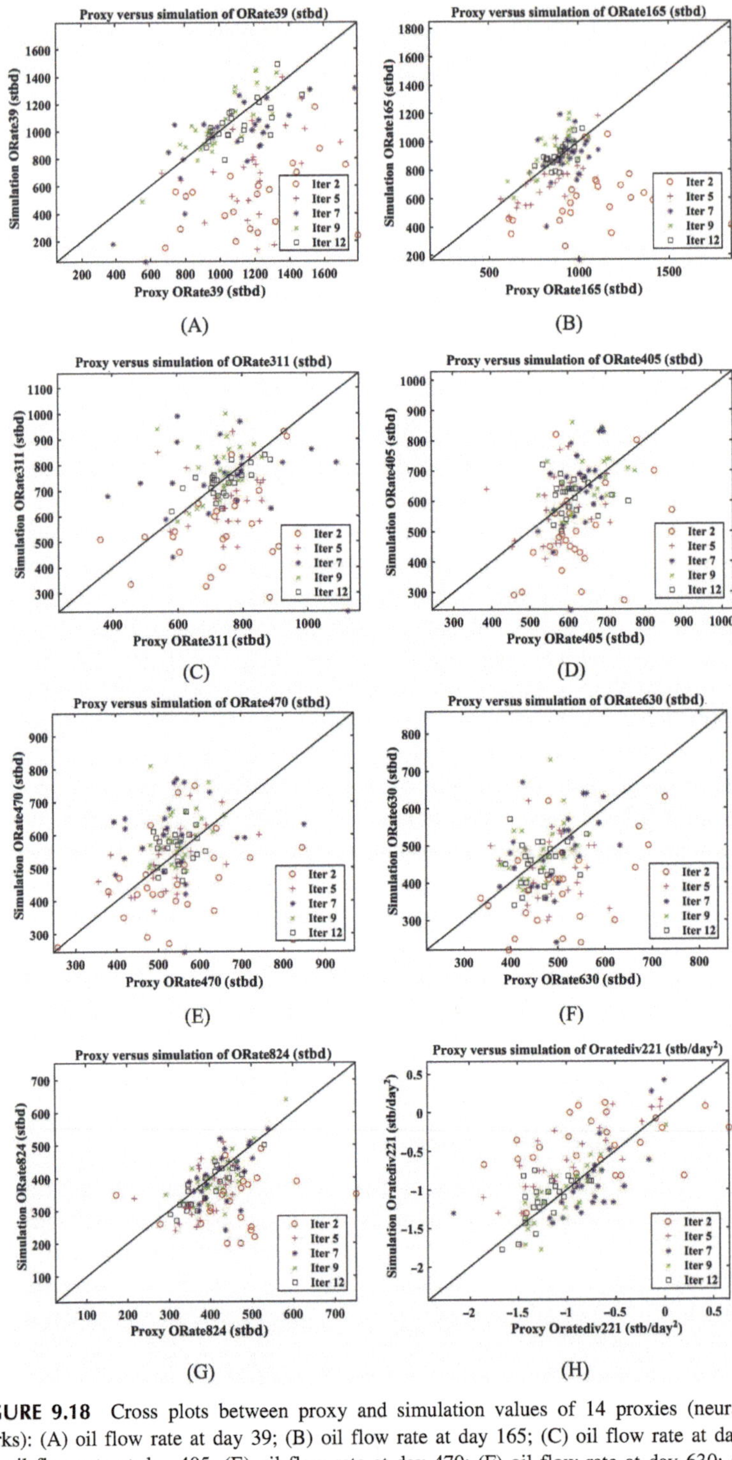

FIGURE 9.18 Cross plots between proxy and simulation values of 14 proxies (neural networks): (A) oil flow rate at day 39; (B) oil flow rate at day 165; (C) oil flow rate at day 311; (D) oil flow rate at day 405; (E) oil flow rate at day 470; (F) oil flow rate at day 630; (G) oil flow rate at day 824; (H) derivative of oil flow rate at day 221; (I) derivative of oil flow rate at day 420; (J) derivative of oil flow rate at day 763; (K) water cut at day 8; (L) water cut at day 200; (M) water cut at day 478; and (N) water cut at day 800 (Tripoppoom et al., 2020).

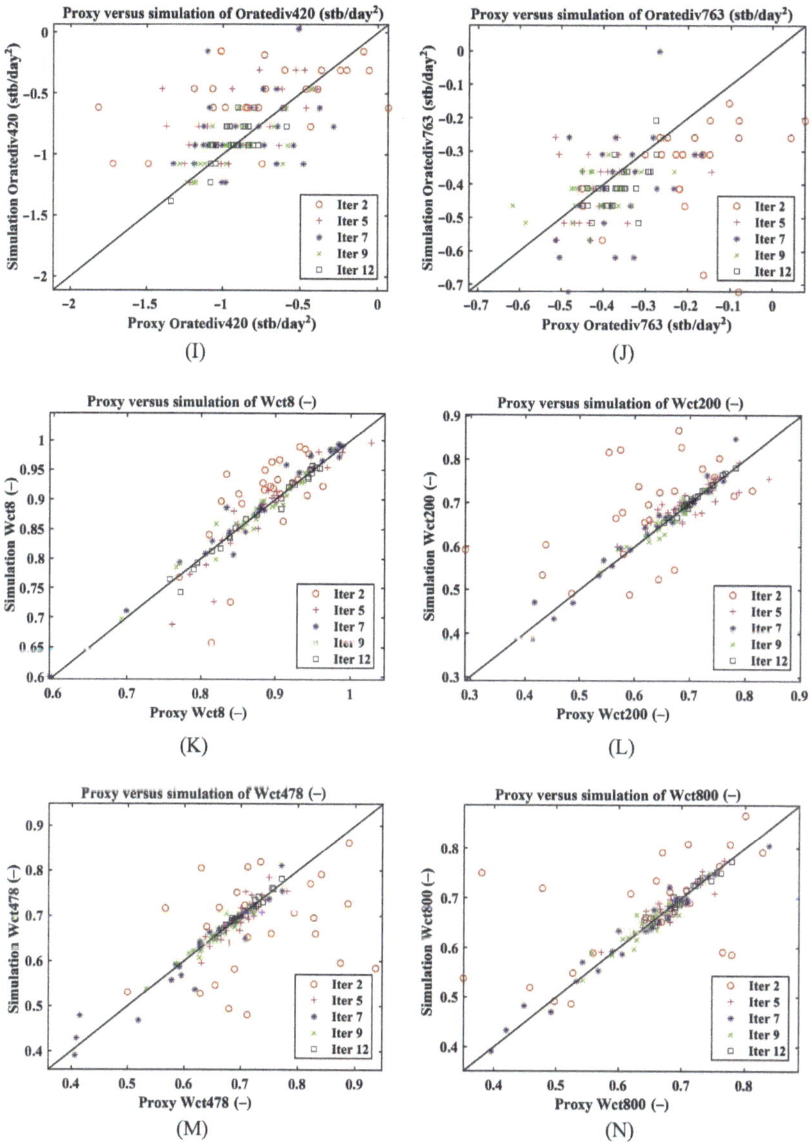

FIGURE 9.18 Continued

Another benefit is that we can obtain dimensionless fracture conductivity (F_{CD}), as shown in Fig. 9.24, from this AHM workflow by calculating from fracture conductivity, fracture half-length, and matrix permeability values of each realization. The results of F_{CD} could be used to reflect the effectiveness

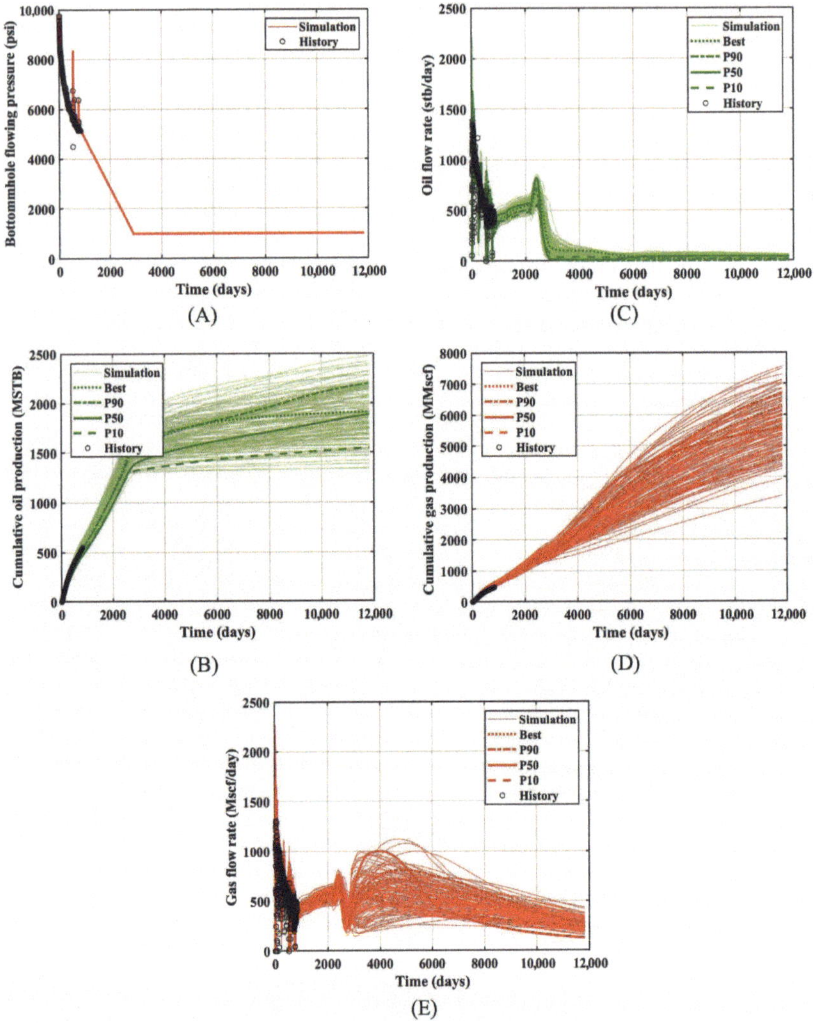

FIGURE 9.19 Production forecast of history-matching solutions of case 1 HFs only: (A) bottomhole pressure; (B) cumulative oil production; (C) oil flow rate; (D) cumulative gas production; and (E) gas flow rate (Tripoppoom et al., 2020). *HF*, Hydraulic fracture.

of hydraulic fracturing and proppant transport performance for this well. As the F_{CD} values are different between these two scenarios, we found that case 3 is 20 times higher than case 1 at P50 value. This means if we simplify the system of HFs with NFs to include only HFs and increased matrix permeability, we may misunderstand that the effectiveness of HFs is still low and can be improved by varying parameters regarding proppants during fracturing operation.

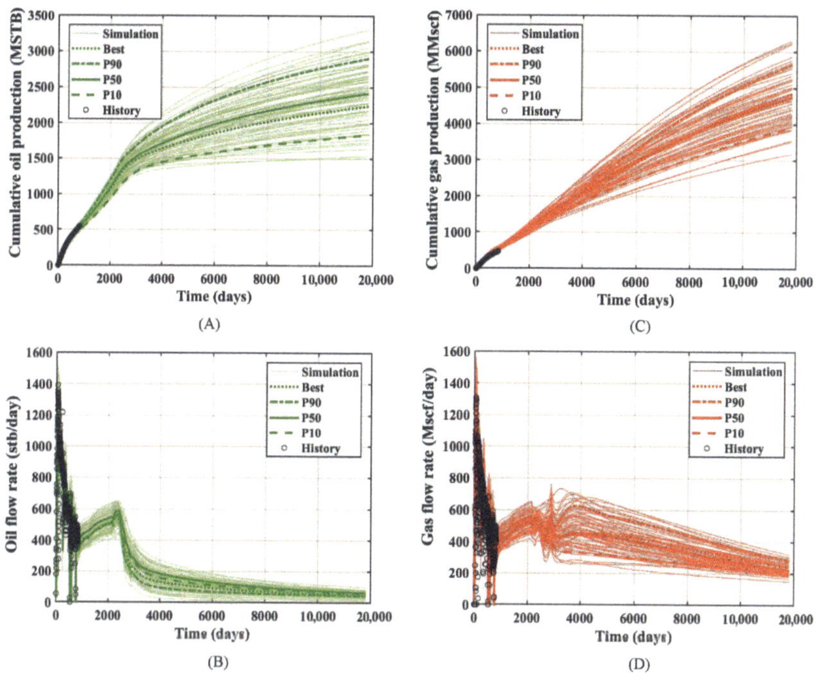

FIGURE 9.20 Production forecast of history-matching solutions of case 3 HFs and full realization of NFs (HF + full NF): (A) cumulative oil production; (B) oil flow rate; (C) cumulative gas production; and (D) gas flow rate (Tripoppoom et al., 2020). *HF*, Hydraulic fracture; *NF*, natural fracture.

Besides, we can characterize reservoir properties and their uncertainties such as matrix permeability, matrix water saturation, and relative permeability which are sometimes challenging to acquire by other methods for shale reservoirs. To illustrate, the relative permeability curves for shale reservoirs are sometimes infeasible to perform in the laboratory. However, with the AHM workflow, the relative permeability curves can be characterized and their uncertainty can be captured. As shown in Fig. 9.25, relative permeability curves of all 71 HM solutions together with the best case are plotted. Furthermore, in cases where we can characterize those reservoir properties by other methods, the reservoir property information from the workflow can be integrated with other techniques such as petrophysical interpretation and core data to produce a more comprehensive analysis. Lastly, we can reduce the uncertainty of NFs from the AHM workflow and include them as the part of realization uncertainty as NFs play an important role in the long-term production forecast, as illustrated in the production forecast. Furthermore, the probabilistic production forecast, capturing the uncertainty during HM, can be obtained from the physically-based numerical model rather than using

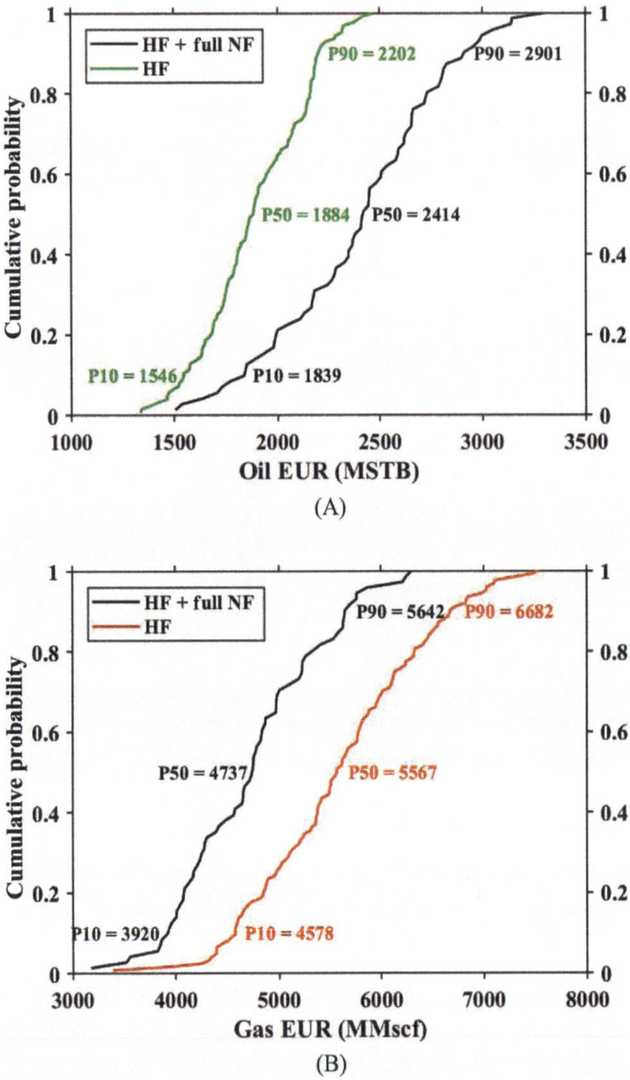

FIGURE 9.21 Comparison of ECDF between case 1 HFs only and case 3 HFs and full realization of NFs (HF + full NF): (A) oil EUR and (B) gas EUR (Tripoppoom et al., 2020). *ECDF*, Empirical cumulative distribution function; *EUR*, estimated ultimate recovery; *HF*, hydraulic fracture; *NF*, natural fracture.

the empirical method such as decline curve analysis alone. Finally, the drainage area over time, as illustrated in Figs. 9.26 and 9.27, can be obtained for the case with only HFs and the case with HFs and NFs. The information of drainage area and volume could help engineers to optimize well spacing in shale-oil reservoirs.

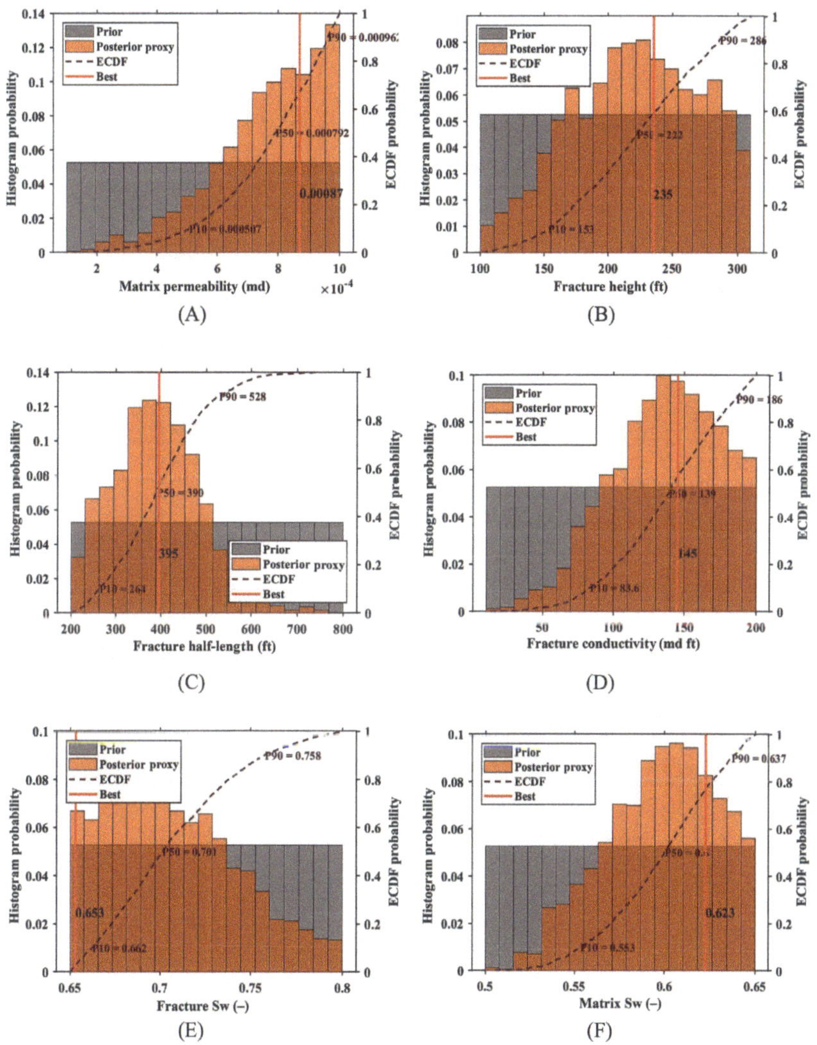

FIGURE 9.22 Prior and posterior distribution of 1000 history-matching solutions from the last proxy of case 3 HFs and full realization of NFs (HF + full NF) with the best case: (A) matrix permeability; (B) fracture height; (C) fracture half-length; (D) fracture conductivity; (E) fracture water saturation; (F) matrix water saturation; (G) exponent of k_{ro}; (H) endpoint of k_{rw}; (I) exponent of k_{rw}; (J) number of NF; (K) NF length; and (L) NF conductivity (Tripoppoom et al., 2020). *HF*, Hydraulic fracture; *NF*, natural fracture.

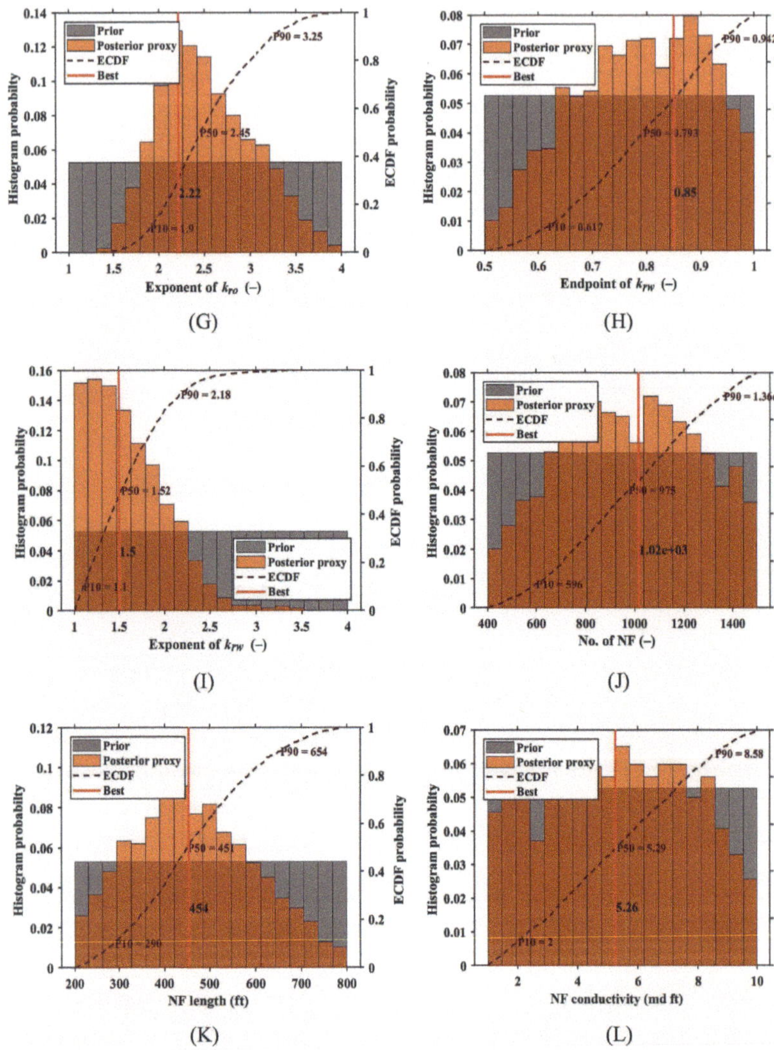

FIGURE 9.22 Continued

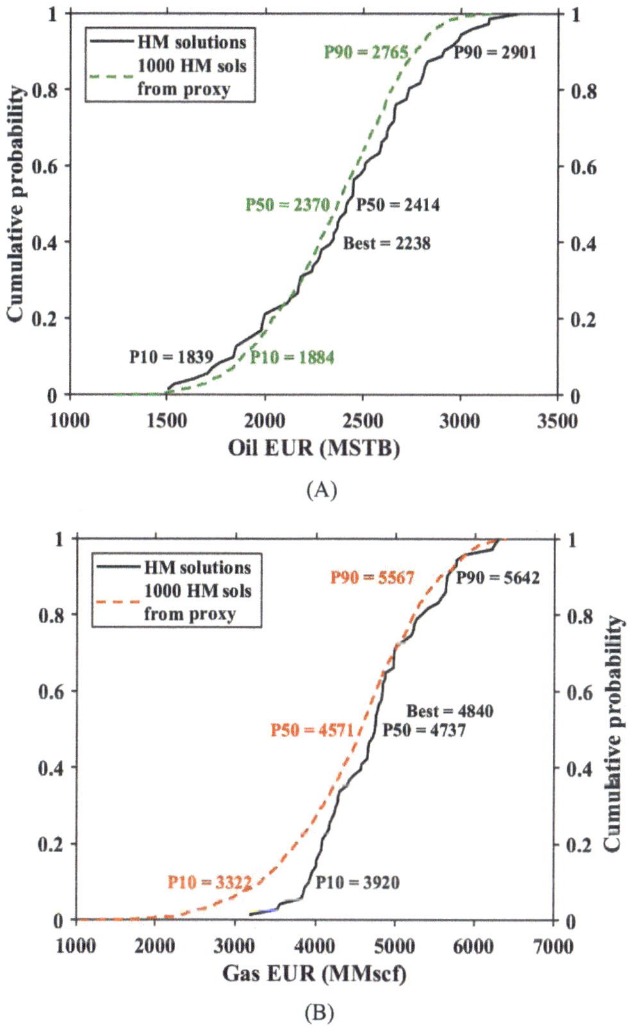

FIGURE 9.23 Comparison of ECDF of case 3 HFs and full realization of NFs (HF + full NF) between 71 history-matching solutions from the AHM workflow and 1000 history-matching solutions generated by proxy (neural networks): (A) oil EUR and (B) gas EUR (Tripoppoom et al., 2020). *AHM*, Assisted history matching; *ECDF*, empirical cumulative distribution function; *EUR*, estimated ultimate recovery; *HF*, hydraulic fracture; *NF*, natural fracture.

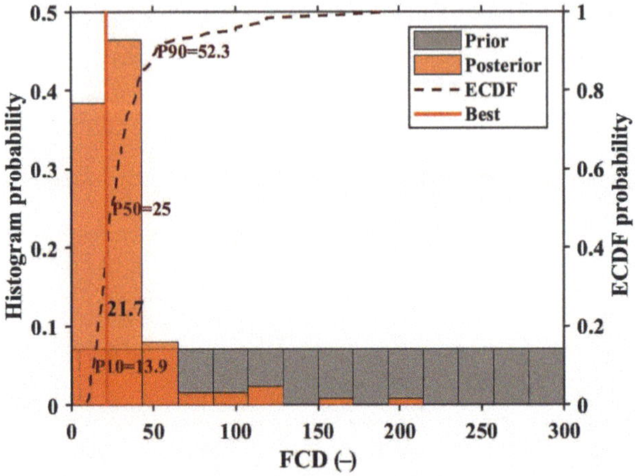

(A)

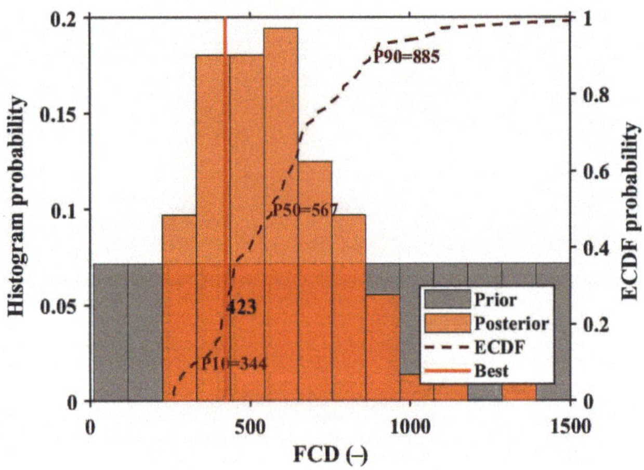

(B)

FIGURE 9.24 F_{CD} of 71 history-matching solutions: (A) case 1 HFs only and (B) case 3 HFs and full realization of NFs (HF + full NF) (Tripoppoom et al., 2020). F_{CD}, Dimensionless fracture conductivity; *HF*, hydraulic fracture; *NF*, natural fracture.

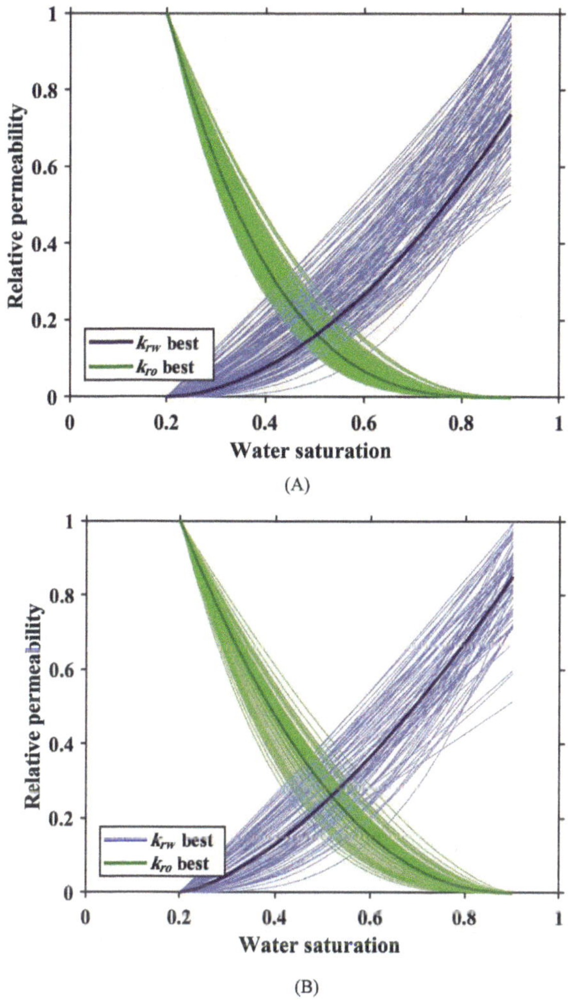

FIGURE 9.25 Relative permeability curves of all history-matching solutions with the best case of (A) case 1 HFs only and (B) case 3 HFs and full realization of NFs (HF + full NF) (Tripoppoom et al., 2020). *HF*, Hydraulic fracture; *NF*, natural fracture.

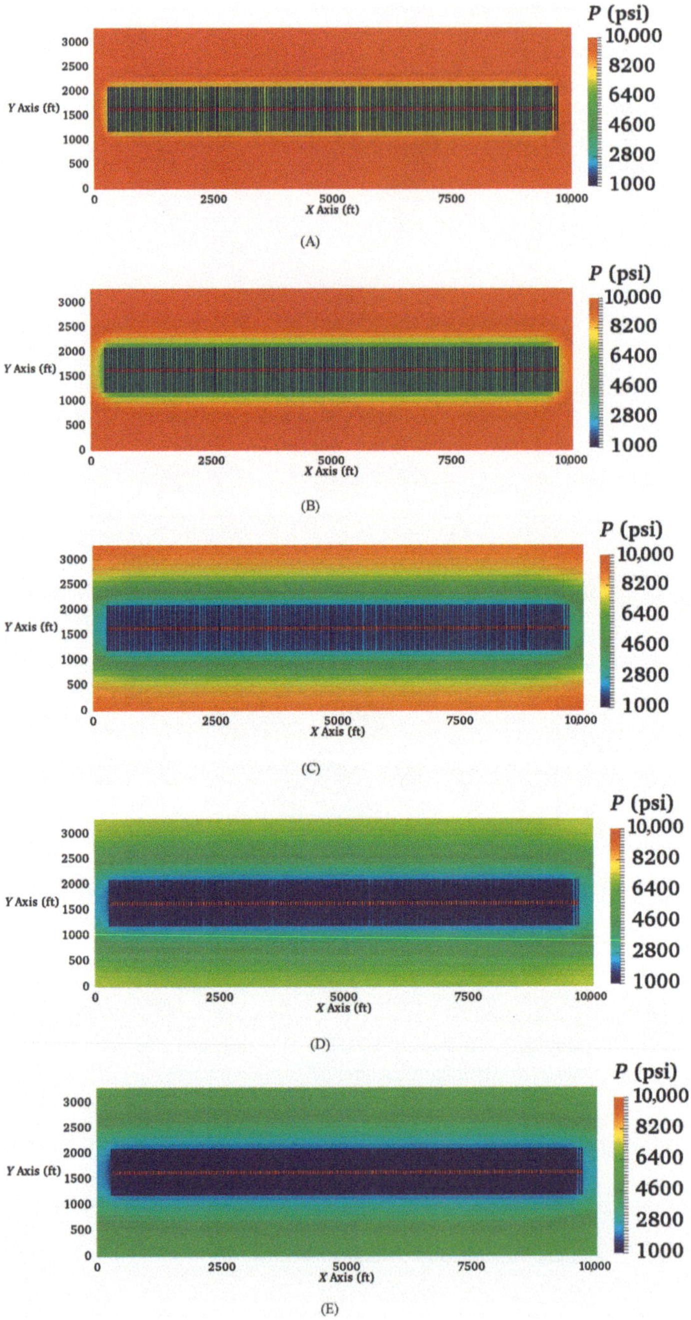

FIGURE 9.26 Pressure distribution of the best case for case 1 HFs only: (A) after 365 days; (B) after 834 days (end of history); (C) after 10 years; (D) after 20 years; and (E) after 30 years (Tripoppoom et al., 2020). *HF*, Hydraulic fracture.

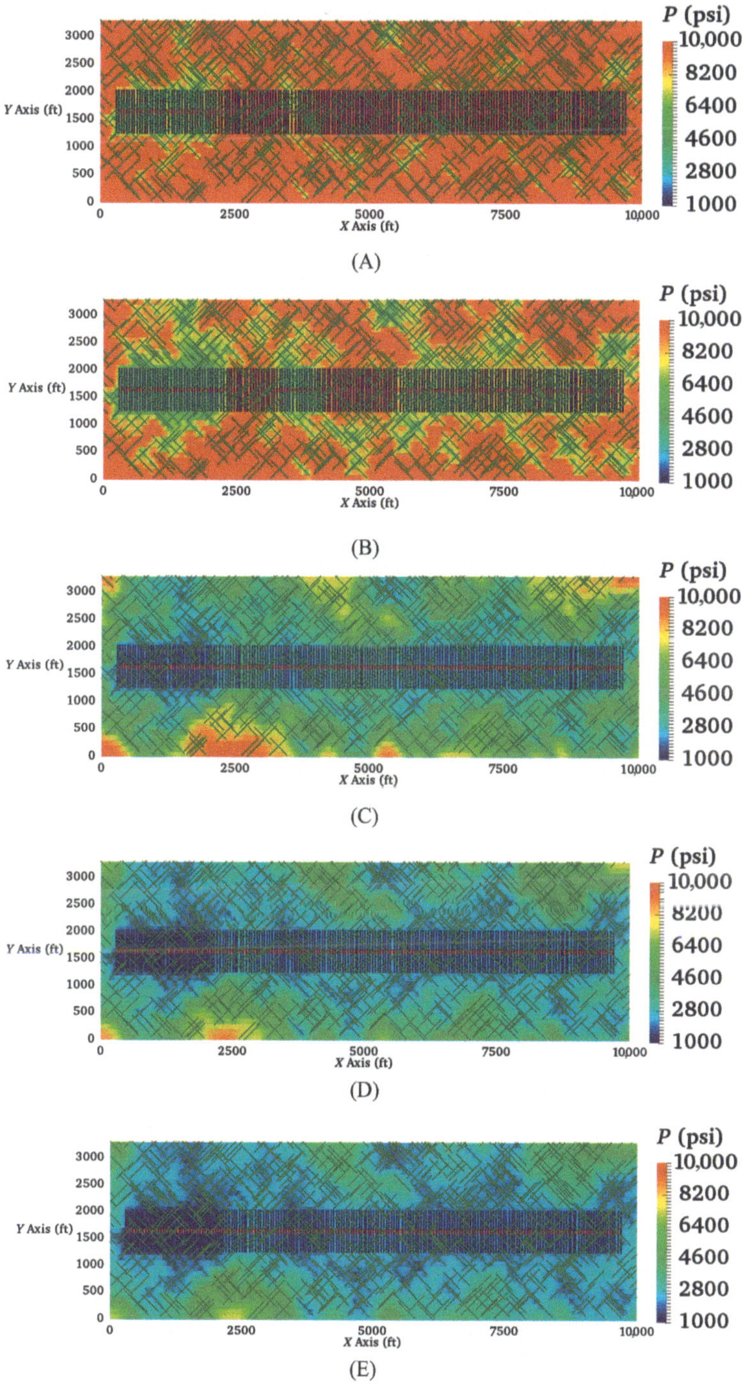

FIGURE 9.27 Pressure distribution of the best case for case 3 HFs and full realization of NFs only (HF + full NF): (A) after 365 days; (B) after 834 days (end of history); (C) after 10 years; (D) after 20 years; and (E) after 30 years (Tripoppoom et al., 2020). *HF*, Hydraulic fracture; *NF*, natural fracture.

9.8 Remarks

We used the NN-MCMC algorithm for AHM in shale-oil reservoirs. We performed automatic HM and production forecasting with a real shale-oil well in Permian Basin. Three scenarios including HFs only, HFs with single realization of NFs, and HFs with a full realization of NFs were investigated. From the HM results, we can evaluate the effect of NF existence on the posterior distribution of uncertain parameters. Moreover, we can obtain multiple HM solutions and characterize HFs, NFs, and reservoir properties in the probabilistic manner. This information can be used further for hydraulic fracturing design improvement and well placement optimization. After that, we performed production forecasts between case 1 HF and case 3 HF + full NF and found that there is a significant difference in the production forecast of both oil and gas. If we simplified the system with NFs by only modeling HFs, we would have underpredicted oil and overpredicted gas production.

References

Li, B., Bhark, E.W., Gross, S.J., Billiter, T.C., Dehghani, K., 2019. Best practices of assisted history matching using design of experiments. SPE Journal 24 (4), 1435–1451.

Tripoppoom, S., Wang, X., Liu, Z., Yu, W., Xie, H., Sepehrnoori, K., et al., 2020. Characterizing hydraulic and natural fractures properties in shale oil well in Permian Basin using assisted history matching. Fuel 275, 117950.

Chapter 10

Investigation of different production performances in multiple shale-gas wells

10.1 Introduction

We sometimes observe different production performances from different shale-gas wells drilled in the same formation. There are many reasons that could cause this difference including different hydraulic fracture geometries and properties, and matrix properties such as porosity and water saturation.

The log data and core data, if they are available, can help characterize matrix or reservoir properties. However, the major challenges are fracture geometries and their properties, such as fracture half-length, fracture height, and fracture conductivity. There is no direct measurement of these fracture properties. One technique that can help quantify fracture properties is diagnostic fracture injection testing (DFIT) analysis. However, the interpretation of DFIT may be limited due to a short cycle of injection data. In some wells, microseismic monitoring data is available and can be integrated to characterize fracture geometry. Nevertheless, one limitation of microseismic data is that it cannot provide the effective fracture geometry that contributes to the production. In other words, the fracture conductivity and propped/unpropped fracture area cannot be extracted from microseismic data (Maxwell et al., 2013). Therefore, in many cases, the fracture geometry, after calibrating with the actual production data, is less than the fracture geometry from microseismic data (Sharma et al., 2004). Cipolla and Wallace (2014) also emphasized that stimulated reservoir volume provided from microseismic data may not be a good indication of the effectiveness of fractures. They also supported using a fracture model to estimate the propped and unpropped fracture area by performing production history matching. Shakiba and Sepehrnoori (2015) also demonstrated that microseismic data cannot capture the small fissures due to microseismic resolution. Therefore, to determine effective fracture geometries and their properties, calibrating with historical production data, which is always available at no additional cost, is necessary.

Therefore, we utilized the assisted history-matching (AHM) workflow to perform automatic history matching with three shale-gas wells to

Assisted History Matching for Unconventional Reservoirs. DOI: https://doi.org/10.1016/B978-0-12-822242-3.00003-1
© 2021 Elsevier Inc. All rights reserved.

characterize reservoir and fracture properties of each well. We had production data of three wells from the same shale formation in China. The historical production data of Wells A, B, and C are 132, 152, and 511 days, respectively, as shown in Fig. 10.1. The direct measurements of gas flow rate, water flow rate, and wellhead pressure (WHP) are available. Each well has a different fracturing treatment and completion design, as summarized in Table 10.1. The cluster spacing of the three wells is the same at 67 ft, while the stage spacing in Well A is the highest one. In terms of well performance, peak gas rate, peak gas rate per cluster, and peak gas rate per drawdown pressure per cluster are shown in Table 10.1. In brief, considering the peak gas rate per drawdown pressure per cluster, Well C has the best production performance among the three wells, while the performances of Wells A and B yield the same gas production performance. For water flow rate the water−gas ratio (WGR) of the three wells is depicted in Fig. 10.1D. The WGR trends from high to low values are Wells A, B, and C, respectively. The bottomhole pressure (BHP) was derived from WHP by using a simple well model considering well depth or hydrostatic pressure, gas flow rate, and WGR values.

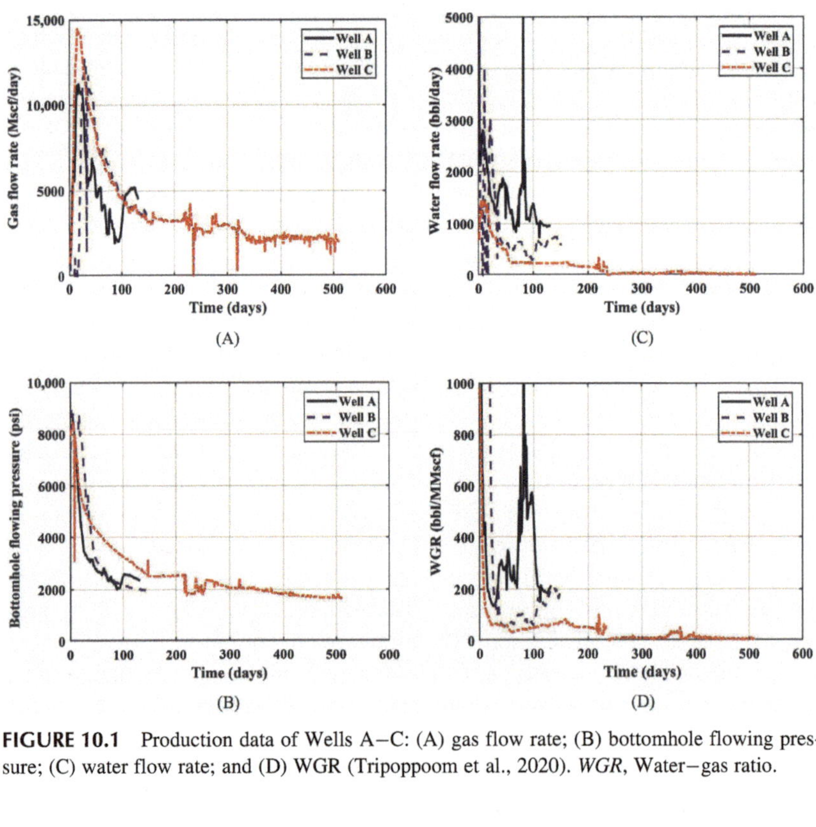

FIGURE 10.1 Production data of Wells A–C: (A) gas flow rate; (B) bottomhole flowing pressure; (C) water flow rate; and (D) WGR (Tripoppoom et al., 2020). *WGR*, Water–gas ratio.

TABLE 10.1 Summary of basic well and fractures information of Wells A, B, and C (Tripoppoom et al., 2020).

Well parameters	Well A	Well B	Well C	Unit
Well length	6234	5577	4921	ft
Number of stages	24	28	18	–
Number of cluster per stage	3	3	3	–
Total cluster number	72	84	54	–
Cluster spacing	67	67	67	ft
Stage spacing	78	65	67	ft
Peak gas rate	11.2	12.7	14.4	MMscf/day
Peak gas rate per cluster	156	151	267	Mscf/day per cluster
Peak gas rate/drawdown pressure per cluster	0.041	0.040	0.106	Mscf/(psi day)

10.2 Reservoir model

We constructed reservoir models, as summarized in Table 10.2. We used the same gridblock size of 20 × 50 × 65 ft in x, y, and z directions for the three wells. As shown in Fig. 10.2, the model dimension in the x direction has an additional length of 250 ft on each side from the horizontal section of the well. For the y direction, the model covers the expected fracture half-length used during history matching. Lastly, the model thickness was constructed as the actual shale reservoir thickness in this basin. Other reservoir properties are similar, so we used the same values for all three wells, as shown in Table 10.2. In shale-gas reservoirs the gas desorption phenomenon exists so we included gas desorption in the model using the Langmuir isotherm, as summarized in Table 10.3 (Yu and Sepehrnoori, 2014, 2018; Yu et al., 2016). When reservoir pressure is depleted after producing, we capture the permeability reduction in both matrix and fractures using permeability modulus, as in Eq. (10.1) and Fig. 10.3. The information regarding gas desorption and permeability modulus were from laboratory results; therefore we did not include them as uncertain parameters for history matching.

$$k = k_i \exp^{-\gamma \cdot (P_i - P)}, \tag{10.1}$$

where k is the pressure-dependent permeability (md), k_i the permeability at initial reservoir pressure (md), γ the permeability modulus (1/psi), P_i the initial reservoir pressure (psi), and P the reservoir pressure (psi).

TABLE 10.2 Summary of basic reservoir inputs for reservoir models of Wells A–C (Tripoppoom et al., 2020).

Reservoir description	Well A	Well B	Well C	Unit
Model dimension ($x \times y \times z$)	$6740 \times 2000 \times 65$	$6060 \times 2000 \times 65$	$5500 \times 2000 \times 65$	ft
Number of gridblocks ($x \times y \times z$)	$337 \times 40 \times 1$	$303 \times 40 \times 1$	$275 \times 40 \times 1$	–
Gridblock dimension ($x \times y \times z$)	$20 \times 50 \times 65$			ft
Initial reservoir pressure	8847			psi
Reservoir temperature	215			°F
Residual water saturation	20%			–
Matrix water saturation	39%			–
Total compressibility	3×10^{-6}			psi^{-1}
Reservoir depth	10,500			ft

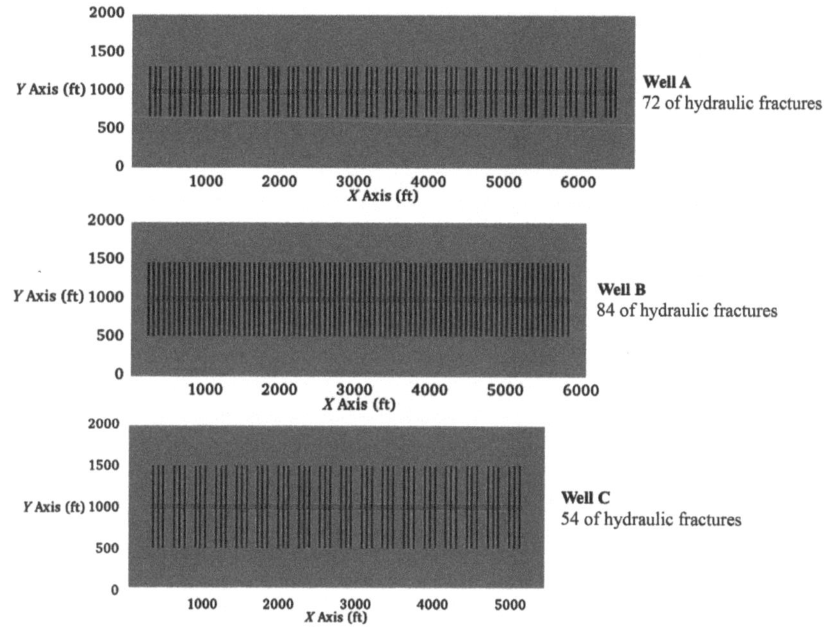

FIGURE 10.2 Reservoir models (top view) of Wells A–C (Tripoppoom et al., 2020).

TABLE 10.3 Langmuir isotherm model parameters for the gas desorption effect (Tripoppoom et al., 2020).

Langmuir isotherm parameters	Value	Unit
Inverse-pressure parameter for the Langmuir Isotherm model, CH_4	0.000534	1/psi
Maximum moles of adsorbed component per unit mass of rock, CH_4	0.06	gmole of component/ lb of rock
Rock density	162	lb/ft^3

As mentioned in section 2.2 in chapter 2, we modeled hydraulic fractures by embedding the fractures into the reservoir grid using the embedded discrete fracture model (EDFM) method. The fracture geometries of the three wells, representing actual completion, as summarized in Table 10.1, are illustrated in Fig. 10.2.

10.3 Automatic history matching

After constructing reservoir models, we performed history matching automatically. First, we selected parameters that are to be uncertain parameters for

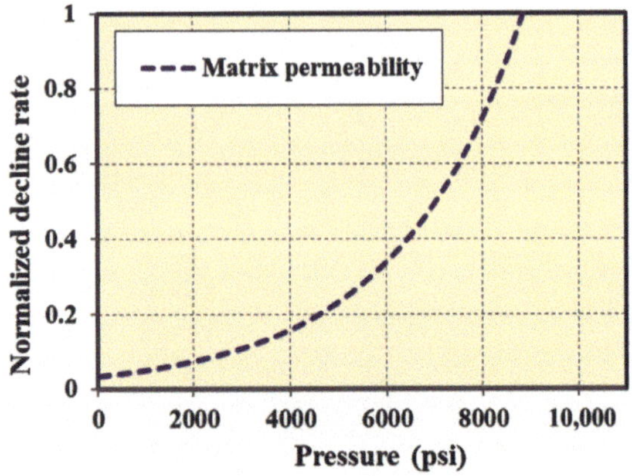

(A)

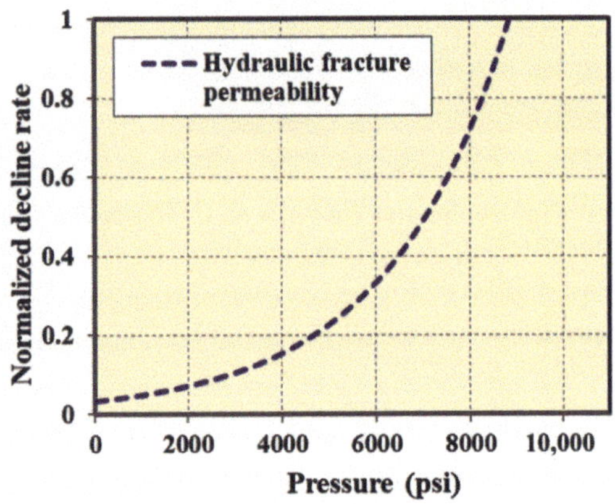

(B)

FIGURE 10.3 Pressure-dependent permeability curve: (A) matrix permeability and (B) hydraulic fracture permeability (Tripoppoom et al., 2020).

history matching. The criteria to select uncertain parameters depends on the certainty of the information we had for each parameter. Information that is still uncertain, especially hydraulic fracture geometry and properties, and some reservoir properties, should be used as uncertain parameters during history matching. In case the number of uncertain parameters is too many, we have two options for acquiring more information regarding those specific parameters to reduce their uncertainty. Alternatively, we can calibrate those parameters with production data during history matching. Normally, in shale-gas reservoirs, hydraulic fracture geometry and properties are the most important and also have high uncertainty even with the acquisition of microseismic data or DFIT analysis.

TABLE 10.4 Summary of 10 uncertain parameters and their prior distributions (Tripoppoom et al., 2020).

Uncertain parameters	Unit	Min value	Max value	Sources
Matrix permeability	md	0.00001	0.0001	• Nearby well core data • DFIT analysis
Fracture height	ft	25	65	• Historical microseismic data • History matched parameters from other wells and • Reservoir thickness
Fracture half-length	ft	200	780	• Historical microseismic data • History matched parameters from other wells
Fracture conductivity	md ft	10	200	• DFIT analysis
Fracture water saturation	–	0.5	0.9	• History-matching experience
Exponent of k_{rg}	–	1	4	
Endpoint of k_{rw}	–	0.5	1	
Exponent of k_{rw}	–	1	4	
Fracture width	ft	0.1	4	
Porosity	–	0.038	0.083	• Nearby well core data • Petrophysical interpretation

DFIT, Diagnostic fracture injection testing.

The more parameters we include during history matching, the more difficult it becomes for the algorithm to find history-matching solutions because we involve more dimensions in the problem – the curse of dimensionality. Other presteps of parameter screening, such as full or partial two-factorial design, can be performed to screen only parameters important to the history-matching problem (Tripoppoom et al., 2019a). The parameter screening step is not discussed in detail in this study but is based on many actual applications. The most important parameters for history matching in shale-gas wells are normally hydraulic fracture geometry and properties and these should be included every time during history matching. In this study, we selected 10 uncertain parameters for history matching, including matrix properties such as matrix permeability, porosity, and relative permeability, as shown in Table 10.4. For relative permeability, we use Corey's equation parameters including the curvature of both gas and water curves and the relative permeability endpoint for water curve. Then, we specified the initial range of each uncertain parameter based on available information and experience, as shown in Table 10.4. This is the integration of core data, log data, DFIT analysis, microseismic data, and engineering experience in this reservoir. Based on our experience, the step of specifying a minimum and a maximum is the most important step in the whole AHM workflow and needs an engineer's expertise and knowledge of the well and reservoir. This step will certainly determine whether the history matching will be successful or not. To illustrate, initially, we tried a maximum fracture width of 2 ft for all wells. This maximum worked for Wells B and C but not for Well A. The workflow rarely finds the history-matching solutions for Well A. Later, we revised the range of maximum fracture width to 4 ft. The initial iteration results were better and a much higher number of history-matching solutions were found. In sum, the ranges specified for each uncertain parameter should be from the knowledge and the integration of information including experience of history matching. The adjustment of each uncertain parameter can occur after evaluating the initial iteration results or even after finishing the whole history-matching workflow attempt.

After finishing the parameter identification step, we selected proxy points to be used in the neural network-Markov chain Monte Carlo (NN-MCMC) algorithm. We chose 12, 11, and 12 proxy points for Wells A, B, and C, respectively, as depicted in Figs. 10.4–10.6. The criteria to select proxy points is that we would like to capture the major historical production data trend by not including unnecessary points (Tripoppoom et al., 2019a, 2019b). Then, we generated an initial 50 cases by using Latin hypercube design. We ran these 50 cases with EDFM and a reservoir simulator by constraining the model with gas flow rate. The response parameters to calculate the objective function are BHP and WGR. The example of simulation results from 50 cases of Well C is illustrated in Fig. 10.7, while simulation results of Wells A and B are shown in Fig. A1 and A2, respectively in Appendix A. In this step, we checked whether first, they cover the historical production data and second, the simulation results create a similar production trend or not. Not achieving these two criteria can be caused by many things, such as model assumption, but the most likely reason is an inappropriate initial range identified in the previous step. Therefore, engineers have to go back and review the uncertain parameter information. As shown in Fig. 10.7, the simulation results of BHP cover the production data and their trends are similar to the actual historical data. However, for WGR, all 50-case simulation results do not yield the actual WGR trend during days 50–250. The reason could be that the current model assumption cannot capture some complicated water flow back behavior in the early period or it could be the measurement uncertainty itself. There is no perfect model and increasing model complexity without more information creates unnecessary challenges. To understand fracture and reservoir properties for the three wells from history matching, we decided to continue performing history matching for all three wells by keeping the current model assumptions.

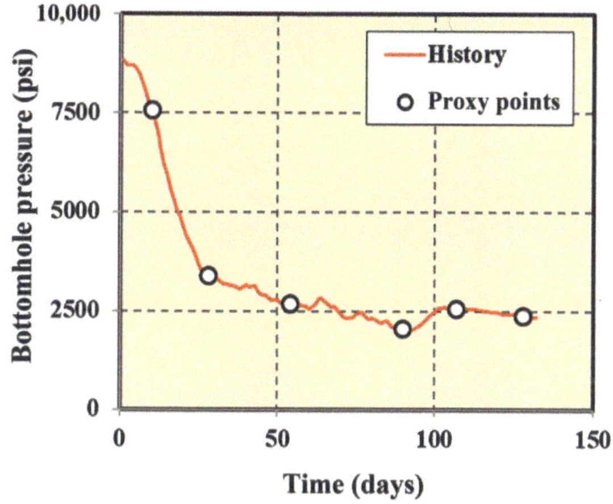

(A)

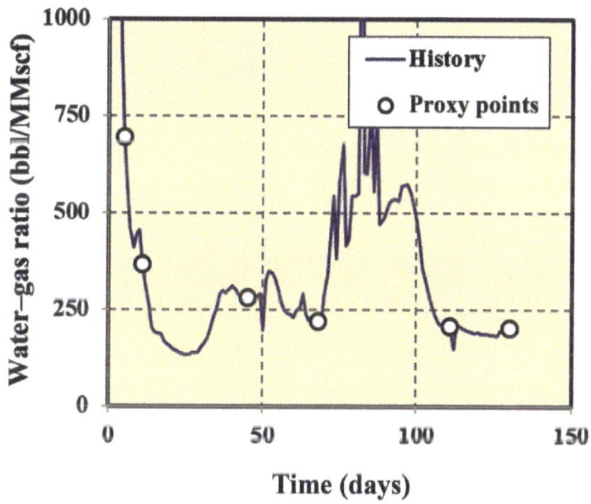

(B)

FIGURE 10.4 12 Multiple proxy points using neural networks for Well A production data: (A) 6 proxies of BHP and (B) 6 proxies of WGR (Tripoppoom et al., 2020). *BHP*, Bottomhole pressure; *WGR*, water–gas ratio.

We used initial simulation cases to train NNs for all proxy points that we specified earlier. Then, we used the NN-MCMC algorithm in the iterative AHM workflow to perform history matching automatically. To determine the misfit value of each realization, the objective function formulation we used is shown in Eq. (10.2). This formulation couples different selected proxy points to represent multiple objective functions such as WGR and BHP into one global value or, as we call it, global error.

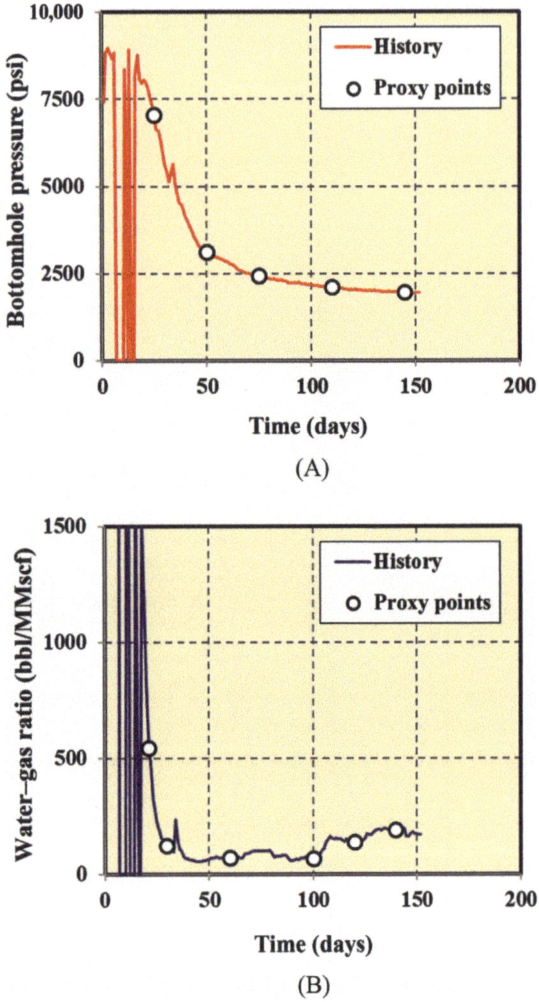

FIGURE 10.5 11 Multiple proxy points using neural networks for Well B production data: (A) 5 proxies of BHP and (B) 6 proxies of WGR (Tripoppoom et al., 2020). *BHP*, Bottomhole pressure; *WGR*, water–gas ratio.

$$\text{Objective function} \quad = \text{global error} \quad (\varepsilon) = \frac{\sum_{j=1}^{p}\sum_{i=1}^{q}\left|\frac{x_{ij,\text{model}} - x_{ij,\text{history}}}{x_{ij,\text{history}}} \times 100\right| \cdot w_{ij}}{\sum_{j=1}^{p}\sum_{i=1}^{q} w_{ij}} \tag{10.2}$$

where i is an index of data points, j is an index of response parameters, q is a number of selected data points, p is the number of response parameters, $x_{ij,}$

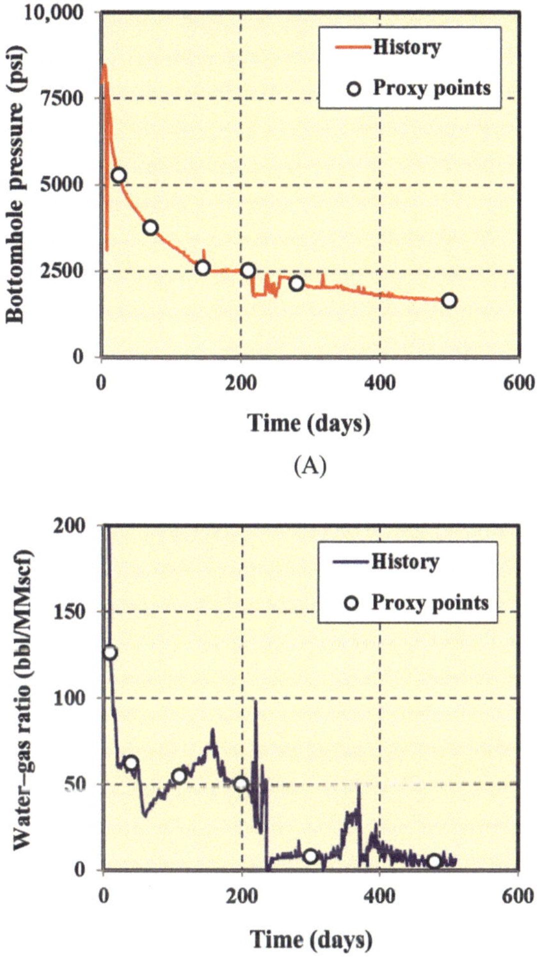

(A)

(B)

FIGURE 10.6 12 Multiple proxy points using neural networks for Well C production data: (A) 6 proxies of BHP and (B) 6 proxies of WGR (Tripoppoom et al., 2020). *BHP*, Bottomhole pressure; *WGR*, water–gas ratio.

$_{\text{model}}$ is the value from the model (simulation run or proxy model) at index i of response parameter j, $x_{ij,\text{history}}$ is the historical data at index i of response parameter j, and w_{ij} is the weight of data points i of response parameter j.

For example, for Well C, six proxy points of BHP and six proxy points of WGR were used to calculate the misfit value as global error using Eq. (10.2). The weights used in this study are 2 and 1 for BHP and WGR, respectively, because we would like to focus more on BHP rather than WGR when we consider the global error.

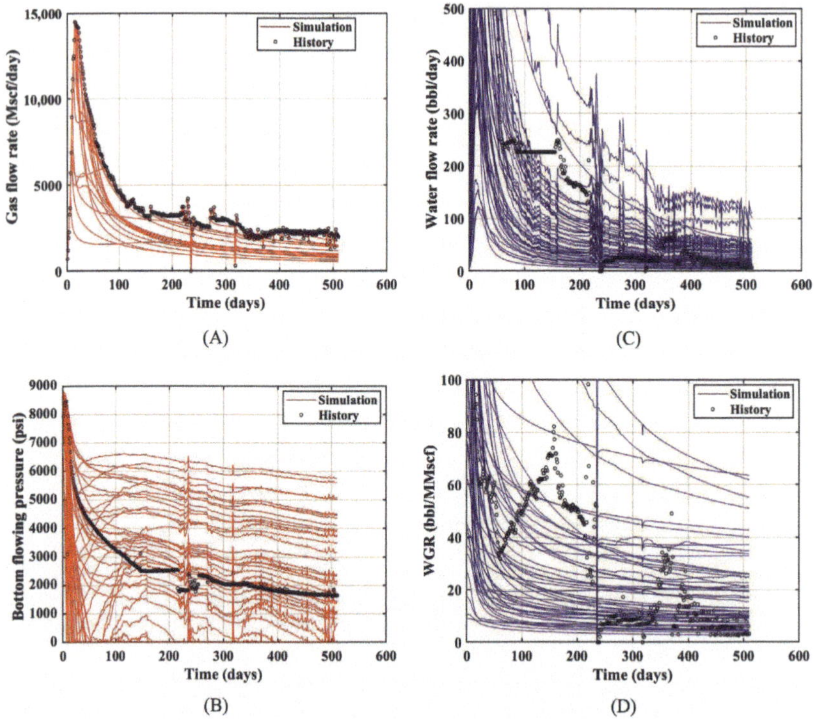

FIGURE 10.7 Comparison between simulation results (50 cases from Latin hypercube design) and historical data of Well C: (A) gas flow rate; (B) flowing BHP; (C) water flow rate; and (D) WGR (Tripoppoom et al., 2020). *BHP*, Bottomhole pressure; *WGR*, water–gas ratio.

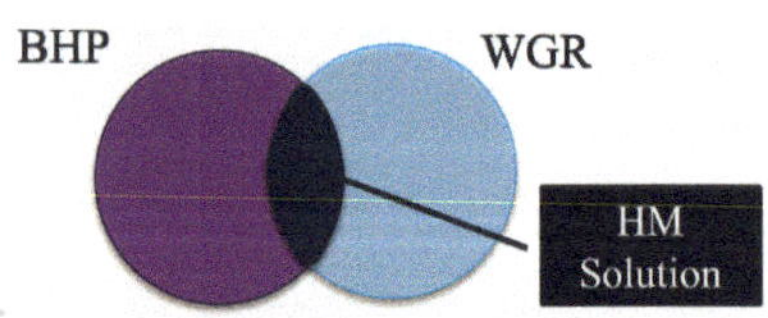

FIGURE 10.8 History-matching solutions criterion of BHP error and WGR error (Tripoppoom et al., 2020). *BHP*, Bottomhole pressure; *WGR*, water–gas ratio.

Next, the NN-MCMC proposes history-matching solution candidates to be validated with the reservoir simulator in each iteration. This continues iteratively, first by NN-MCMC proposing the next 25 cases, then running simulations, and finally, incorporating these new 25 simulation results into a training data set for re-training NNs. The iterative workflow stops either when we reach the maximum limit we set for number of iterations or the NNs are not significantly different between consecutive iterations. In this study, the iterative workflow stopped because the workflow reached the maximum of 12 iterations or 325 simulation runs. With a 3.7 GHz, 4-core

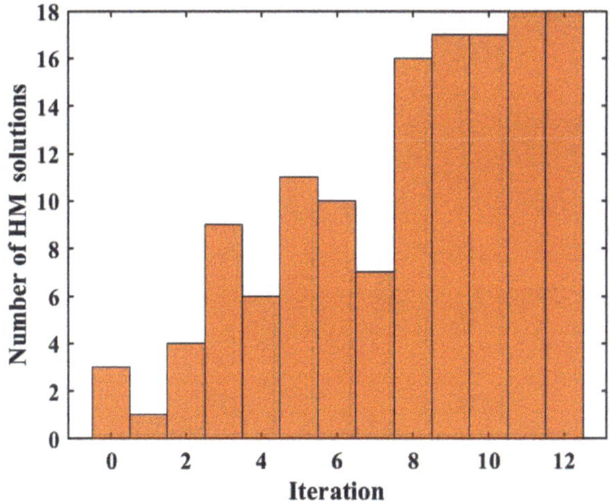

FIGURE 10.9 Number of history-matching solutions of Well C versus iteration number (Tripoppoom et al., 2020).

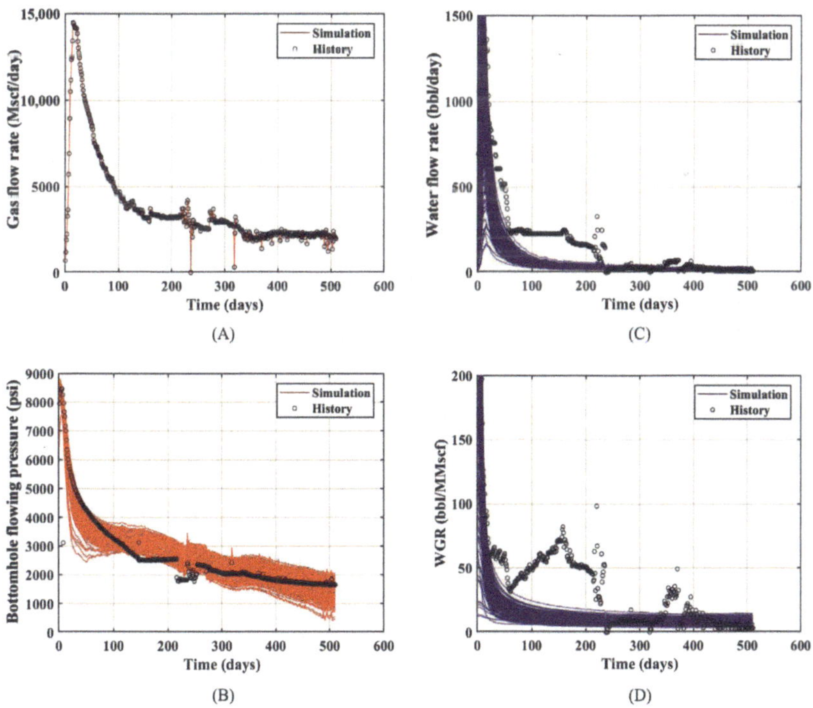

FIGURE 10.10 Comparison between simulation results of 137 history-matching solutions and historical data of Well C: (A) gas flow rate; (B) flowing BHP; (C) water flow rate; and (D) WGR (Tripoppoom et al., 2020). *BHP*, Bottomhole pressure; *WGR*, water–gas ratio.

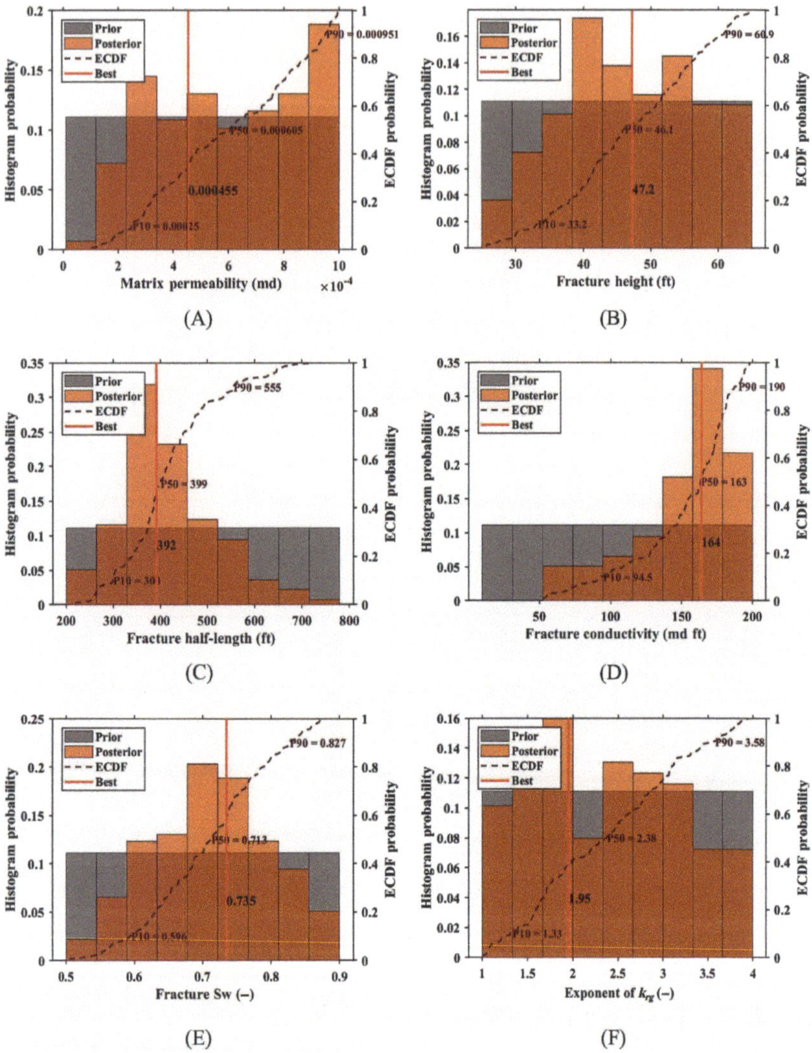

FIGURE 10.11 Continued.

processor computer and 4-parallel computation for simulator, the workflow spent around 32 hours or less than 1.5 days for Well C. For Wells A and B, the workflow spent around 30 hours each due to a shorter period of historical data. This elapsed time includes case auto-generation, EDFM preprocessor, NN-MCMC algorithm, and validation with reservoir simulator. Around 40% of the total elapsed time was dedicated to the NN-MCMC algorithm, another 40% to the reservoir simulator, and 20% was used by the EDFM preprocessor.

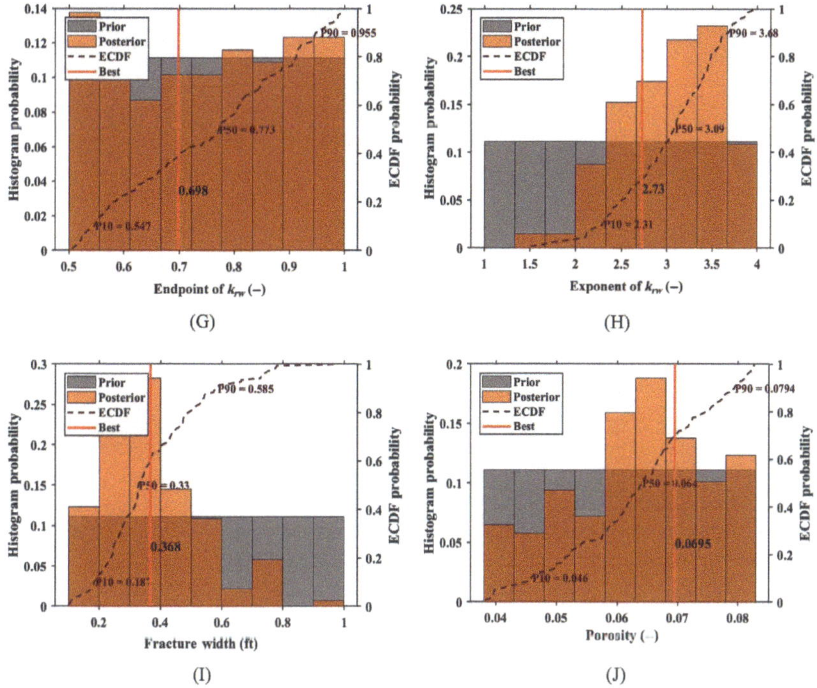

FIGURE 10.11 Comparison between prior and posterior distribution of uncertain parameters (137 solutions) of Well C: (A) matrix permeability; (B) fracture height; (C) fracture half-length; (D) fracture conductivity; (E) fracture water saturation; (F) exponent of k_{rg}; (G) endpoint of k_{rw}; (H) exponent of k_{rw}; (I) fracture width; and (J) porosity (Tripoppoom et al., 2020).

Next, we screened all simulation runs (history-matching solution candidates from the NN MCMC algorithm from all iterations) with the history-matching solution threshold as shown in Fig. 10.8. We identified the threshold by reviewing the actual simulation results compared with the historical production data. This is to see which cases were the least acceptable and what the BHP error and WGR error threshold should be (Li et al., 2019). In this study, for Well C, the history-matching solutions must have a BHP error and a WGR error of less than 20% and 70%, respectively. For Wells A and B, we used 25% and 20% for BHP error and 60% and 45% for WGR error, respectively. The higher threshold value of WGR error was used due to the fact that the model results cannot mimic the WGR trend during days 50–250 for Well C, as discussed in the initial simulation cases shown in Fig. 10.7. The higher values of the WGR threshold were also applied to Wells A and B with a similar reason. With this threshold, the workflow found 137 solutions from 325 simulation runs, or around 42%. The number of history-matching solutions versus iteration number for Well C is illustrated in Fig. 10.9. For other wells, the figures are shown in Fig. A3 and A4

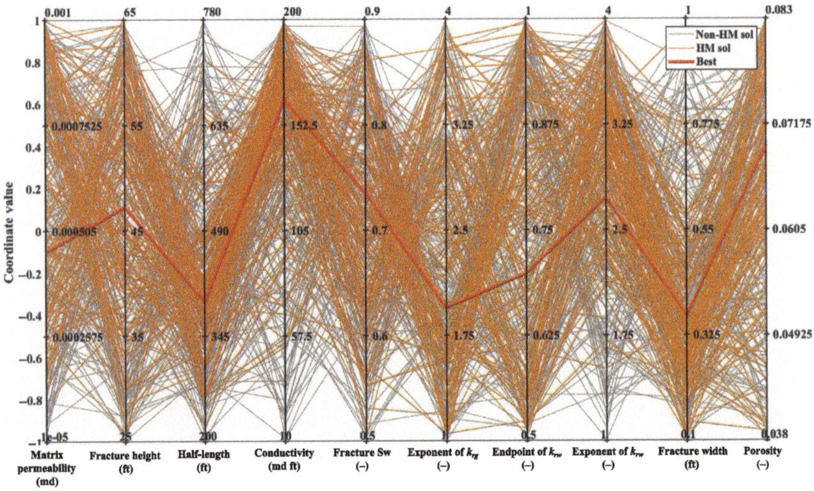

FIGURE 10.12 Parallel coordinate plot showing all simulation runs of Well C during history matching, including nonhistory-matching solutions, history-matching solutions, and best case (Tripoppoom et al., 2020).

for Well A and B, respectively, in Appendix A. We can observe that the later iterations found more solutions than the early ones. This is because the NNs' accuracy is improved in every iteration by adding more measured points from actual simulation results into the interesting area near the history-matching solution domain. All 137 simulation results of the history-matching solutions are shown in Fig. 10.10. The significant improvement in matching quality can be observed by comparing with the initial iteration results in Fig. 10.7. Wells A and B history-matching results are also shown in Fig. A5 and A6, respectively, in Appendix A.

Next, we obtained a posterior distribution of each uncertain parameter from all 137 history-matching solutions. Fig. 10.11 illustrates the prior and posterior distribution of all 10 uncertain parameters of Well C with the best case. For the best case, another iteration was performed to generate realizations around P50 values (plus and minus 10% of P50 values). This is the same as Chapter 9, Shale Oil Well With and Without Natural Fractures, and the best case was selected from the realization with the lowest global error in this iteration. This was done to ensure the best case to be aligned with P50 values. The figures of Wells A and B can be found in Fig. A7 and A9, respectively,in Appendix A. You can observe that the best case values align with P50 values in the posterior distribution.

From the posterior distribution of Well C, we obtained the uncertainty range and a representative value, such as P50, for each parameter. For example, the fracture half-length range falls in the lower range of our initial input (P50 is at 399 ft) while fracture conductivity implies a higher value, near a

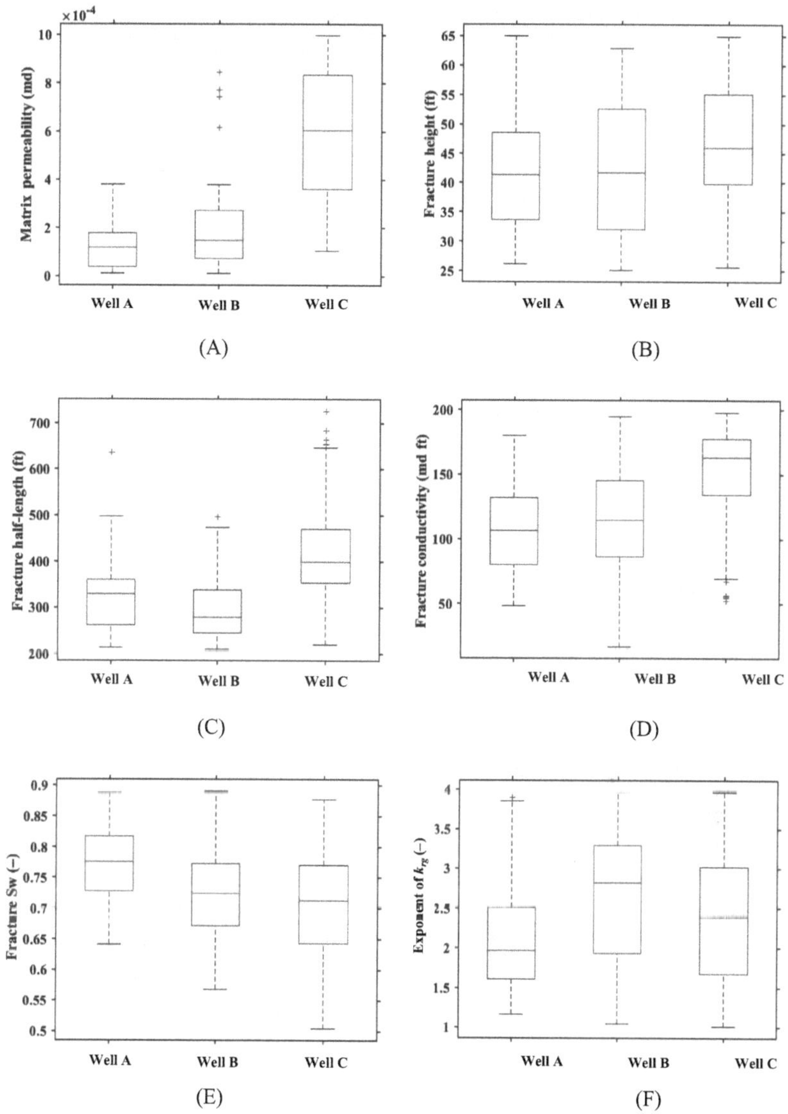

FIGURE 10.13 Continued.

maximum of 200 md ft. One interesting point to be highlighted is that the posterior distribution of uncertain parameters such as exponent of k_{rg}, end-point of k_{rw}, and porosity do not change significantly from prior distribution and any values within the specified range found history-matching solutions. This is because these uncertain parameters have a lower impact on the objective function within the specified value ranges compared with other uncertain

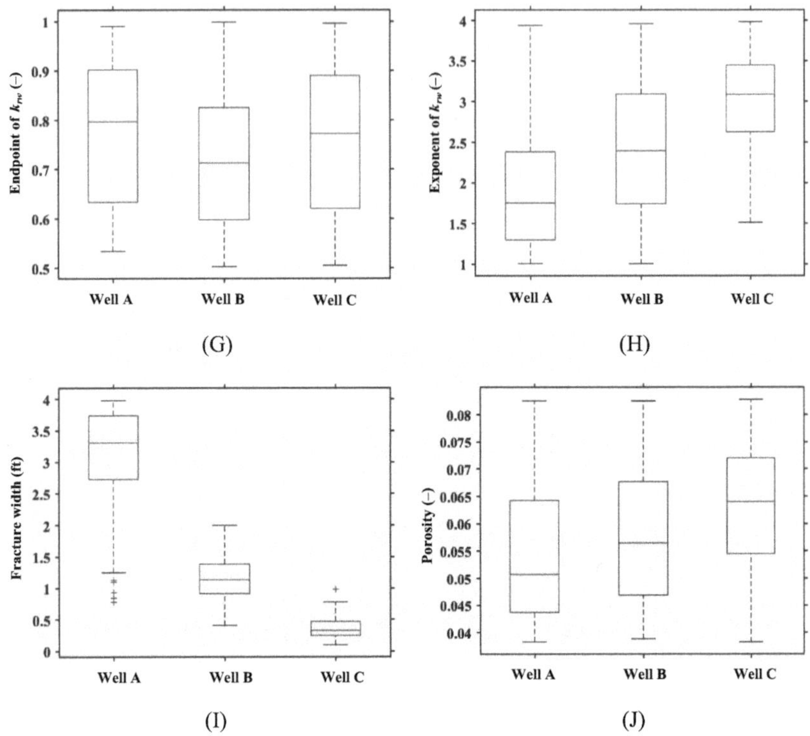

FIGURE 10.13 Box plot comparison of uncertain parameter posterior distribution of Wells A–C: (A) matrix permeability; (B) fracture height; (C) fracture half-length; (D) fracture conductivity; (E) fracture water saturation; (F) exponent of k_{rg}; (G) endpoint of k_{rw}; (H) exponent of k_{rw}; (I) fracture width; and (J) porosity (Tripoppoom et al., 2020).

parameters. All simulation runs during the iterative AHM workflow are illustrated in Fig. 10.12 for Well C. Each line represents one realization of fracture and reservoir properties. As you can see, only some certain combinations and value ranges provide history-matching solutions. For Wells A and B, similar parallel coordinate plots are shown in Fig. A8 and A10, respectively, in Appendix A.

For comparison, a box plot comparison of posterior distribution of each uncertain parameter for Wells A–C is shown in Fig. 10.13. We found that uncertain parameters such as fracture half-length, fracture conductivity, fracture width, and matrix permeability were significantly different among the three shale-gas wells while fracture height, fracture water saturation, porosity, and relative permeability curves were less different. For fracture half-length, Well C has the highest value followed by Wells A and B. For fracture conductivity, Well C is also the most conductive well followed by Wells B and A. On the other hand, Well C has the lowest effective fracture width, while Well A has the highest effective fracture width. For fracture height, there is no significant

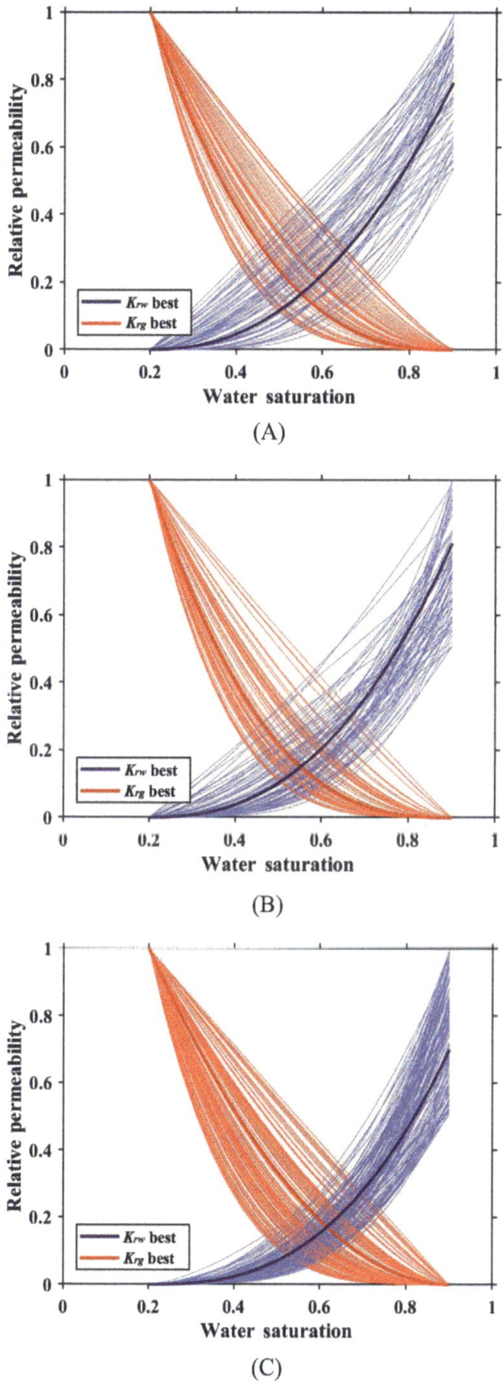

FIGURE 10.14 Relative permeability curves of history-matching solutions and best case: (A) Well A; (B) Well B; and (C) Well C (Tripoppoom et al., 2020).

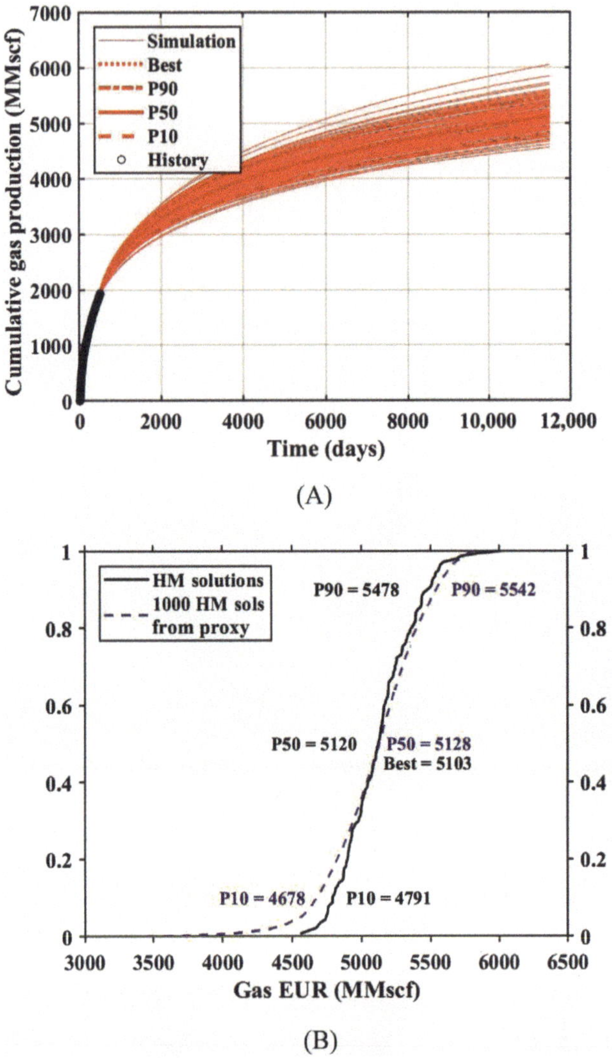

FIGURE 10.15 Production forecast of Well C: (A) cumulative gas production and (B) ECDF of gas EUR for history-matching solutions from AHM workflow and 1000 history-matching solutions generated by proxy model (Tripoppoom et al., 2020). *AHM*, Assisted history matching; *ECDF*, empirical cumulative distribution function; *EUR*, estimated ultimate recovery.

difference between the three wells. For matrix permeability, we observed significantly higher matrix permeability from Well C, while the values of Wells A and B are close to each other. The matrix permeability found from history-matching represents the combination of shale permeability and the induced unpropped fractures around main hydraulic fractures. The same trend was also

TABLE 10.5 Summary of gas estimated ultimate recovery (EUR) forecast of Wells A–C (Tripoppoom et al., 2020).

Well parameters	Well A	Well B	Well C	Unit
20-year gas EUR	3507	3261	4669	MMscf
30-year gas EUR	3859	3552	5103	MMscf
20-year gas EUR per cluster	49	39	86	MMscf per cluster
30-year gas EUR per cluster	54	42	95	MMscf per cluster

observed for matrix porosity—Well C has the highest value. For relative permeability curves, we observed more curvature on k_{rw} in Well C than Wells B and A, respectively, while there is no significant difference in the k_{rg} curvature endpoint of k_{rw}, as depicted in Fig. 10.14.

With this information from history matching, the difference in fracture half-length, fracture conductivity, and fracture width can be used to investigate new fracture treatment designs to improve any future fractured wells. In addition, it can be useful to integrate information of matrix properties such as permeability, porosity, and relative permeability with other methods.

10.4 Probabilistic production forecast

We obtained multiple history-matching solutions for all three wells. Then, we utilized the matched models to perform probabilistic production forecasting. We continued running the reservoir simulation for another 30 years after the history period by constraining the model with BHP at 1000 psi. The production forecast of Well C is illustrated in Fig. 10.15. The production forecast's uncertainty can be quantified by utilizing multiple matched models, and the P10–P50–P90 of gas estimated ultimate recovery (EUR) can be retrieved, as illustrated in Fig. 10.15B. The figures illustrate both the solutions from the AHM workflow and 1000 solutions generated by using the latest iteration of NNs as proxy models. The 1000 solutions from NNs have the benefit of generating a smoother distribution and can be used to compare with the actual history-matching solutions from the AHM workflow. However, one limitation of these 1000 solutions is that they have not been validated with a reservoir simulator, as it may be infeasible to perform this validation. Therefore, we should be aware of the limitations of these 1000 solutions for a later interpretation.

As shown in Fig. 10.15B, both the empirical cumulative distribution function from the AHM workflow and proxy models are quite in line to each other for Well C. The production forecasts for Wells A and B are also illustrated in Fig. A11 and A12, respectively, in Appendix A. The 20- and 30-year gas EUR of the best case is summarized in Table 10.5. Well C has the

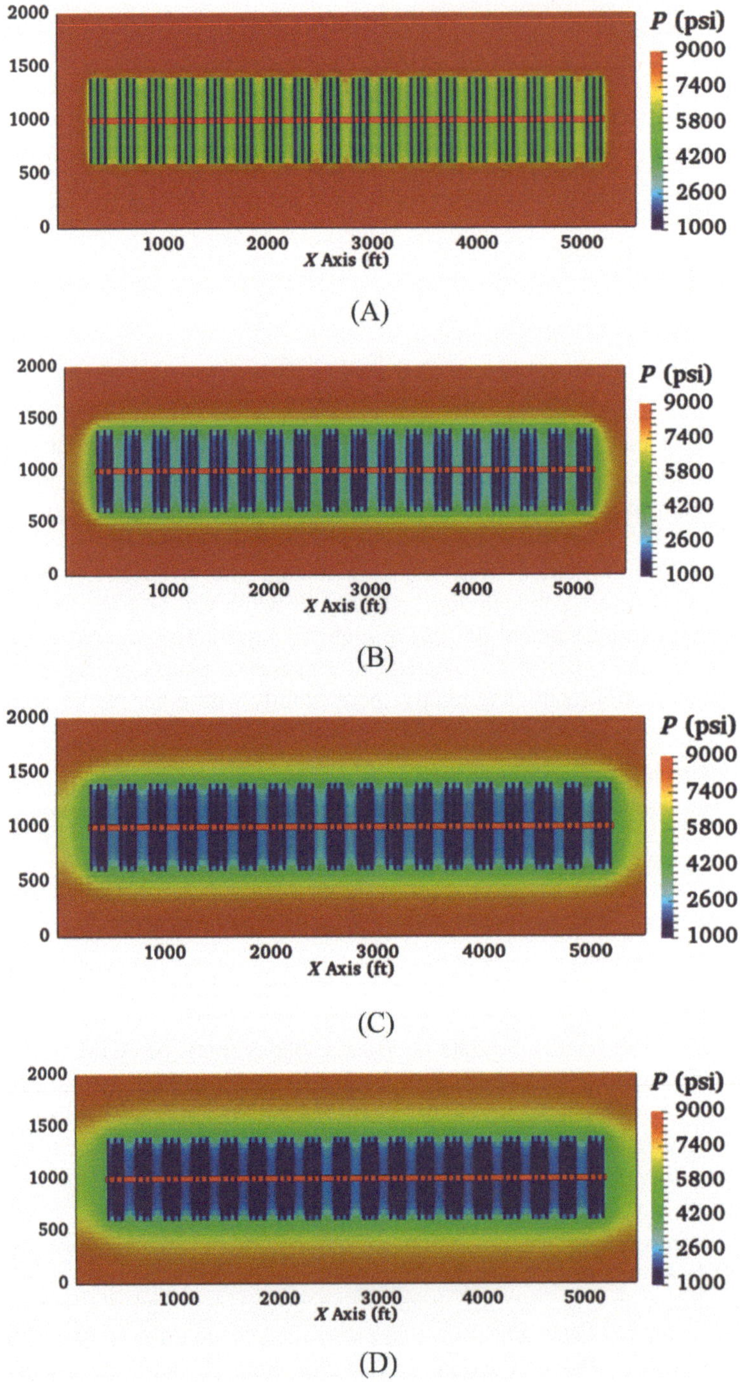

FIGURE 10.16 Pressure distribution of the best case for Well C: (A) after 511 days (end of history); (B) after 10 years; (C) after 20 years; and (D) after 30 years (Tripoppoom et al., 2020).

highest gas EUR and gas EUR per cluster, followed by Wells A and B. The pressure distribution at different points of time in the best case for Well C is also animated in Fig. 10.16. For Wells A and B the pressure distributions are illustrated in Fig. A13 and A14, respectively in Appendix A.

10.5 Remarks

We characterized the hydraulic fracture and reservoir properties of three different shale-gas wells by using historical production data. We performed automatic history matching using the AHM workflow, NN-MCMC algorithm, and EDFM. Two-phase flow of gas and water were history matched by using ten uncertain parameters. We found that the major uncertain parameters that cause production performance difference are fracture half-length, fracture conductivity, fracture width, and matrix permeability. Less significant are fracture height, fracture water saturation, porosity, and relative permeability curves. The information from uncertain parameter posterior distributions of these three wells will be used to help engineers better understand fracture geometry and properties related to fracturing design and completion. We also forecasted production probabilistically and compared gas EUR of three shale-gas wells.

References

Cipolla, C., Wallace, J., 2014. Stimulated reservoir volume: a misapplied concept?. In: Paper SPE 168596, Presented at SPE Hydraulic Fracturing Technology Conference, The Woodlands, TX, February 4–6, 2014.

Li, B., Bhark, E.W., Gross, S.J., Billiter, T.C., Dehghani, K., 2019. Best practices of assisted history matching using design of experiments. SPE Journal 24 (4), 1435–1451.

Maxwell, S.C., Weng, X., Kresse, O., Rutledge, J., 2013. Modeling microseismic hydraulic fracturing deformation. In: Paper SPE 166312, Presented at SPE Annual Technical Conference and Exhibition, New Orleans, LA, September 30–October 2, 2013.

Shakiba, M., Sepehrnoori, K., 2015. Using embedded discrete fracture model (EDFM) and microseismic monitoring data to characterize the complex hydraulic fracture networks. In: Paper SPE 175142, Presented at SPE Annual Technical Conference and Exhibition, Houston, TX, September 28–30, 2015.

Sharma M.M., Gadde P.B., Sullivan, R., Sigal, R., Fileder, R., Copeland, D., et al., 2004. Slick water and hybrid fracs in the Bossier: some lessons learnt. In: Paper SPE 89876, Presented at SPE Annual Technical Conference and Exhibition, Houston, TX, September 26–29, 2004.

Tripoppoom, S., Yu, W., Huang, H., Sepehrnoori, K., Song, W., Dachanuwattana, S., 2019a. A practical and efficient iterative history matching workflow for shale gas well coupling multiple objective functions, multiple proxy-based MCMC and EDFM. Journal of Petroleum Science and Engineering 176, 594–611.

Tripoppoom, S., Yu, W., Sepehrnoori, K., Miao, J., 2019b. Application of assisted history matching workflow to shale gas well using EDFM and neural network-Markov chain Monte

Carlo algorithm. In: Paper URTeC 2019-659, Presented at Unconventional Resources Technology Conference, Denver, CO, July 22–24, 2019.

Tripoppoom, S., Xie, J., Yong, R., Wu, J., Yu, W., Sepehrnoori, K., et al., 2020. Investigation of different production performances in shale gas wells using assisted history matching: hydraulic fractures and reservoir characterization from production data. Fuel 267, 117097.

Yu, W., Sepehrnoori, K., 2014. Simulation of gas desorption and geomechanics effects for unconventional gas reservoirs. Fuel 116, 455–464.

Yu, W., Sepehrnoori, K., 2018. Shale Gas and Tight Oil Reservoir Simulation. Gulf Professional Publishing, Elsevier.

Yu, W., Sepehrnoori, K., Patzek, T.W., 2016. Modeling gas adsorption in Marcellus shale with Langmuir and BET isotherms. SPE Journal 21 (2), 589–600.

Appendix A Wells A and B results

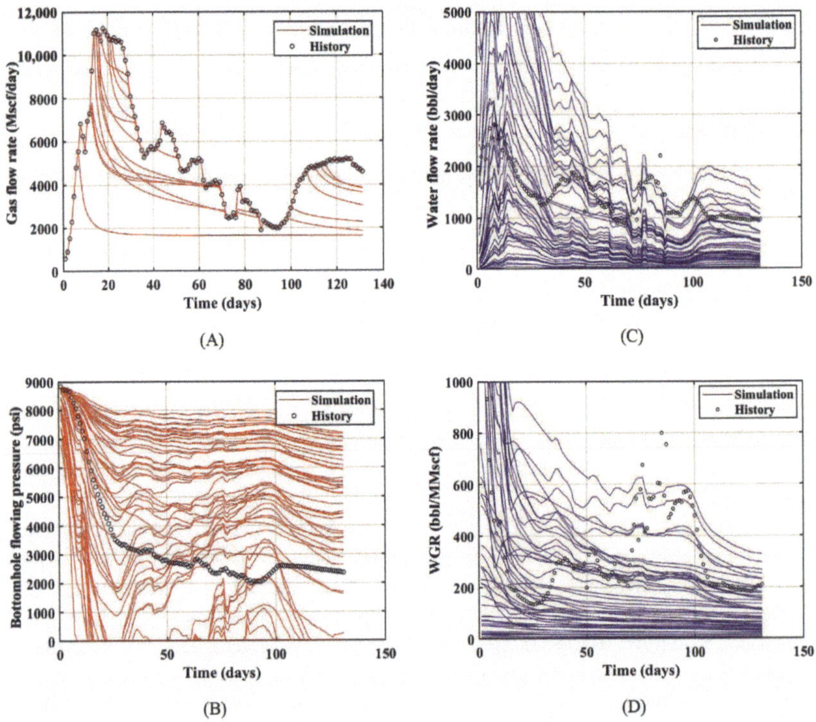

FIGURE A1 Comparison between simulation results (50 cases from Latin hypercube design) and historical data of Well A: (A) gas flow rate; (B) flowing BHP; (C) water flow rate; and (D) WGR (Tripoppoom et al., 2020). *BHP*, Bottomhole pressure; *WGR*, water–gas ratio.

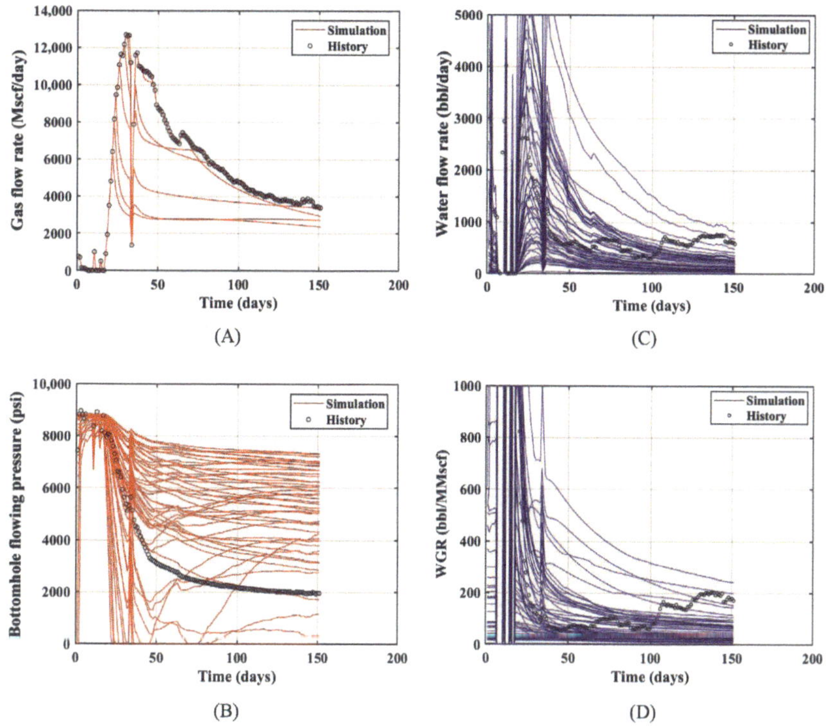

FIGURE A2 Comparison between simulation results (50 cases from Latin hypercube design) and historical data of Well B: (A) gas flow rate; (B) flowing BHP; (C) water flow rate, and (D) WGR (Tripoppoom et al., 2020). *BHP*, Bottomhole pressure; *WGR*, water–gas ratio.

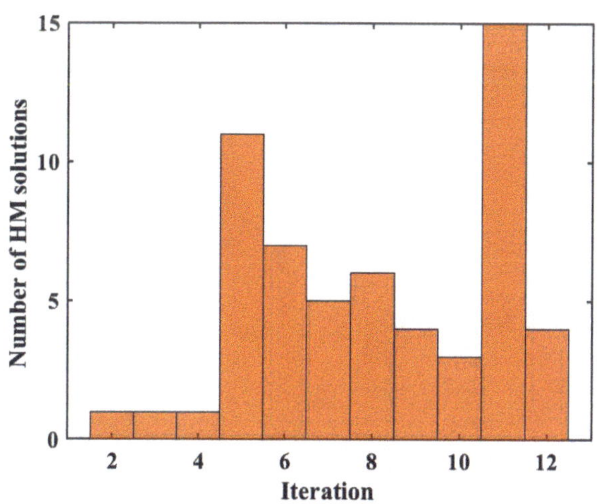

FIGURE A3 Number of history-matching solutions of Well A versus iteration number (Tripoppoom et al., 2020).

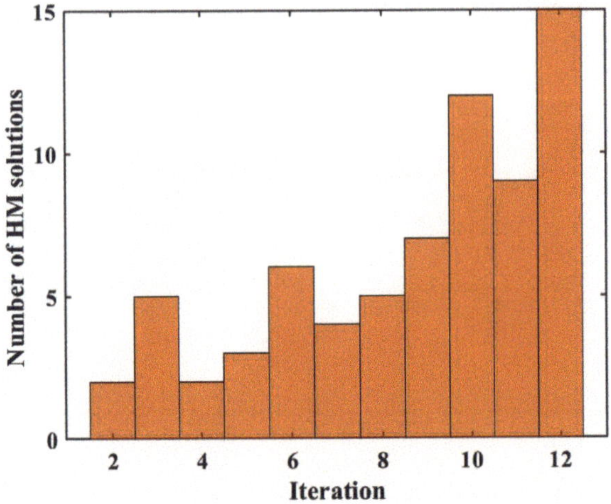

FIGURE A4 Number of history-matching solutions of Well B versus iteration number (Tripoppoom et al., 2020).

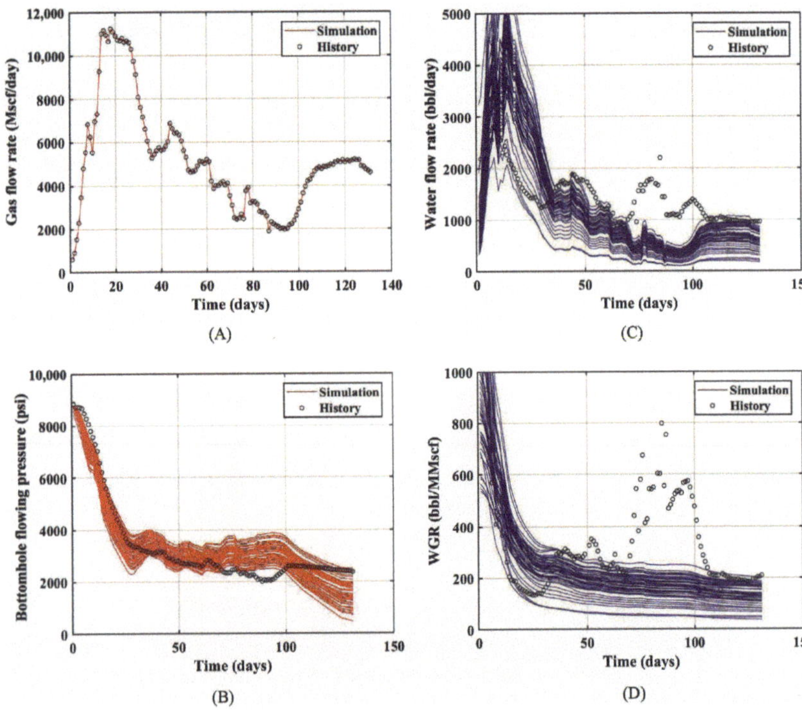

FIGURE A5 Comparison between simulation results of 58 history-matching solutions and historical data of Well A: (A) gas flow rate; (B) flowing BHP; (C) water flow rate; and (D) WGR (Tripoppoom et al., 2020). *BHP*, Bottomhole pressure; *WGR*, water–gas ratio.

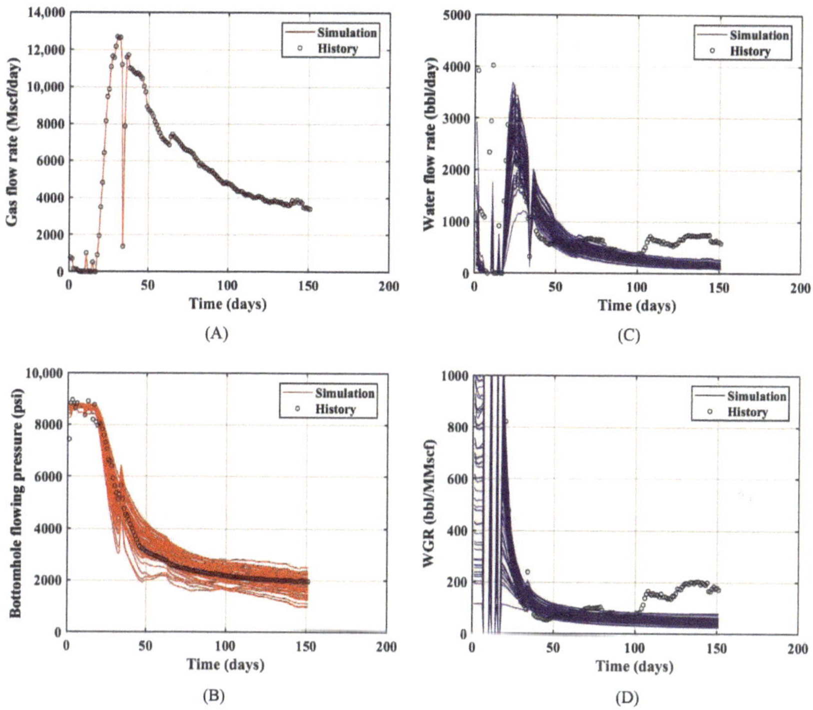

FIGURE A6 Comparison between simulation results of 70 history-matching solutions and historical data of Well B; (A) gas flow rate; (B) flowing BHP; (C) water flow rate; and (D) WGR (Tripoppoom et al., 2020). *BHP*, Bottomhole pressure; *WGR*, water–gas ratio.

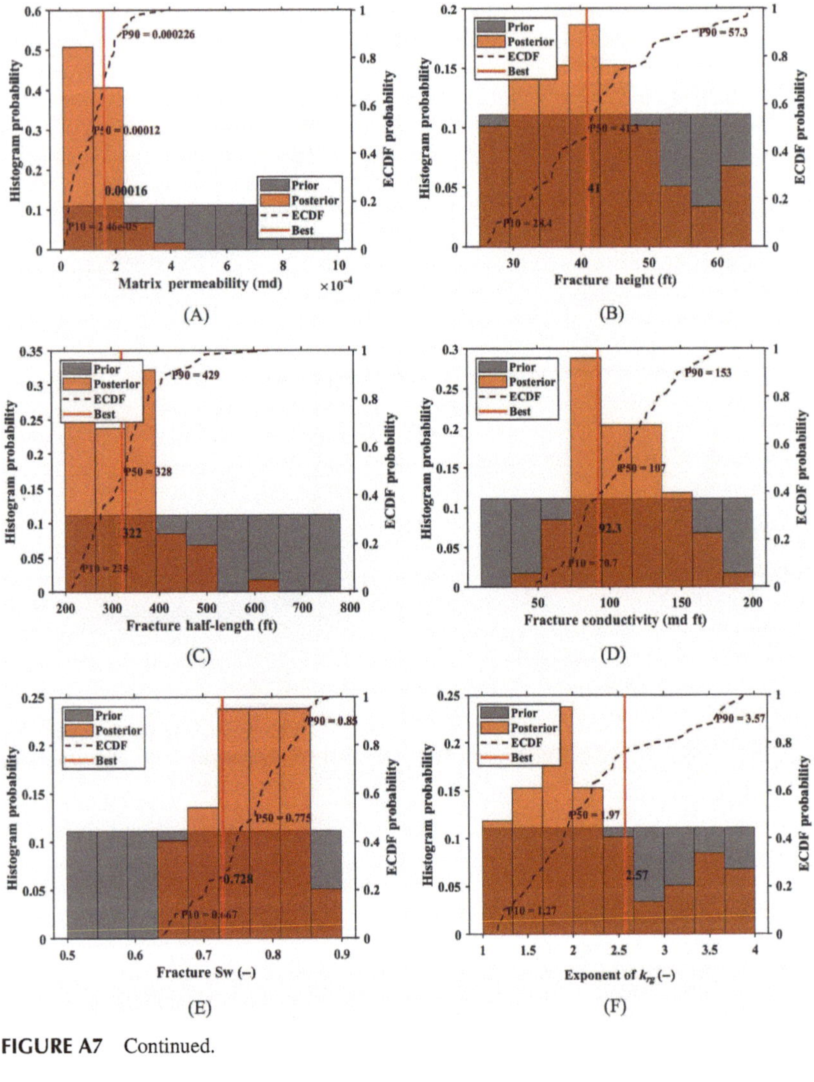

FIGURE A7 Continued.

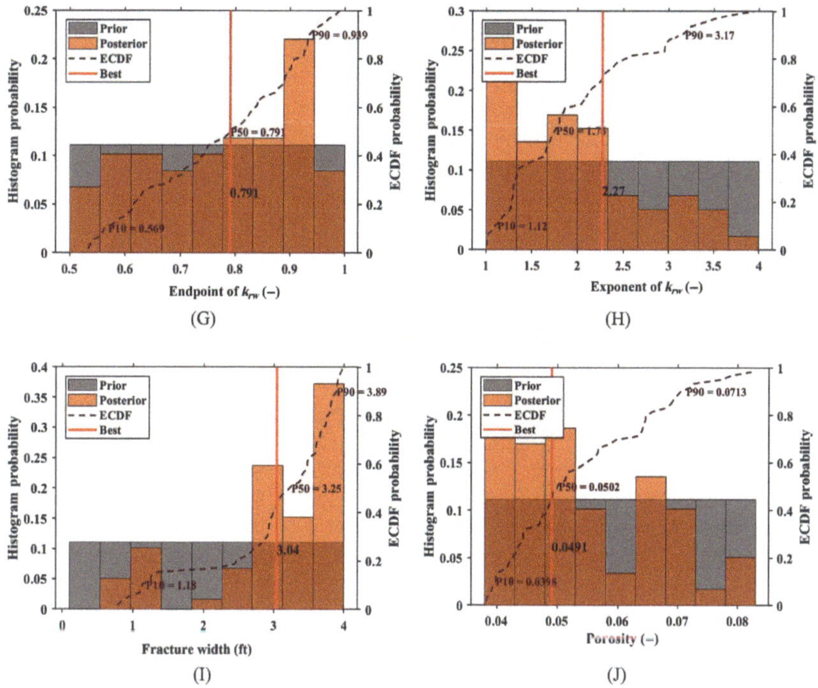

FIGURE A7 Comparison between prior and posterior distribution of uncertain parameter (58 solutions) of Well A: (A) matrix permeability; (B) fracture height; (C) fracture half-length; (D) fracture conductivity; (E) fracture water saturation; (F) exponent of k_{rg}; (G) endpoint of k_{rw}; (H) exponent of k_{rw}; (I) fracture width; and (J) porosity (Tripoppoom et al., 2020).

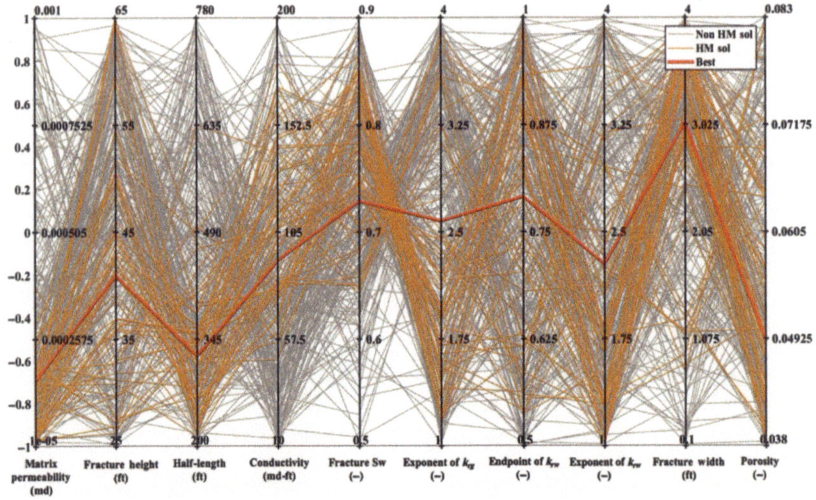

FIGURE A8 Parallel coordinate plot showing all simulation runs of Well A during history matching including nonhistory-matching solutions, history-matching solutions and the best case (Tripoppoom et al., 2020).

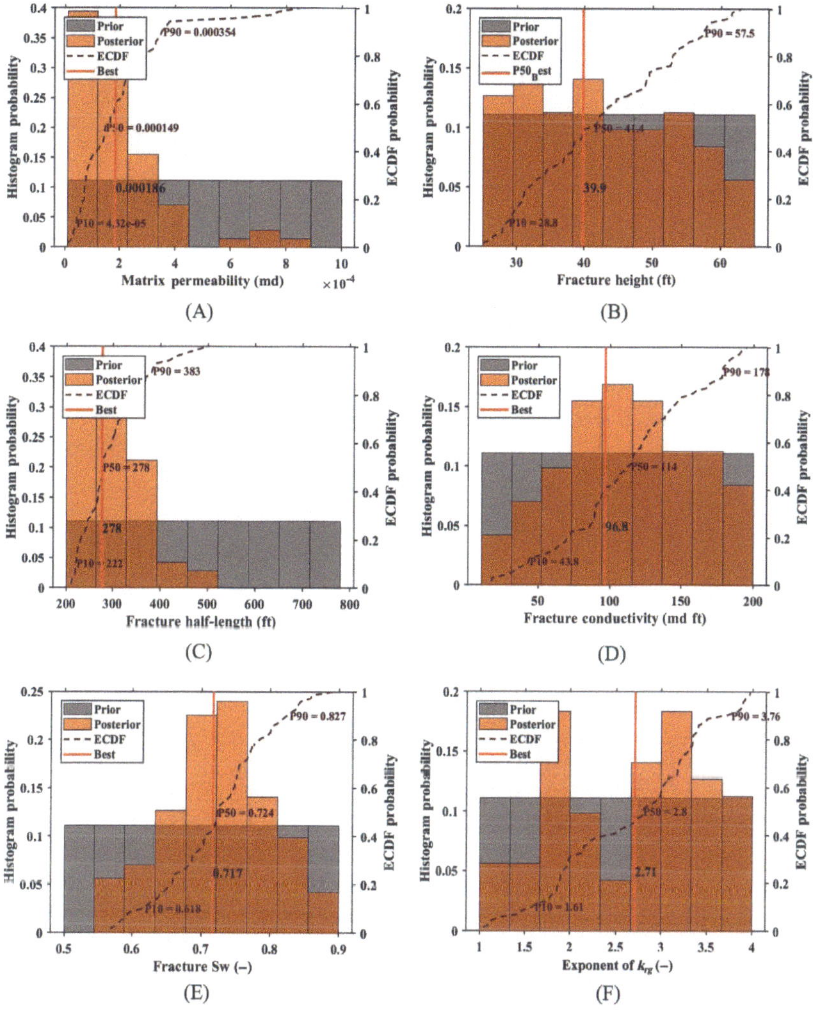

FIGURE A9 Continued.

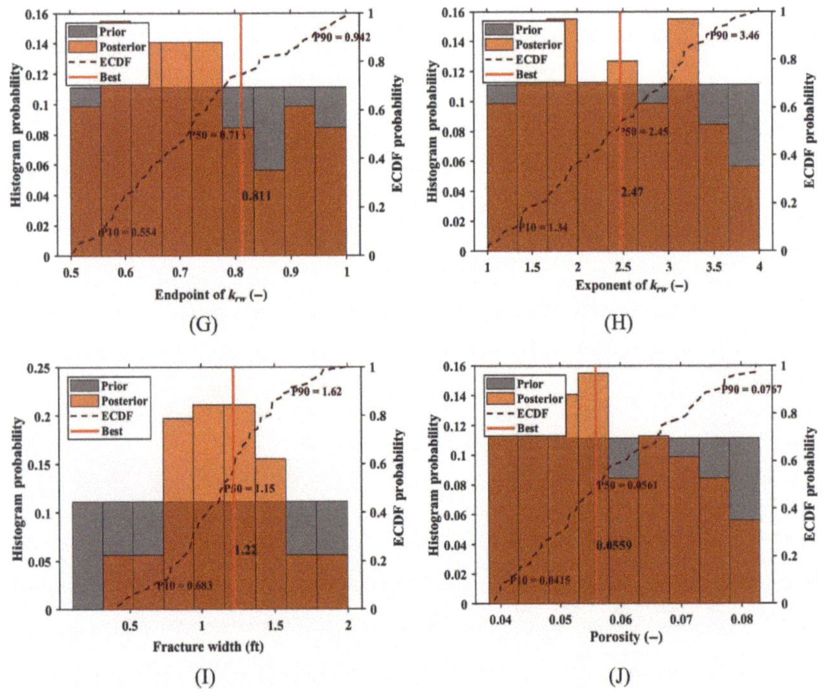

FIGURE A9 Comparison between prior and posterior distribution of uncertain parameter (70 solutions) of Well B: (A) matrix permeability; (B) fracture height; (C) fracture half-length; (D) fracture conductivity; (E) fracture water saturation; (F) exponent of k_{rg}; (G) endpoint of k_{rw}; (H) exponent of k_{rw}; (I) fracture width; and (J) porosity (Tripoppoom et al., 2020).

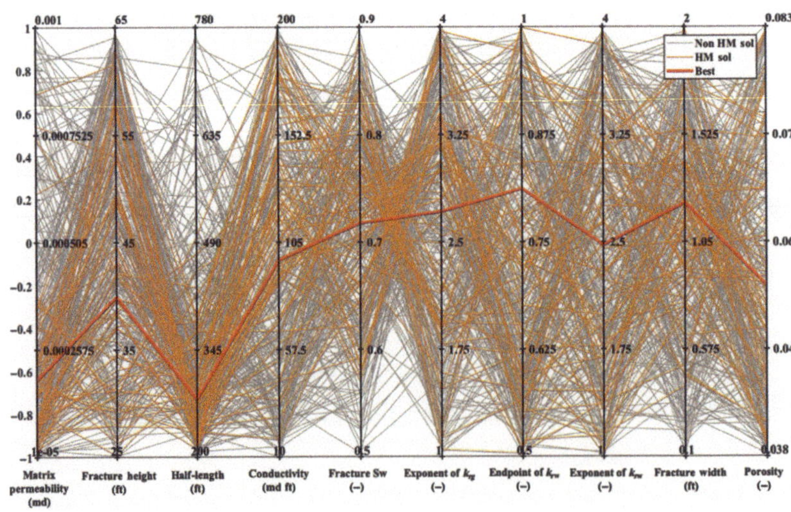

FIGURE A10 Parallel coordinate plot showing all simulation runs of Well B during history matching including nonhistory-matching solutions, history-matching solutions and the best case.

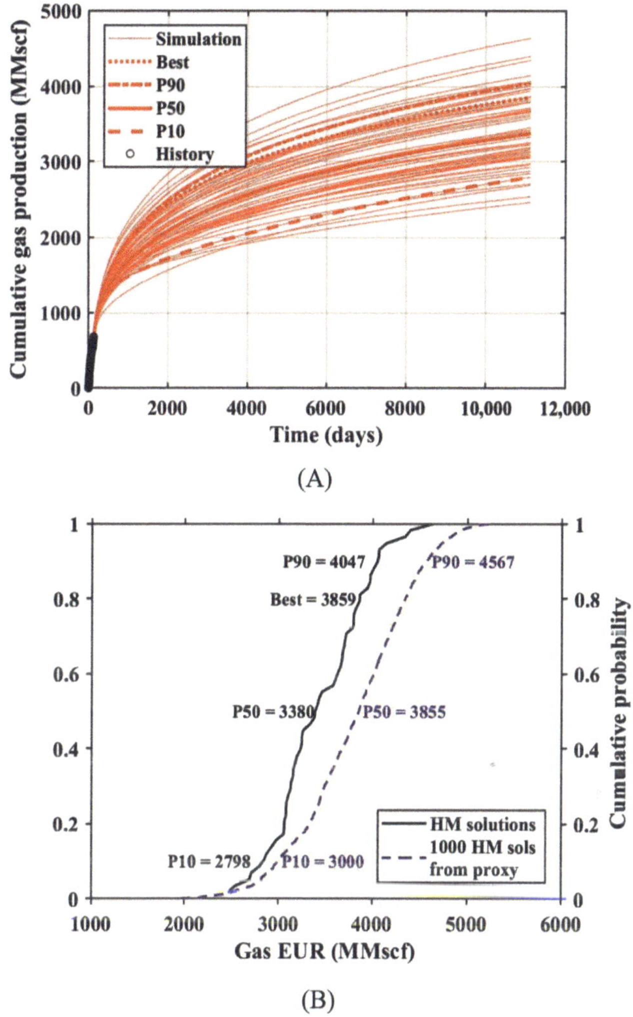

FIGURE A11 Production forecast of Well A: (A) cumulative gas production and (B) ECDF of gas EUR for history-matching solutions from AHM workflow and 1000 history-matching solutions generated by proxy model (Tripoppoom et al., 2020). *AHM*, Assisted history matching; *ECDF*, empirical cumulative distribution function; *EUR*, estimated ultimate recovery.

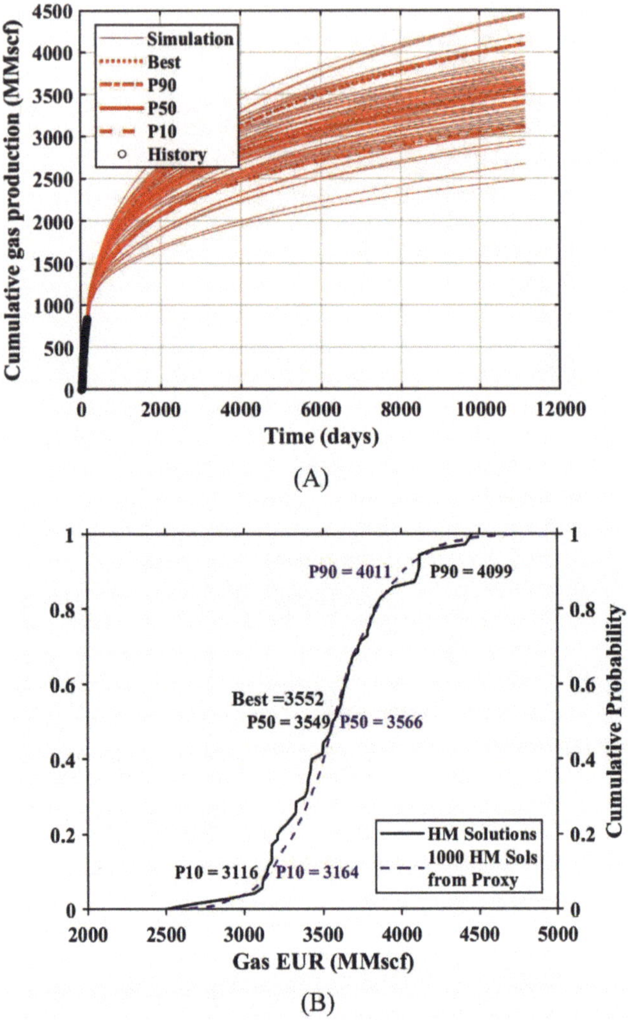

FIGURE A12 Production forecast of Well B: (A) cumulative gas production and (B) ECDF of gas EUR for history-matching solutions from AHM workflow and 1000 history-matching solutions generated by proxy model (Tripoppoom et al., 2020). *AHM*, Assisted history matching; *ECDF*, empirical cumulative distribution function; *EUR*, estimated ultimate recovery.

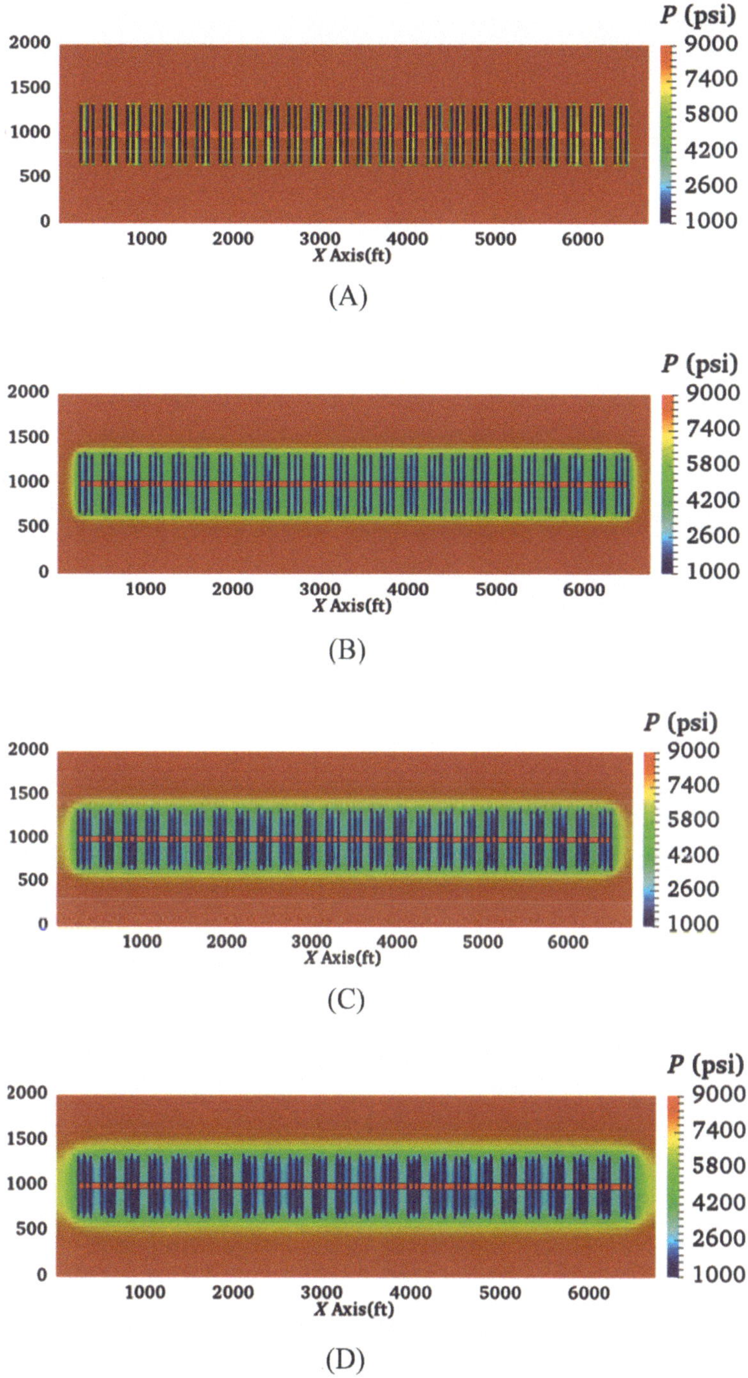

FIGURE A13 Pressure distribution of the best case for Well A: (A) after 132 days (end of history); (B) after 10 years; (C) after 20 years; and (D) after 30 years (Tripoppoom et al., 2020).

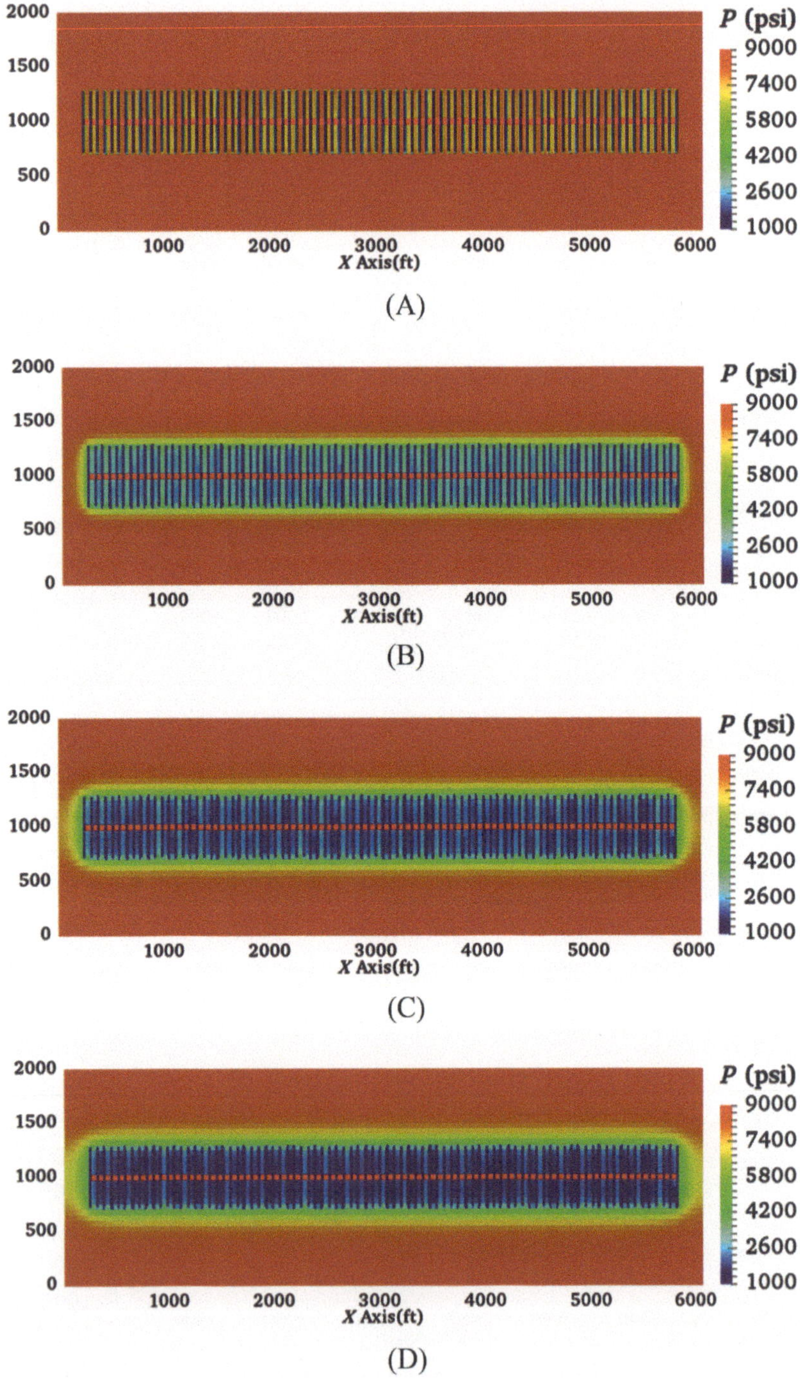

FIGURE A14 Pressure distribution of the best case for Well B: (A) after 152 days (end of history); (B) after 10 years; (C) after 20 years; and (D) after 30 years (Tripoppoom et al., 2020).

Chapter 11

Concluding remarks

11.1 Key conclusions

We developed the assisted history matching (AHM) workflow for shale reservoirs with hydraulic fractures and natural fractures and adopted a practical, efficient, and accurate AHM workflow by using embedded discrete fracture model (EDFM), in conjunction with a commercial simulator and multiple proxy-based Markov chain Monte Carlo (MCMC) algorithm. Also, instead of retrieving one matched model, we automatically obtained multiple solutions and their posterior distributions of uncertain parameters including fracture geometry, fracture properties, and reservoir properties from the AHM workflow.

In Chapter 3, Validation of Assisted History Matching for a Synthetic Shale Gas Well, we demonstrated the validation of the proposed AHM workflow for both scenarios, with and without natural fractures. The history-matching solutions found from the workflow were in line with the true solution. Also, the production forecast of history-matching solutions and true solution were similar. These findings ensure the robustness, reliability, and accuracy of the workflow for applying our approach to real cases.

Then, several applications to shale-gas and shale-oil wells were demonstrated. In Chapter 4, Shale-Gas Well in Longmaxi Shale With Bi-Wing Hydraulic Fractures, and Chapter 5, Shale-Gas Well in Marcellus Shale With Bi-Wing Hydraulic Fractures, we demonstrated the application of the AHM workflow to real shale-gas wells in Longmaxi shale and Marcellus shale, respectively. First, the comparison between local grid refinement and EDFM was performed and confirmed the accuracy and efficiency of using EDFM to model fractures nonintrusively in the AHM workflow. Then, parameter identification and screening was performed to determine the uncertain value ranges and optimize the number of uncertain parameters to be later used during the history-matching step. Subsequently, we used the multiple proxy-based MCMC algorithm with *k*-nearest neighbors as proxy model to perform history matching automatically and then obtained the posterior distributions of uncertain parameters. Lastly, we conducted a probabilistic production forecast by using multiple matched models to account for the nonuniqueness of history matching.

Next, in Chapter 6, Proxy Comparison Between Neural Network and *k*-Nearest Neighbors, we applied the AHM workflow to another real

 DOI: https://doi.org/10.1016/B978-0-12-822242-3.00002-X
© 2021 Elsevier Inc. All rights reserved.

shale-gas well. We investigated the performance of using different proxy models in the proxy-based MCMC algorithm including neural networks and *k*-nearest neighbors. The effects of pressure-dependent permeability and gas desorption were included in the shale-gas reservoir modeling. We demonstrated a more complex case with a greater number of uncertain parameters. We found that neural network performs better in terms of accuracy and requires a smaller number of simulations than *k*-nearest neighbors in order to obtain the same number of history-matching solutions. However, *k*-nearest neighbors has the benefits of requiring lower running time, almost half, compared to neural networks. Therefore, which proxy model to be used depends on whether the number of simulations or the running time is more important. Also, the history-matching solutions from the two different proxy models were aligned confirming the robustness of the history-matching algorithm.

In Chapter 7, Shale-Gas Well With and Without Enhanced Permeability Area, we demonstrated a more realistic, but simple, concept of modeling hydraulic fractures with complex fractures as enhanced permeability area (EPA). We compared the cases with and without EPA around main hydraulic fractures in shale-gas well. By performing AHM with the neural network-MCMC algorithm, we found that the existence of EPA would affect the history-matching solutions. The posterior distributions of fracture geometries and properties and matrix permeability shifted to the lower values. These changes in history-matched solutions have an impact on the probabilistic production forecast. In this case study, we observed that the case with EPA has lower gas estimated ultimate recovery (EUR), around 25%–30%, compared to the case without EPA. This means if we simplify the simulated rock volume (SRV) created during hydraulic fracturing by modeling only hydraulic fractures, it could overestimate fracture geometry and gas EUR prediction.

In shale reservoirs, natural fractures also often exist. Therefore, in Chapter 8, Shale-Gas Well With and Without Natural Fractures, we continued applying the AHM workflow to a shale-gas well to investigate the impact of natural fractures existence on the history-matching solutions and production forecast. The effects of gas desorption and pressure-dependent permeability were also included in shale-gas reservoir modeling. Then, we performed two scenarios of with and without natural fractures as uncertain parameters. The neural network-MCMC algorithm was still adopted. With the proposed workflow, we can handle not only reservoir and hydraulic fractures properties, but also natural fractures, as uncertain parameters during history matching.

We found that the scenario with natural fractures has posterior distributions significantly lower in fracture height and fracture conductivities, and higher in fracture water saturation than the case without natural fractures. Also, we observed less curvature on relative permeability curves of k_{rg} and

k_{rw}. All changes due to the existence of natural fractures are to compensate more flow contributed by natural fractures and changes in other uncertain parameters. Moreover, uncertainty of natural fractures was reduced by calibrating with production data using the AHM workflow. For production forecast, in this case, we found that the scenario with natural fractures has lower gas EUR, around 10%, compared with the scenario without natural fractures due to lower fracture height. With this effect on production forecast, this means a decision to include natural fractures modeling in the actual system of natural fractures and hydraulic fractures will depend on the natural fractures and reservoir properties and whether they have a significant effect on history-matching solutions and production forecast or not. However, these effects can only be quantified after we included natural fractures as uncertain parameters in AHM, and the proposed AHM workflow is capable to handle both hydraulic and natural fractures for history matching.

Then, in Chapter 9, Shale-Oil Well With and Without Natural Fractures, we applied the AHM workflow with another actual case of a shale-oil well in Permian basin. Similarly, we performed scenarios with and without natural fractures. From the workflow, we automatically characterized hydraulic fractures, natural fractures, and shale properties in forms of posterior distributions. The impact of ignoring natural fractures in the system was highlighted and production forecast comparison is demonstrated. We found that if we assume the system of hydraulic fractures and natural fractures with the scenario of hydraulic fractures and increased matrix permeability, we would underestimate the oil production and overestimate gas production.

Lastly, in Chapter 10, Investigation of Different Production Performances in Multiple Shale-Gas Wells, we utilized the proposed workflow to solve actual issue in shale-gas wells. As we observed different production performance among shale-gas wells, we leveraged the proposed AHM workflow to investigate this issue by performing history matching automatically in three shale-gas wells. In this case study, from the AHM work, key parameters that cause different production performances among wells are fracture half-length, fracture conductivity, fracture width, and matrix permeability. We also performed a probabilistic production forecast of three shale-gas wells. The findings from the study can assist engineers to understand fracture geometry and properties relating to fracturing design and completion.

To sum up the benefits of the proposed AHM workflow, we can characterize fracture geometries and properties, reservoir properties, and natural fracture uncertainties in a probabilistic manner from the production data which is always available at no additional cost. This information from history matching can be used to calibrate fracture propagation models to optimize any future hydraulic fracturing design and infill well placement decision. Lastly, we can forecast production in a probabilistic way to capture subsurface uncertainties and the nonuniqueness of history-matching problems.

11.2 Recommendations

11.2.1 Proxy-based MCMC algorithm and AHM workflow codes

- In this book, we only used one long MCMC chain during the proxy-based MCMC algorithm. The effect of using multiple short MCMC chains has not been studied. Therefore, the investigation of using multiple short MCMC chains versus one long MCMC chain could determine whether there is any significant difference or not and which option gives a better efficiency of proxy-based MCMC algorithm.
- The current study handled the convergence of MCMC by using a high number of MCMC steps. However, the further development of evaluating MCMC chain convergence specifically for each case rather than one constant number of MCMC steps could improve the efficiency of the algorithm.
- The number of initial points in the first iteration of this study is based on experience and empirical method, which is five times the number of uncertain parameters. However, a study on varying number of initial points could help determine the optimum number of initial points and could improve the efficiency and accuracy of the proxy-based MCMC algorithm.
- The current study only adopted Metropolis–Hastings (MH) MCMC, which is a random walk with probability. Future work could evaluate the performance comparison between MH MCMC and other types of MCMC, such as Gibbs and Hamiltonian MCMC, in terms of the number of history-matching solutions found and computational efforts.
- In Chapter 6, Proxy Comparison Between Neural Network and *k*-Nearest Neighbors, we compared the performance of neural networks versus *k*-nearest neighbors and found that neural networks provide a lot more accuracy in terms of number of history-matching solutions found but with a higher elapsed time. Therefore, to speed up neural networks related calculation, the study to find alternative ways of training neural networks could be beneficial to the history-matching algorithm. This is to improve the computational time of neural networks while accuracy should be the same or better. Also, the architecture of neural networks we used in this study is the fully connected neural networks and has one hidden layer of 10 neurons. While many literatures recommend focusing on the quality of training data rather than finding the optimized neuron network architecture, another study on different architecture of neural network such as partially connected structure, number of hidden layers, and number of hidden neurons can be performed to observe whether they could improve the efficiency and accuracy of the history-matching algorithm or not.

11.2.2 Fracture modeling

- The current study modeled hydraulic fractures either by planar fractures or planar fractures with enhanced permeability representing SRV. The comparison study of other SRV modeling techniques such as discrete fracture network model around main hydraulic fractures with the current methods could be useful to determine whether there is any necessity to model fractures with a more complicated concept or not and may provide more realistic fracture flow behavior.

- The study to integrate the current AHM workflow with fractures output from fractures propagation model can provide more realistic and complex fractures geometry. In addition, more constraints for history matching from other diagnostic tools and analyses such as microseismics, fiber optic distributed acoustic sensing/distributed temperature sensing data, and diagnostic fracture injection test analysis can help narrowing down the parameter uncertainty ranges during AHM workflow. A more representative case of nonuniform fractures with different cluster efficiency can be performed if related information is available.

11.2.3 Applications of AHM workflow

- All current studies were performed for a single well model. The study of multiwell history matching can be useful to characterize the multiwell performance in a more holistic view and to account for their interaction and inference between wells.
- In the current study, the geological and petrophysical inputs were not available to provide a static model that represents heterogeneity of shale reservoirs; only a homogeneous model was assumed to perform AHM. Therefore, the study to integrate the AHM workflow with multidisciplinary inputs to obtain a realistic static model accounting for shale heterogeneity can extend and illustrate the benefits of AHM workflow in a more comprehensive point of view. Also, the grid type used for modeling in this book is Cartesian grid. EDFM can be used with any grid types, such as corner point grid, which is more convenient when the static model is available as an initial input.
- Another interesting study is to investigate the effect of different days of historical production data to the history-matching solutions, posterior distribution of uncertain parameters, and production forecast. The findings from the study may help determine the results validity of history matching and what concerns to be aware of when using history-matching results at different stages of well life.
- Cluster efficiency can be used as another uncertain parameter for history matching in case other fracture geometry and properties are more certain and solid.

Index

Note: Page numbers followed by "*f*" and "*t*" refer to figures and tables, respectively.

F

G

H

I

K

L

M

N

O

P

FSC
www.fsc.org
MIX
Papier | Fördert
gute Waldnutzung
FSC® C083411

Kolibri 360 GmbH
Ferdinand-Jühlke-Straße 7
99095 Erfurt, Deutschland
produktsicherheit@kolibri360.de